Circuit Modeling for
Electromagnetic Compatibility

Other titles in the series

Designing Electronic Systems for EMC (2011)
by William G. Duff

Electromagnetic Measurements in the Near Field, Second Edition (2012)
by Pawel Bienkowski and Hubert Trzaska

Circuit Modeling for Electromagnetic Compatibility (2013)
by Ian B. Darney

The EMC Pocket Guide (2013)
by Kenneth Wyatt and Randy Jost

Forthcoming titles in the series

EMC Essentials (2014)
by Kenneth Wyatt and Randy Jost

Electromagnetic Field Standards and Exposure Systems (2014)
by Eugeniusz Grudzinski and Hubert Trzaska

Guide to EMC Troubleshooting and Problem-solving (2014)
by Patrick G. André and Kenneth Wyatt

Designing Wireless Communication Systems for EMC (2014)
by William G. Duff

Circuit Modeling for Electromagnetic Compatibility

EMC Series

Ian B. Darney

SCITECH
PUBLISHING
an imprint of the IET

Edison, NJ
scitechpub.com

SciTech
PUBLISHING
an imprint of the IET

Published by SciTech Publishing, an imprint of the IET.
www.scitechpub.com
www.theiet.org

10 9 8 7 6 5 4 3 2 1

ISBN 978-1-61353-020-7 (hardback)
ISBN 978-1-61353-028-3 (PDF)

Typeset in India by MPS Limited
Printed in the USA by Sheridan Books, Inc.
Printed in the UK by Hobbs the Printers Ltd

The SciTech Series on Electromagnetic Compatibility

The SciTech Series on Electromagnetic Compatibility provides a continuously growing body of knowledge in the latest developments and best practices in electromagnetic compatibility engineering. EMC is a subject that has broadened its scope in the last 20 years to include effects associated with virtually all electronic systems, ranging from the nanoscale to large installations and from physical devices to distributed communications systems. Similarly, EMC knowledge and practices have spread beyond the EMC specialist to a much wider audience of electronic design engineers. No longer can ESD/EDI problems be addressed as a solution to an unforeseen problem in a reactive response. Rather, design engineers can model and simulate systems specifically to root out the potential for such effects. Similarly, knowledge and practice from other engineering disciplines have become an integral part of the subject of electromagnetic compatibility. The aim of this series is to provide this broadening audience of specialist and non-specialist professionals and students books by authoritative authors that are practical in their application but thoroughly grounded in a relevant theoretical basis. Thus, series books have as much relevance in a modern university curriculum as they do on the practicing engineer's bookshelf.

Circuit Modeling for Electromagnetic Compatibility, EMC Series
Ian B. Darney

Understanding a problem often means focusing on the heart of the issue. That is what this book does: it strips away the clutter in order to help develop an appreciation and understanding of some of the core issues for EMC. *Circuit Modeling for Electromagnetic Compatibility* demonstrates how powerful the simple models for lumped parameter, transmission line, and the antenna can be. The origins of this book go back over 40 years and emphasize the huge amount that can be garnered from simplified analytical approaches. Ian Darney's clear approach is that if you can simulate the observed response, you are a long way toward solving the problem.

Ian and I first spoke about this book about a year and a half ago, and it was apparent that, having spent a successful career as an electronic systems designer, he had a firm intention to share his career's learning in a distilled and accessible book. Some people may feel that too much of the detail has been stripped away, but the vast majority of the engineers I have shared this with have enjoyed both the technical underpinnings and Ian's approach to communicating it.

I think this is a great companion book for any electronic engineer's bookshelf. It will help non-EMC engineers get to grips with the core technology challenges and help EMC engineers visualize the driving mechanisms for some of the phenomena they are working with on a daily basis.

Alistair Duffy – Series Editor
2013

Contents

Preface   xiii
Acknowledgments   xvii

1 Introduction   1

  1.1 Background   1

    1.1.1 The need for EMC   1
    1.1.2 Pragmatic approach   1
    1.1.3 Academic approach   2
    1.1.4 Managerial approach   2
    1.1.5 Misleading concepts   2
    1.1.6 Circuit modeling   3
    1.1.7 Computations   3
    1.1.8 Testing   3
    1.1.9 Essence of the approach   4

  1.2 Developing the model   4

    1.2.1 Basic model   4
    1.2.2 Parameter types   5
    1.2.3 Derivation process   6
    1.2.4 Composite conductors   7
    1.2.5 Proximity effect   8
    1.2.6 Electrical length   8
    1.2.7 Distributed parameters   9

  1.3 Intra-system interference   11

    1.3.1 The signal link   11
    1.3.2 Simulating the structure   11
    1.3.3 Equivalent circuits   12
    1.3.4 Conducted emission   12
    1.3.5 Conducted susceptibility   13

	1.3.6	Voltage transformer	14
	1.3.7	Current transformer	14
	1.3.8	Representative circuit model	14
1.4	Inter-system interference		15
	1.4.1	Dipole model	15
	1.4.2	The virtual conductor	16
	1.4.3	The threat voltage	17
	1.4.4	Worst-case analysis	18
1.5	Transients		19
1.6	The importance of testing		20
1.7	Practical design techniques		21
1.8	System design		22
	1.8.1	Guidelines	22
	1.8.2	Top-down approach	23
	1.8.3	Formal EMC requirements	23

2   Lumped parameter models   **25**

2.1	Primitive capacitance		27
2.2	Primitive inductance		30
2.3	Duality of L and C		34
2.4	Loop parameters		35
2.5	Circuit parameters		38
	2.5.1	Inductance	38
	2.5.2	Capacitance	39
	2.5.3	Maintaining duality	40
	2.5.4	Resistance	41
	2.5.5	Basic assumption	42
2.6	Twin-conductor model		42
2.7	Three-conductor model		45
2.8	Optimum coupling		49
2.9	Transfer admittance		52
2.10	Co-axial coupling		55
2.11	The ground plane		57

3   Other cross sections   **61**

3.1	Single composite conductor	62
3.2	The composite pair	67
3.3	The screened pair	74

4   Transmission line models   **81**

    4.1   Single-T model   82

    4.2   Triple-T model   86

    4.3   Cross-coupling   89

    4.4   Bench test models   94

5   Antenna models   **101**

    5.1   The half-wave dipole   102

          5.1.1   Radiated power   102

          5.1.2   Power density   104

          5.1.3   Field strength   105

          5.1.4   Power received   106

    5.2   The virtual conductor   107

    5.3   The threat voltage   113

    5.4   The threat current   117

    5.5   Coupling via the structure   122

    5.6   Radiation susceptibility   130

    5.7   Radiated emission   132

6   Transient analysis   **135**

    6.1   Time-step analysis   137

          6.1.1   Basic concept   137

          6.1.2   Basic equations   137

          6.1.3   Series LCR circuit   138

          6.1.4   Parallel LCR circuit   140

    6.2   Delay-line model   143

    6.3   Line characteristics   149

    6.4   Antenna-mode current   155

    6.5   Radiated emission   161

          6.5.1   Current linking the transformer   161

          6.5.2   Line voltage   164

          6.5.3   Source current and voltage   164

          6.5.4   Radiated current   165

          6.5.5   Cable losses   166

          6.5.6   Line parameter measurements   167

    6.6   Transient emission model   168

7   Bench testing   175

7.1   Voltage transformer   176

7.2   Current transformer   183

7.3   Triaxial cable   188

7.4   The isolated conductor   189

7.5   Cable characterization   197

7.6   Cable transients   206

7.7   Capacitor characterization   213

8   Practical design   217

8.1   Grounding   218

8.2   Conductor pairing   221

8.3   Ground loops   224

8.4   Common-mode rejection   226

    8.4.1   Differential amplifier   226

    8.4.2   Differential logic driver   227

    8.4.3   Differential analogue driver   228

    8.4.4   Common-mode choke   228

    8.4.5   Transformer coupling   229

    8.4.6   Center-tapped transformer   230

    8.4.7   Opto-isolator   230

8.5   Differential-mode damping   231

    8.5.1   Transient damping   231

    8.5.2   Mains filtering   232

    8.5.3   Solenoid switching   233

    8.5.4   Commercial filters   234

    8.5.5   Use of carbon   235

8.6   Common-mode damping   235

    8.6.1   Common-mode resistor   235

    8.6.2   Minimizing pickup   236

    8.6.3   Triaxial cable   237

    8.6.4   Transformer inter-winding   238

    8.6.5   Common-mode filter   238

    8.6.6   Transformer-coupled resistors   239

8.7   Shielding   240

    8.7.1   Equipment shielding   241

    8.7.2   Shielding of buildings   242

    8.7.3   Use of carbon   244

9   System design   **245**

  9.1   Design guidelines   246

    9.1.1   Structure as a shield   246

    9.1.2   Return conductors   247

    9.1.3   Ground loops   247

    9.1.4   Current balance   248

    9.1.5   Differential-mode damping   248

    9.1.6   Common-mode damping   248

    9.1.7   System assessment   249

    9.1.8   Bench tests   249

  9.2   Relating the diagrams   249

    9.2.1   Circuit diagrams   249

    9.2.2   Wiring diagrams   251

    9.2.3   Block diagrams   252

    9.2.4   Interface diagrams   252

    9.2.5   Circuit models   253

    9.2.6   Deriving component values   254

    9.2.7   Analyzing the signal link   255

    9.2.8   Testing the link   256

  9.3   Printed circuit boards   256

  9.4   Susceptibility requirements   257

  9.5   Emission requirements   260

  9.6   Planning   263

    9.6.1   Performance requirements   263

    9.6.2   Bench test equipment   263

    9.6.3   Software   264

    9.6.4   Critical signal links   264

    9.6.5   Critical frequencies   265

    9.6.6   Characterization   265

    9.6.7   General approach   266

Appendix A   Mathcad worksheets   269

Appendix B   MATLAB®   271

Appendix C   The hybrid equations   273

Appendix D   Definitions   277

References   281

Index   285

Preface

Back in the 1960s, the author was a member of a team designing a Flight Trainer. In this equipment, an analog computer generated a set of waveforms which resulted in a trapezoidal raster being displayed on the screen of a flying-spot scanner. The light from the screen illuminated a continuously moving film of a five-mile wide strip of terrain. The light which penetrated the film was focused by a collimator lens onto a photomultiplier. The video output simulated that of a camera mounted on a low-flying missile.

The system worked reasonably well, but was plagued by a wide range of interference problems which were never satisfactorily solved. The underlying reason was the fact that, at the outset of the project, the customer insisted that a single-point ground terminal be located at the bottom of the rack of equipment and that three wire-braids be connected to that point. These were designated the 'analog ground', 'logic ground', and 'power ground', and the star-point was specified as the only place where ohmic contact was allowed. At all other locations in the equipment, the reference 'grounds' were isolated from each other. This was, and is, the worst possible configuration to adopt.

Even so, the concept of the 'star-point ground' has gained widespread acceptance by the engineering community. Other misleading concepts in vogue are the 'equipotential ground' and the need to 'avoid earth loops'.

This book started out as a study report that advocated the use of a set of guidelines which could replace these misleading concepts. Since engineers are skeptical individuals, there was always someone who could point out a defect in the reasoning. So more background material was gathered, tests were carried out, and further analyses performed. It eventually became clear that circuit modeling could be used to analyze the coupling mechanisms.

But there were still critics who pointed out that such an approach could not be used to handle high-frequency simulations. So the modeling technique was developed further to cater for transmission-line effects and to take into account the action of cables as antennae. The end result is a technique that can be used to assess and analyze the mechanisms usually associated with electromagnetic interference (EMI). That is

- common impedance,
- electric fields (capacitive induction),
- magnetic fields (magnetic induction), and
- electromagnetic fields (plain waves).

The following pages provide many circuit models which can simulate the various conducted EMI and radiated EMI problems. The approach is unique in that it uses simple analytical methods. It is easy to implement.

The contents are useful to practical design engineers at various levels, such as circuit designers, printed circuit board designers, electronic system engineers, power system engineers, EMC engineers, and EMC consultants. Time is precious to such individuals, so it is recommended that the busy designer first reads Chapter 9, which describes a top-down approach and provides a set of simple guidelines. If this systematic approach is implemented, then the design can be made fundamentally sound. Then it is worth reading Chapter 8, which identifies most of the techniques which reduce the level of EMI coupling and describe the mechanisms involved. The preceding chapters can be regarded as material which justifies the detailed recommendations.

Lecturers who teach subjects such as electronic circuit design (analogue, digital, switched-mode, radio-frequency, etc.) should find it useful, since it relates fundamental concepts to the considerations of practical design.

Students of electrical engineering will benefit from this book, since EMC is no longer an optional topic and the approach described in the following pages is the simplest possible.

It should also be useful to universities who provide special courses on the subject of EMC, since it identifies a different approach to the analysis of EMI. Since it does not require an ability to manipulate the mathematics of electromagnetic field theory, it is understandable to a wider range of engineers.

One of the tests in Chapter 7 identifies the fact that antenna-mode current propagates faster than differential-mode current, and shows how the two velocities can be measured. This should be of interest to researchers.

There are many books which describe the various interference coupling mechanisms, and which identify practical design solutions. Others delve into the analysis of electromagnetic field propagation. Since these aspects are well-covered elsewhere, there is no need to reprise their contents. Such a policy keeps this book relatively short.

The first chapter identifies the underlying concepts and summarizes the approach.

Chapter 2 defines the building blocks of all circuit models and derives simple models of familiar configurations such as the coupling between common-mode circuits and differential-mode circuits. These models are useful in providing an insight into the coupling mechanisms. They are amenable to analysis using SPICE software. The simulated response is reasonably accurate up to the frequency at which the wavelength of the signal is one-tenth the length of the cable.

Chapter 3 develops the process to allow the electromagnetic coupling in complex assemblies such as aircraft wings or multilayer boards to be simulated. Although the frequency response of such models is subject to the same limitation as that of the simpler configurations, the range of possible applications is vastly extended.

An open-circuit line will resonate at a frequency where the quarter wavelength of the signal is equal to the length of the line. A short-circuited line will resonate at the half-wave frequency. At resonance the level of interference will reach a peak value. If it is hoped to simulate the interference-coupling characteristics of any signal link, then the model should be capable of handling signals up to, and beyond, the half-wave frequency of the line. Chapter 4 achieves this objective by invoking the relationships of transmission-line theory.

Chapter 5 takes the process one step further, to simulate the behavior of cables as antennae.

Chapter 6 derives a circuit model that can replicate the transient behavior of a twin-conductor cable as an antenna.

Chapter 7 shows how circuit models can be used to simulate the response of bench tests on actual hardware. This establishes the all-important connection between theory and practice.

Chapter 8 describes a number of techniques that have been used by engineers to improve EMC, and relates these designs to the interference coupling mechanisms identified in the previous chapters.

Chapter 9 outlines a systematic method of analyzing the EMC characteristics of the system-under-development. It establishes a clear link between the formal EMC design requirements and the performance of the equipment. The S.I. system of units is used throughout the book.

Although the analytical process is dramatically simpler than one based on the use of electromagnetic field theory, the calculations still require the use of a computer. Simulation Programs with Integrated Circuit Emphasis (SPICE) can deal with the simple configurations described in Chapter 2, but cannot handle the computations described in the later chapters. Mathematical software is needed.

It was found that Mathcad software was ideal for the purpose since it can combine the equations of circuit analysis with those of electromagnetic field propagation. It can accept input data in the form of geometrical measurements of the hardware-under-review and combine this with data derived from tests on that hardware. Appendix A provides a brief introduction to this software. Copies of the worksheets can be downloaded from the website www.designemc.info.

Subsequent to the completion of the first draft of this book, an exercise was carried out to translate the Mathcad worksheets into MATLAB® m-files. These also can be downloaded from the website. Appendix B identifies the relationships between the two software packages which help engineers who are familiar with MATLAB to read and understand the contents of the Mathcad worksheets.

One of the key features of the analysis is the use of a transformation formula derived from the equations of transmission-line theory. Appendix C provides a succinct introduction to the concept of distributed parameters and derives the hybrid equations used as a starting point in Chapter 4.

Although many of the concepts used in this book are familiar to electrical and electronic engineers, some are new. So a set of definitions is provided in Appendix D.

Reports on further tests and analyses will be filed at www.designemc.info as and when they are completed. The website also has a page for feedback from readers.

Acknowledgments

Thanks are due to the many old colleagues in British Aerospace who provided encouragement and criticism in equal measure, to Alistair Duffy who promoted the book, to Dudley Kay who agreed to publish it, and to my wife Frances for her patience and support.

CHAPTER 1

Introduction

1.1 Background

1.1.1 The need for EMC

The development of electronic equipment has come a long way since the invention of valves and transistors, to the extent that modern society is highly dependent on the smooth functioning of the myriad systems that myriad systems now in operation.

Concurrently with that development, Electromagnetic Interference (EMI) has also increased, both in the number of daily incidents and in the severity of the possible consequences. Initially, most of the effects were annoying; for example, crackles on the radio due to a nearby thunderstorm were something one learned to accept. Latterly, some of the effects could be life threatening. The phenomenon described as 'sudden unintended acceleration' could be a case in point.

A succinct definition of Electromagnetic Compatibility (EMC) is 'the ability of a device, unit of equipment, or system to function satisfactorily in its electromagnetic environment without introducing intolerable electromagnetic disturbances to anything in that environment' [1.1].

1.1.2 Pragmatic approach

There is no shortage of regulatory requirements. In fact, the continuous stream of new or revised regulations is enough to keep a team of researchers occupied full time [1.2].

However, the task of designing equipment to meet the requirements does not seem to be amenable to a simple, systematic approach. Normal practice is to break the phenomenon down into four distinct types: common impedance, electric fields, magnetic fields, and electromagnetic fields [1.3]. Each type of coupling is treated separately, and examples are provided of the effects which can be expected with different design fixes. The objective is to help readers to achieve an overall understanding of the physics involved. Armed with such an understanding, the designer should be able to assess the EMC characteristics of the equipment-under-review.

This pragmatic approach tends to focus on the multitude of methods by which interference can be coupled from one system to another, on layout and grounding, printed circuit layout, cables and connectors, filtering, transient suppression, and shielding. Many useful design techniques are identified in books that adopt this approach [1.4]. This being so, there is no need to reprise the material they provide.

While such an approach leads to many useful design techniques, it is essentially hit or miss. A technique that works well in one application can cause disastrous effects in another. Since it involves subjective judgment, there is still plenty of scope for disagreement between designers.

This is in marked contrast to requirements such as functional performance, frequency response, power consumption, reliability, mass, and size, which are all amenable to rigorous analysis. Care is taken at every stage of the design process to ensure that these other requirements will be met. Bench testing is carried out on prototype equipment. In situ tests are carried out on assembled systems. Test results are compared with predicted performance. Regular design reviews are carried out. If there are any problems, modifications are implemented.

1.1.3 Academic approach

Another approach to the subject is based on the use of computational electromagnetics. There is at least one book which compiles the results of research carried out on this topic [1.5].

This is essentially a review of the various techniques used in computational electromagnetics, followed by a selection of an appropriate method of analysis. Such techniques are: radiation models for wire antennas, diffraction and scattering models for apertures, field coupling using transmission line theory, and shielding models. The problem with this approach is that the mathematical processes require an expertise which is well beyond the capability of the average equipment designer. Moreover, the focus is mostly on the behavior of electromagnetic fields. This being so, the functional performance of the equipment-under-review is not an aspect that is analyzed.

1.1.4 Managerial approach

In large organizations, the approach to the achievement of EMC is to create a set of Design Rules which are updated periodically by standing committees. The nature of such a management system is that rules which have been formulated decades in the past come to be regarded as sacrosanct. Development of new equipment involves design processes which follow these rules, the equipment is subjected to performance testing, environmental testing, reliability analyses, design reviews, and is eventually submitted to the EMC Test House for final testing. If it fails this final test, this fact is discussed at the Critical Review meeting, emergency measures are invoked, design fixes are implemented, and the equipment resubmitted for EMC testing. Eventually, the equipment muddles through.

1.1.5 Misleading concepts

There is a widespread belief among engineers that the structure can be represented by an equipotential surface. Such an assumption is inherent in every circuit diagram which contains the ubiquitous 'ground' symbol. Attempts to reconcile this concept with the observed interference in the system have led at least one engineer to declare that the subject is a 'black art'.

Another disturbing concept which has been given formal blessing by successive teams of experts is that of the 'single ground point'. Some point on the structure is designated as the reference point for all signals in the system. Guidance for the design of UK military equipment formalizes this concept and provides detailed requirements for its implementation [1.6]. If this concept is implemented into the design of any electronic equipment, it can be guaranteed that the equipment will suffer from intractable interference problems.

Closely related to the concept of the 'single-point ground' is the stricture to 'avoid earth loops'.

Although none of these concepts is to be found in any textbook on electromagnetic theory, they have become firmly entrenched beliefs in the engineering community. There has to be a better way of approaching the task of achieving EMC.

1.1.6 Circuit modeling

The approach described in the following pages enables electronic equipment to be designed to meet the EMC requirements using mathematics which is understandable to every circuit designer. It establishes a clear relationship between the performance of the system-under-review and the requirements placed on the system. The ability to analyze the coupling mechanisms leads to a better understanding of the underlying physics.

Given a better understanding, the designer is able to appreciate the value of the detailed advice provided by the existing books and to reject the misleading concepts prevalent in the workplace.

The simplifications inherent in the use of circuit theory inevitably mean that some fidelity is lost in the simulation. If high fidelity is required, then the results obtained by the use of circuit modeling can be refined by the techniques of computational electromagnetics.

1.1.7 Computations

Although the simplifications achieved through the use of circuit modeling provide a dramatic reduction in the complexity of the mathematics, it is still necessary to employ the use of personal computers. Since many of the calculations are beyond the capability of Simulation Programs with Integrated Circuit Emphasis (SPICE), another type of general-purpose software is needed – mathematical software.

A Mathcad worksheet is used to illustrate the details of each computation. Since mathematical notation is used throughout in these programs, they are much easier to follow than ones written in, say, the JAVA language. A few special features of Mathcad are described in Appendix A. After reading this appendix, it should not be difficult for any reader to understand the contents of the worksheets. Since every worksheet is fully explained, the reader is not left to formulate his or her own program from a set of mathematical relationships.

In the following pages, the parameter dimensions adhere to those of the SI System.

1.1.8 Testing

The ability to analyze the mechanisms involved in coupling interference into and out of the signal link-under-review leads to the ability to devise tests which measure the actual coupling parameters and correlate them with those of the relevant circuit model. A connection has been established between test and analysis.

Since circuit theory is capable of analyzing signals in either the time domain or the frequency domain, it becomes possible to identify the most probable cause of unexpected interference during product development. It is but a small step to modify the design of the signal link, build a prototype of the new link, check its performance using bench test equipment, create a representative circuit model, and prepare a progress report. More than that, it becomes possible to check functional performance against formal requirements.

That is, EMC can be subjected to the same design process that applies to every other performance requirement.

1.1.9 Essence of the approach

The essence of the approach used by the author has always been to carry out tests on a representative assembly, search for a circuit model which allows the results to be simulated, and then to review available literature to relate that model to accepted theory. Having identified a clear relationship between the model and accepted theory, it becomes possible to define that relationship. Since the theory of electromagnetics is based on the use of the Maxwell Equations, it can be said that the model is based on those equations.

Since there are always deviations between the model and the test observations, it becomes possible to focus on the deviations and search for ways of refining the model. It is tempting to continue with this process and develop ever more complex models. However, the objective of the exercise is to enable the engineer at the bench to gain an understanding of the mechanisms underlying electromagnetic interference. Once the cause of a particular interference problem is understood, human ingenuity enables a solution to be found.

The ability to create a circuit model of a particular signal link allows the user to predict the interference characteristics of that link, enabling tests to be defined to measure those characteristics. Ultimately, the accuracy of the technique is determined by the accuracy of those measurements.

1.2 Developing the model

1.2.1 Basic model

Basic assumptions can be clearly stated. Every conductor (including the structure) possesses the properties of inductance, capacitance and resistance. Any circuit model which simulates interference must represent at least three conductors. Only a three-conductor configuration can be used to simulate the coupling between two independent circuit loops. For example, to simulate

- cross-coupling between two signals in a multi-conductor cable or on a printed circuit board (see section 4.3),
- coupling between the differential-mode loop and the common-mode loop (see section 4.4),
- coupling between a transmission line and the environment (see section 5.2).

A circuit model of inductive coupling between three parallel conductors would be as shown on Figure 1.2.1. Similarly, a circuit model of the capacitive coupling would be as shown on Figure 1.2.2

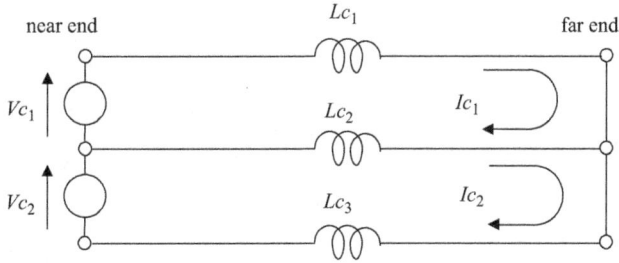

Figure 1.2.1 Inductive coupling between three parallel conductors.

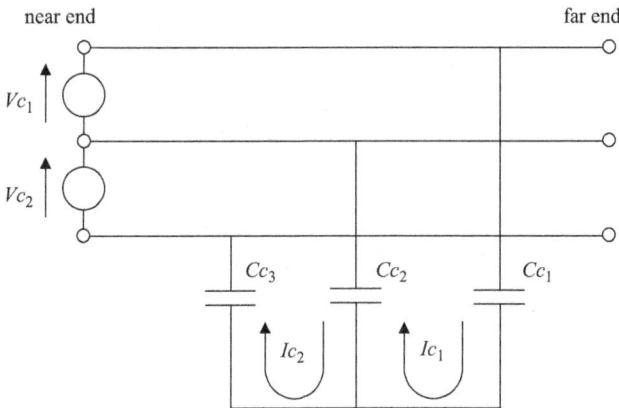

Figure 1.2.2 Capacitive coupling between three parallel conductors.

Since current must flow along a conductor in order to flow into the capacitance, and since a cable behaves in the same manner no matter at which end voltages are applied, the most logical way of simulating the combined effect of inductive and capacitive coupling would be to use the circuit of Figure 1.2.3.

This model includes resistors to represent the effect of the series resistance of each conductor.

1.2.2 Parameter types

Much of the analysis of the coupling parameters deals with the relationships between electromagnetic theory and circuit theory.

The mathematics of the former involves divs, dels, and curls, as well as the relationships between rectangular, circular, and spherical co-ordinate systems. A basic assumption is that currents and voltages anywhere in the system affect currents and voltages everywhere else in the system.

Circuit theory involves a totally different set of concepts. The system is depicted as a network of nodes and branches and the voltage across each branch is a unique function of the current flowing in that branch. Although circuit theory is much simpler, it still involves a great deal of computation. So it is essential to ensure at the outset that equations derived

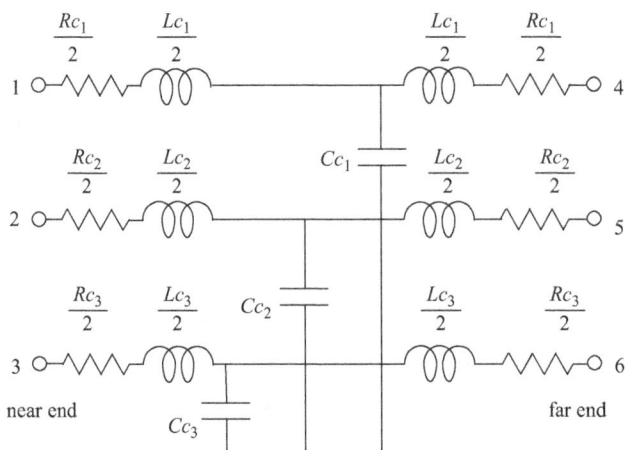

Figure 1.2.3 Combined effect of inductive and capacitive coupling.

using one theory are always distinguishable from equations derived from the other. This can be done by defining different types of parameter and by assigning a unique symbol to each. Four distinct types of parameter can be identified; primitive, partial, loop, and circuit.

- A primitive parameter is one which relates the current in a conductor of circular section to the energy level of the electromagnetic field associated with that current.
- A partial parameter is one which relates the current in a conductor of any cross-section to the energy level of the electromagnetic field associated with that current.
- Loop parameters are derived from primitive parameters and are used in equations which relate loop voltages to loop currents. They are the parameters which can be measured directly by electronic test equipment.
- Circuit parameters are those which appear in circuit diagrams.

These parameters are described more fully in Chapters 2 and 3.

Electromagnetic theory invokes the concept of distributed parameters; resistance per meter, capacitance per meter, inductance per meter, and conductance per meter. Chapter 4 identifies a simple transformation formula which obviates the need to use these parameters.

Reflections can occur at transmission line terminations. Incident current flows in one direction along a conductor while reflected current flows in the opposite direction. Total current is the sum of these two partial currents. The term 'partial' is also used to identify the associated voltages.

1.2.3 Derivation process

The process used to derive circuit models is based on the method described in textbooks to derive formulae for the equivalent phase inductance [1.7] and the equivalent phase capacitance [1.8] of a three-phase power line. It is assumed that three conductors are routed in parallel. It is also assumed that the waveforms of currents and voltages are sinusoidal.

The process can be summarized:

1. Define the length of the assembly, the radii of the conductors, and the spacing between the centers.
2. Establish a set of three primitive equations relating the voltage on each conductor to the current in all three conductors.
3. Define the loops in terms of conductor pairs.
4. Derive a set of loop equations.
5. Define the loop inductors and loop capacitors in terms of the primitives.
6. Postulate the existence of a circuit model which creates two mesh equations.
7. Relate the components of the circuit model to the constants in the loop equations.
8. Relate the components of the circuit model to the primitive parameters.

The key feature to note in the above process is that it contains a discontinuity. Step (6) does not follow logically from step (5). Lateral thinking is needed. The purpose of the circuit model is to create a set of mesh equations which correlate precisely with those of the loop equations. Mesh equations are derived using the rules of circuit theory. Loop equations are derived from the relationships of electromagnetic field theory.

Setting up such a relationship is conceptually the same as defining x as an unknown variable. If circuit theory is treated in this way, it can be utilized to simulate all types of electromagnetic coupling.

It is useful to always bear in mind the fact that circuit theory does not define the physical mechanisms. Concepts such as the 'equipotential ground plane' and the 'single point reference' are convenient assumptions which allow circuit models to simulate the behavior of complex printed circuit boards in signal processing equipment. In so doing, they carry with them the unstated assumption that there is no such thing as electromagnetic interference.

Since mesh analysis caters for the fact that partial currents can flow in both directions along a conductor at the same time, this form of analysis is used in the following pages.

1.2.4 Composite conductors

In practice, there are a great many conductors which are not circular in section. Conduits and cable trays fall into this category. Figure 1.2.4 illustrates one way of dealing with conductors of any cross section; simulate the section as a set of elemental conductors which are short-circuited at each end.

This is a development of a technique pioneered by researchers at Culham Laboratories (as it then was) to analyze the magnetic field distribution in an aircraft wing or fuselage during the occurrence of a direct lightning strike [1.9].

Since a composite conductor is built up from an array of elemental conductors and since it has inductive and capacitive properties akin to that of $Lp_{i,j}$ and $Cp_{i,j}$, it is necessary to invoke the use of partial parameters: partial inductance $Lq_{m,n}$ and partial capacitance $Cq_{m,n}$, where m and n identify the composite conductors.

This allows a computer program to be compiled which makes a clear distinction between partial parameters and primitive parameters. Chapter 3 describes this technique in greater detail.

(a) section of conduit

(b) array of elemental conductors

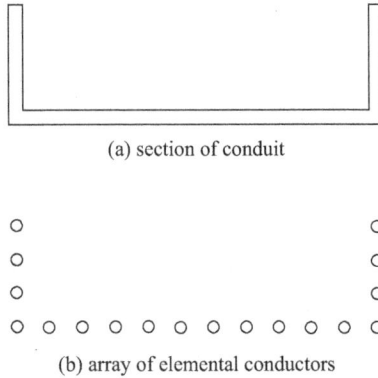

Figure 1.2.4 Concept of the composite conductor.

1.2.5 Proximity effect

The original purpose of elemental conductors was to analyze the indirect effects of a lightning strike on aircraft cables. By determining the magnetic field distribution inside the wing or fuselage when lightning current was flowing in the structure, it was possible to identify regions where the field was less concentrated. This helped in the task of deciding where to route cables.

It is also possible to use the technique to analyze the current distribution in the surface of cables, ducting and conduits, as illustrated by Figures 3.2.10 and 3.3.5. These depict the current distribution on the surface of conductors, and show that when loop current is flowing, the current concentrates on adjacent surfaces. These pictures could equally well illustrate the distribution of charge on adjacent charged surfaces.

The technique can be used to simulate the current distribution on the conducting surfaces of printed circuit boards, as illustrated by Figure 8.2.1. Such a depiction is probably more meaningful to circuit designers than color images created by full-field simulation techniques. As well as providing results which are easier to interpret, the technique does not require the use of special-purpose software or the ability to understand the underlying mathematics of such software.

Normal analysis of skin effect, as described in section 2.5.4, assumes that the current is evenly distributed over the surface of the conductor. Proximity effect indicates that, although current concentrates on the surface, the distribution is not at all uniform. Even so, it does not alter the fact that conductor resistance increases as the frequency increases.

1.2.6 Electrical length

The wavelength λ represents the distance an electromagnetic wave must travel in order to change phase by 360 degrees. If the frequency of the wave is f and the velocity of propagation is v, then:

$$\lambda = \frac{v}{f}$$

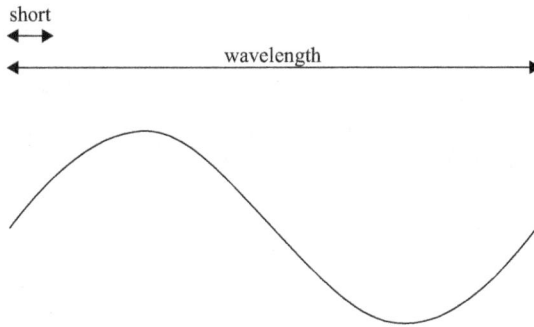

Figure 1.2.5   Electrical lengths.

The velocity of propagation of a signal along the conductor of a transmission line is the same as that of the associated electromagnetic wave. The electrical length k of a conductor of length l can be defined as:

$$k = \frac{l}{\lambda}$$

The usual textbook definition of a short electrical length is that:

$$k < 0.1$$

Figure 1.2.5 illustrates the relationship between short electrical length and wavelength. Experience has shown that the circuit model of Figure 1.2.3 is reasonably accurate, provided the conductor is electrically short. Equally, it can be said that the model is reasonably accurate up to a frequency at which the wavelength is ten times the length of the assembly-under-review.

1.2.7 Distributed parameters

Transmission line theory caters for the fact that currents and voltages vary along the length of a conductor by invoking the concept of distributed parameters. That is, derivation of formulae is based on the parameters:

$$\frac{Rc}{l}, \frac{Lc}{l}, \frac{Cc}{l}, \quad \text{and} \quad \frac{Gc}{l}$$

where l is the length of the conductor and Rc, Lc, Cc, and Gc are the series resistance, inductance, capacitance, and conductance of the line. The analysis results in a pair of hybrid equations which relate the current and voltage at the near end of the line to the current and voltage at the far end. If it is postulated that the complete length of transmission line can be represented by the circuit model of Figure 1.2.6, then it is possible to define the impedances $Z1$ and $Z2$ in terms of the lumped parameters; Rc, Lc, Cc, Gc.

Extending the concept to deal with a three-conductor transmission line gives the circuit model of Figure 1.2.7. This can be described as a distributed parameter model since all the impedance values are derived via the use of distributed parameters in the derivation process.

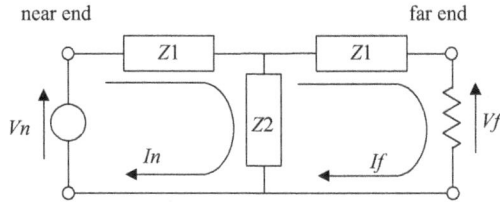

Figure 1.2.6   Transmission line model.

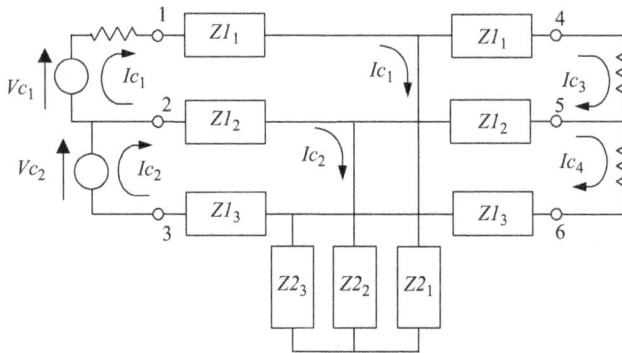

Figure 1.2.7   Distributed parameter model.

Although these impedances are derived in an unconventional way, they still evaluate to the form $R + j \cdot X$, where X is the value of the reactance at frequency f.

Since there is a clear correlation between the models of Figures 1.2.3 and 1.2.7, the transformation process is extremely simple. The three-conductor line can be defined in terms of the lumped parameter model and analyzed using the distributed parameter model. This means that the designer retains visibility of the properties of the model throughout the analytical process. This method of simulation is much simpler than stringing dozens of lumped parameter models in series.

Using the distributed parameter model, the maximum frequency of the simulation is no longer limited by the length of the cable. But there is still a limit. It is assumed that action and reaction between adjacent conductors in any cross section of the cable is instantaneous. By analogy with the limitation defined in section 1.2.6, the maximum frequency is that at which the wavelength is ten times the maximum spacing between conductors.

Best accuracy is obtained if the cross section of the cable is exactly uniform along its entire length and that the lengths of the three conductors are identical. If these requirements are not met because different types of cable are used, or because intermediate connectors are included in the wiring harness, then the accuracy of the simulation will be reduced. However, accuracy can be restored by carrying out bench testing on a representative assembly and creating a circuit model which replicates the test results.

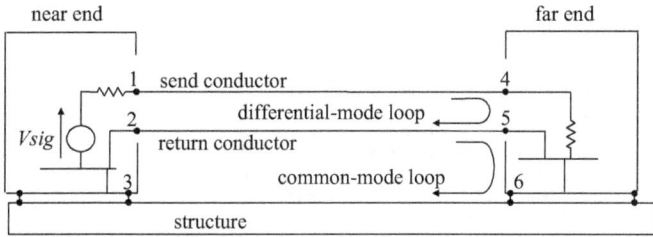

Figure 1.3.1 Signal link wiring.

1.3 Intra-system interference

1.3.1 The signal link

Having derived a method of simulating the coupling between three separate conductors, it becomes possible to relate the model of Figure 1.2.3 to actual hardware. Figure 1.3.1 illustrates the general method of transmitting an electrical signal from one unit of equipment to another.

Comparing Figure 1.3.1 with Figure 1.2.3 establishes a clear correlation. Terminals 1, 2, and 3 at the near end of the signal link can be correlated with terminals 1, 2, and 3 of the circuit model. Terminals 4, 5, and 6 bear a similar relationship.

1.3.2 Simulating the structure

The simplest way of simulating the structure is to treat it as a plane surface and employ the method of images to calculate the relevant parameter values. This technique involves treating the plane surface as a perfectly reflecting mirror, then using the properties of the image conductors to represent the effect of the actual surface as illustrated by Figure 1.3.2.

A set of four primitive equations is created to relate voltages on the four conductors (two real, two image; the structure is transparent) to the currents in those conductors. These can be simplified to two loop equations. Then the process described in section 1.2.3 is used to derive formulae for the circuit inductors and capacitors of the triple-T model.

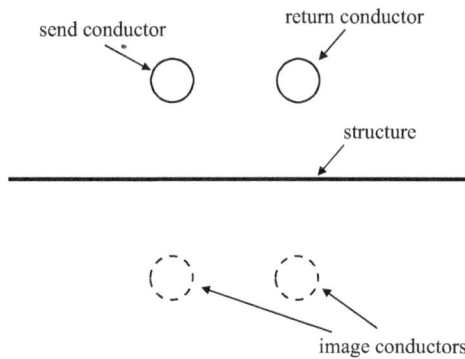

Figure 1.3.2 Simulating the structure.

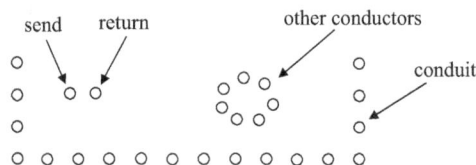

Figure 1.3.3   Using method of composite conductors.

A more accurate method of deriving values for the inductance and capacitance of the structure is to use the method of composite conductors, as shown by Figure 1.3.3. In this simulation, all other conductors in vicinity of the send and return conductors are assumed to form a single composite conductor; and the properties of that conductor are assigned to the structure.

As far as this method of simulation is concerned, it does not matter that the other signals in the system are interfering with each other. All that matters is how they affect the link-under-review and how the signal in the link under-review couples with the structure.

Chapter 3 derives computer programs for calculating values for the components of Figure 1.2.3, using data on the geometry of the structure and cables.

1.3.3 Equivalent circuits

One of the concepts inherent in circuit theory is that of the 'equivalent circuit'. Two inductors in series can be represented by a single inductor whose value is the sum of the two inductors it represents. Two capacitors in parallel can be represented by a single capacitor whose value is the sum of the original two. This means that a series of triple-T networks can be simulated by the single network of Figure 1.2.3. That is, a triple-T network can be used to simulate any signal link between two units of equipment, no matter how the cable is routed.

With a cable that follows a complex route along different parts of the structure, the task of calculating accurate values for the components of the model can become quite arduous. It is possible to make an initial estimate; but the more complex the route, the less confidence there is in the final result of the calculations.

This is where test equipment comes into the picture. Given a sure knowledge that the link can be represented by a triple-T network, it is not particularly difficult to devise a set of tests to measure the value of every component of the model. An LCR meter could be used. Other ways are described in Chapter 7.

1.3.4 Conducted emission

Adding components to the near and far ends of the triple-T model to represent the interface circuitry within the two units of equipment results in the model of Figure 1.3.4. The geometry of the signal link provides sufficient data to provide an initial estimate of the value of every component of the triple-T model.

Given this information, values for the four loop currents can be calculated. Current Ic_2 gives a measure of the common-mode current at the near end of the link. Calculating the

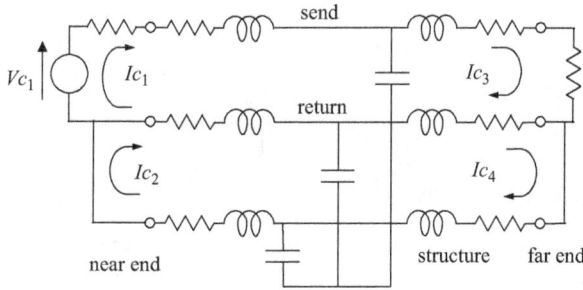

Figure 1.3.4   Circuit model of signal link.

ratio of current in the common-mode loop to the voltage in the differential-mode loop gives a value for the transfer admittance *YT*.

$$YT = \frac{Ic_2}{Vc_1}$$

Calculating the value of *YT* at a number of spot frequencies allows a graph to be created, relating transfer admittance to frequency. This graph gives a clear indication of the conducted emission characteristic of this particular link.

It is equally possible to create a graph which defines the frequency characteristic of the voltage developed from end to end along the structure.

1.3.5 Conducted susceptibility

Given knowledge of the voltages developed along the structure by a number of signal links, the principle of superposition can be used to define the total voltage developed along the structure.

The combined effect of several sources of interference can be represented by a single voltage source, Vc_2, in series with the structure. Figure 1.3.5 shows how the model can be

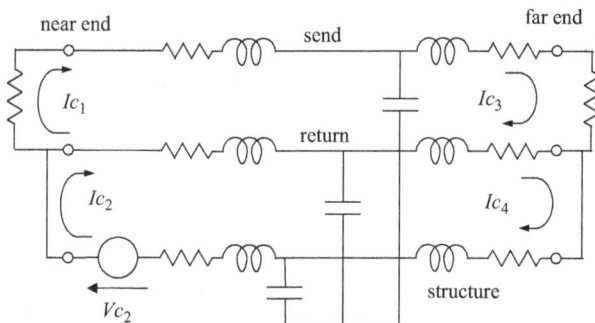

Figure 1.3.5   Circuit model for analysis of conducted susceptibility.

used to simulate the effect of this interference. Again, the transfer admittance characteristic can be used to define the susceptibility of the link:

$$ZT = \frac{Ic_1}{Vc_2}$$

1.3.6 Voltage transformer

Having established a method of simulating the effects of conducted emission and conducted susceptibility, the next step is to carry out bench tests of the signal link. Comparing the results of such tests with the simulation allows a check to be made on the accuracy of the model. But first, it is necessary to have the right sort of test equipment available.

It is reasonable to assume that anyone responsible for developing electronic equipment will already have available a number of items of general-purpose test equipment and that these items include a signal generator and an oscilloscope.

The first requirement is to be able to induce a fairly constant voltage over a wide range of frequencies into the loop-under-test without the need to physically break into the loop. This can be done by using a split-core toroid, similar to those used in EMC test houses. The second requirement is to minimize the impedance reflected into the loop-under-test by the test equipment. This can be done by winding several turns on the primary. Signal generators usually have an output impedance of 50 Ω. With a turns ratio of ten to one, the reflected impedance will be 0.5 Ω. The third requirement is to monitor the output voltage, since the load presented to the transformer can vary dramatically. The addition of a separate monitor turn caters for this requirement.

If the transformer is clamped round both conductors of the signal link, it will induce the same voltage into the send and return conductors. Net voltage induced into the differential-mode loop will be zero. The full voltage is induced into the common-mode loop however. Since current in the common-mode loop flows via the structure, the effect can be simulated by inserting a voltage source in series with the conductor representing the structure.

1.3.7 Current transformer

Current transformers are already in widespread use. However, it is useful to recap on the construction and use of such transformers when used to measure interference. The same core used with the voltage transformer is ideal for the current transformer. The requirement to reflect minimal impedance into the loop-under-test still applies. It is necessary to use a 50-Ω co-axial cable to link the transformer to the monitoring equipment. So a ten-turn secondary will reflect 0.5 Ω into the loop-under-test.

If a 50-Ω resistor is placed in parallel with the secondary winding, the transformer output can be simulated by a current source in parallel with 50 Ω. This can also be simulated by a voltage source in series with 50 Ω, the ideal configuration for interfacing with a co-axial cable.

1.3.8 Representative circuit model

Chapter 7 describes several tests which can be carried out on the signal link to determine the frequency response of the transfer admittance. When these are plotted on the same graph as that of the simulated response it is inevitable that there will be some deviation between the two curves.

However, it will always be possible to modify a few parameters which define the component values of the model to achieve close correlation between the two curves. The number of independent variables is much fewer than the number of components. Experience has shown that it does not take many iterations of the simulation to achieve this objective.

The end result of such an exercise is the creation of a circuit model which defines the coupling characteristics of the signal link. Such a model can be utilized in much the same way as one of the modules in the library of components available in most SPICE software packages.

Advantage can be taken of the fact that every parameter (R, L, C, and G) of the model is a function of length. Tests can be carried out on a long test rig at low frequencies and used to create a model of that rig. Knowledge of the length of that rig can be used to modify the component values in the model, to allow the modified model to simulate the behavior of a much shorter link. Tests at relatively low frequencies can be used to predict performance at much higher frequencies. Such an approach is analogous to the use of wind-tunnel tests on a small-scale model to predict the behavior of the actual aircraft.

This means that a circuit model for a 10-m line, tested at frequencies up to 20 MHz, can be used to create a model for a 1-m line, valid at frequencies up to 200 MHz. Care would need to be taken that the cross section of the test rig is constant over the entire length and that the components at the interfaces are suitable for use at the frequency at which the equipment will operate.

The method can be used to simulate the cross-coupling between two 50 mm lengths of printed circuit track, valid up to 4 GHz. Testing the accuracy of this particular model would require the use of sophisticated test equipment.

Most of the tests described in Chapter 7 are limited to the short-wave band. But this is only because the test equipment available was limited to 20 MHz.

The model itself is limited only by the assumptions inherent in the theory of transmission lines. That is, it is assumed that action and reaction between adjacent conductors is instantaneous. In practice, the upper frequency limit is that at which the wavelength is ten times the maximum dimension of any section of the assembly-under-review.

Section 7.7 shows that the technique of circuit modeling can be used to characterize small components such as capacitors, at frequencies over the range 200 kHz to 1 GHz.

1.4 Inter-system interference

1.4.1 Dipole model

Having developed a model which is not restricted to the simulation of conductors of short electrical length, the way is open to extend the approach to the analysis of antenna-mode coupling. A reading of the textbook analysis of the half-wave dipole reveals that a new component can be added to the existing set of primitive parameters: the radiation resistance.

When a sinusoidal voltage is applied between the two dipoles of the antenna and the frequency adjusted to coincide with the quarter-wave frequency of each dipole, the current distribution along each conductor follows a sinusoidal waveform. Maximum current flows at the center of the antenna while the current at the tips is zero. The current has to go somewhere; it is converted into electromagnetic radiation.

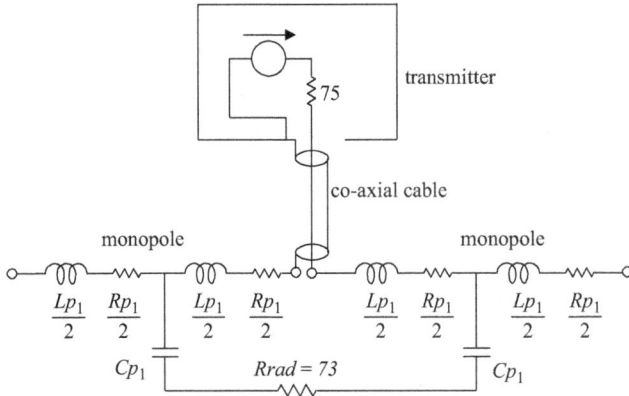

Note. This is a first order approximation only.

Figure 1.4.1   Circuit model of half-wave dipole.

A commonly accepted formula for the average radiated power Pt over a spherical surface is:

$$Pt = \frac{1}{2} \cdot Rrad \cdot Ip^2$$

where Ip is the peak amplitude of the current. The parameter $Rrad$ is not a resistor in the conventional sense of the word, but a mathematical constant derived from a complex process of integration which happens to have the dimensions of resistance. For a half-wave antenna the radiation resistance, $Rrad$, is 73 Ω.

Since the current distribution along a monopole of a half-wave dipole is identical to that along an open-circuit transmission line operating at the quarter-wave frequency and since the T-network model of Figure 1.2.6 simulates the behavior of the line, it is logical to assume that this model can also simulate the behavior of a monopole. This leads to the model of Figure 1.4.1, where Lp and Cp are the primitive inductance and primitive capacitance of a monopole.

Tests on a single length of isolated conductor, when it is acting like a dipole, are described in section 7.4. The fact that there is close correlation between the responses of the model and the hardware provides a high level of confidence in the soundness of the technique.

1.4.2 The virtual conductor

The common factor in every model developed for analysis of intra-system interference is that every conductor is represented by a T-network of inductors, resistors, and a capacitor. The concept can be extended further, to simulate the coupling between a twin-conductor cable and the environment.

The model of Figure 1.4.2 represents a configuration where a voltage source is applied to one conductor of a twin-core cable, at the mid-point along the length of the cable. This conductor acts as a transmitting dipole. The adjacent conductor acts as a receiving dipole.

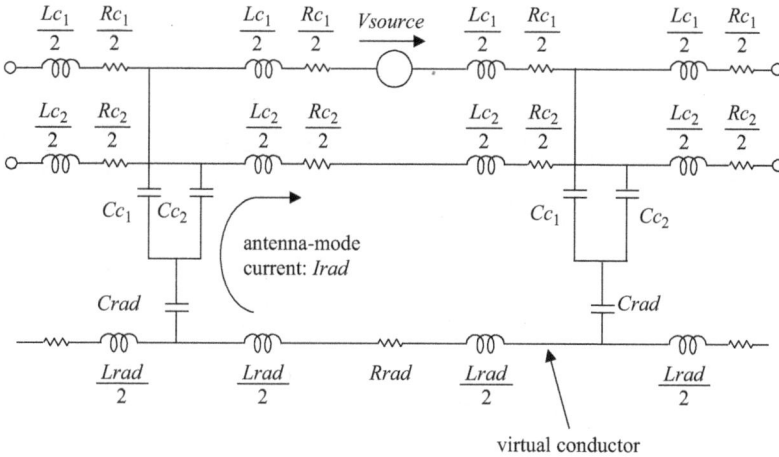

Figure 1.4.2   The virtual conductor.

Although most of the transmitted energy is picked up by the receiving antenna, a significant proportion constitutes radiated interference.

Analysis of the geometry of the twin-conductor cable provides formulae for the inductance $Lrad$ and capacitance $Crad$. The other components remain exactly the same as those of the transmission-line model of the cable.

Since the components $Lrad, Crad$, and $Rrad$ are also configured as a T-network, it is logical to give it the name 'virtual conductor', since it does not simulate an actual piece if hardware. It represents the effect of the environment.

The amplitude of the current radiated into the environment $Irad$ gives a measure of the radiated field strength. The maximum strength of the magnetic field H at a distance r is:

$$H = \frac{Irad}{2 \cdot \pi \cdot r}$$

This formula is recognisable as that derived by integrating the Biot-Savart equation along an infinitely long, straight conductor, as (2.2.1) would indicate. However, it can also be derived from the equations of antenna theory. Section 5.7 shows that it is related to the maximum power delivered to the environment by a wire pair routed along the structure. It can be used to predict the results of formal EMC testing of radiated emission of the equipment-under-review, by indicating the maximum strength of the signal which would be picked up by the monitor antenna.

1.4.3 The threat voltage

The model of Figure 1.4.2 can also be used to simulate the behavior of a twin conductor cable when it is exposed to an external field. The only change necessary is to move the location of the voltage source to that shown on Figure 1.4.3.

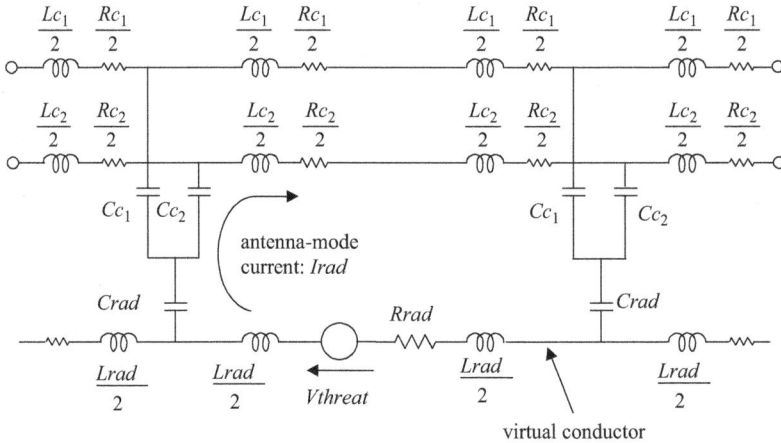

Figure 1.4.3   The threat voltage.

The amplitude of the threat voltage *Vthreat* can be calculated by integrating the value of the electric field strength E over the length l of the cable. In the case where $l = \frac{\lambda}{4}$:

$$Vthreat = \frac{\lambda}{\pi} \cdot E$$

Vthreat is the maximum voltage which can be injected into the antenna-mode loop by an electromagnetic field of electrical field strength E.

Analysis of this model allows the differential-mode current caused by an external field to be calculated.

1.4.4 Worst-case analysis

By assuming that optimum coupling exists between conductors, the end result of any analysis is a calculation of the maximum level of interference. By providing an accurate simulation of the response over the frequency range which includes that of the quarter-wave frequency and the half-wave frequency, the analysis guarantees that the frequency at which interference is a maximum will be predicted.

Skin effect ensures that the amplitude of the interference current reduces with frequency. Also, the amplitude of the threat voltage due to an external field of constant power density decreases with frequency, as illustrated by Figure 5.3.6. These factors ensure that the first resonant peak is always the highest; at least, in a design that is not intended to act as an antenna.

These factors enable the designer to focus on the performance of each signal link at those frequencies where EMI problems are most likely to arise.

By invoking the simple relationships of (5.3.6) and (5.7.3), which define maximum levels, the designer avoids the need to consider the field distribution pattern in the region of the signal link.

By far the majority of signal links in a system are carried by cables which are routed over a conducting structure. If it is assumed that the shielding effectiveness of the structure is zero, then the maximum power which can be delivered to the line can be represented by a voltage source *Vthreat* in series with *Rrad*. One way of simulating the effect of the threat environment is to insert both these components in series with the structure. Section 5.5 illustrates how the response of such a signal link can be analyzed.

Similarly, if the structure provides no shielding, then current in the structure can be regarded as the source of interference. Since this conductor carries the common-mode current, then knowledge of the amplitude of the common-mode current will provide a first estimate of the level of emission which can occur.

1.5 Transients

Transients are an ever-present source of glitches in electronic systems. Sources can be relays, switches, motors, and power supplies. They can easily corrupt the data streams handled by microprocessors. Depending on the criticality of the processing circuitry, such events could be inconsequential, annoying, dangerous, or catastrophic.

Since most signal processing is now carried out by digital signals, it is essential that this topic be included in any analysis of interference. The lumped parameter models developed for frequency response analysis can also be used for analysis in the time domain. The textbook approach to such analysis is to invoke the use of Fourier transforms, Laplace transforms, and even more complex techniques. The approach adopted here follows the example of SPICE programs and uses time-step analysis.

However, the problems encountered with the analysis of the frequency response of a transmission line also appear in the use of transient analysis. It takes a finite time for a signal to traverse from one end of the line to another. With frequency analysis, it was possible to transform the lumped parameter model into a distributed parameter model. With transient analysis, another solution is called for.

In concept, the solution is much simpler; use the computer memory to store the signal applied to the near end of the line for a fixed number of time steps before delivering it to the terminals at the far end.

Books on electromagnetic theory introduce the concept of partial currents and partial voltages to explain the behavior of transient signals at the interface between cables and equipment terminations. Incident current flows toward the interface; some of it is absorbed by the equipment and some is reflected back down the line. The total current at any section of the cable is the sum of the incident and reflected currents at that location. Section 6.2 describes the phenomenon in more detail.

The reflected signal is also delayed a fixed number of time steps before appearing back at the near end. A simple program to simulate the propagation of incident and reflected currents is also described in section 6.2.

An experiment was carried out by applying a square wave to one end of a twin-core cable via a 5-Ω resistance, leaving the far end open-circuit and monitoring the current flowing in the line. Since the impedance at the near end was much lower than the characteristic impedance of the line, it was expected that there would be multiple reflections. Attention was focused on the response to the leading edge; that is, on the response to a step input.

Textbook theory predicted that the incident current would be inverted by the open-circuit terminals at the far end and reflected straight back to the near end. At the near end, the inverted current would almost double in amplitude. The response to a step input was expected to be a square waveform of slowly diminishing amplitude. It wasn't.

During the time the leading edge of the pulse took to make the return trip, the current remained steady. This was in accordance with classical theory. However, the trailing edge was not sharp, as one would expect. It took the form of an exponential decay. This process continued and the waveform underwent a metamorphism, rapidly changing shape into a sine wave.

A trial-and-error process ensued in an effort to create a circuit model which replicated the observed waveform and which could be explained in terms of electromagnetic theory. It was eventually reasoned that, as the step waveform propagated along the line, it left a residual charge on the conductors and that this charge gradually decayed away via current flow into the environment. This charging current could be simulated.

In addition, it proved to be possible to measure and simulate the proportion of the current which departed into the environment. Results from these transient tests could also be correlated with observations and analysis of tests which had been made using sinusoidal waveforms. Section 6.6 explains the reasoning.

1.6 The importance of testing

Chapters 2 to 6 are essentially concerned with methods of developing circuit models which replicate the relationships defined by electromagnetic theory. Since electromagnetic theory relates the electrical parameters to geometric parameters, the result is the establishment of a theoretical relationship between each circuit model and the hardware it represents.

Electrical tests establish a clear relationship between the response of the model and the response of signal link it represents. By correlating the response of the model with that of the test setup, the circuit model can be used to define the characteristics of the link. Electrical tests can be carried out on the configuration-under-review and the results used to define the electromagnetic coupling characteristics of that configuration. The model can be described as a 'representative circuit model'.

Testing and analysis can be carried out concurrently. This eliminates the need for a 'try-it-and-see' approach.

In view of the several chapters of mathematical analysis which precede the introduction of the subject of testing, readers can be led to the belief that theoretical analysis needs to be carried out before testing can begin. This is not strictly true. Testing precedes analysis.

From the point of view of the author, the starting point was the observation of annoying glitches in electronic systems. The sources were easily identifiable, but the coupling mechanisms did not seem to be amenable to analysis. Circuit models were created to simulate the observed phenomena and deviations noted between simulation and observation. These deviations led to speculation as to the probable causes. When a review of electromagnetic effects revealed a plausible explanation, it became possible to establish a firm link between electromagnetic theory and the circuit model. The model was refined and further tests carried out – to reveal other deviations.

For example, tests to characterize a twin-conductor cable reveal the fact that antenna-mode current propagates at a higher velocity than differential-mode current. Section 7.5 describes how this can be demonstrated.

In hindsight, the phenomena can be explained by the fact that the antenna-mode waves propagate mainly in the air, whereas the differential-mode waves are mostly in the cable insulation. From the viewpoint of field propagation, this explanation comes quite readily to mind. In this case, however, the evidence comes from observations on the behavior of currents and voltages rather than the analysis of H-field and E-fields.

To someone who thinks in terms of system function, this is something of a revelation. Although the electromagnetic fields propagate in the insulating medium, the currents which create them flow in the conductors. The antenna-mode current and the differential-mode current are separate entities, just as surely as the incident and reflected currents in transmission lines are independent of each other.

This conclusion is supported by the analysis of the transient tests described in section 7.6. In the setup of Figure 7.6.1, a step voltage is injected into the signal conductor of a conductor pair. This creates an antenna-mode current which propagates in the same direction down the pair of conductors, from near end to far end. Current in the signal conductor also causes a current to flow back along the return conductor.

Just behind the leading edge of the antenna-mode current step, current is flowing in both directions along the return conductor. Current flowing in opposite directions in a pair of conductors constitutes differential-mode current. The leading edge of the differential-mode step follows behind the leading edge of the antenna-mode current, at a lower velocity. From the results of the tests of section 7.5, the two velocities are:

- Antenna-mode current: 230 m/μs
- Differential-mode current: 170 m/μs

That is, the use of circuit modelling techniques provides the user with an ever-improving understanding of the mechanisms involved in interference coupling.

The key point here is that progress can only be made by engineers who carry out measurements for themselves.

An equally important aspect of testing is that it provides firm evidence of the validity of the model. Being able to verify theoretical results by comparing them with other theoretical results is encouraging; but it does not provide the confidence achievable when practical measurements correlate with theoretical predictions.

1.7 Practical design techniques

Since the concepts of the 'equipotential ground', the 'single-point reference', and the advice to 'avoid earth loops' have acquired universal acceptance as critically important guidelines, the first three sections of Chapter 8 are devoted to an explanation as to why they are so misleading. The remaining sections identify many of the techniques employed by generations of designers to improve circuit immunity and reduce the level of unwanted emissions.

More than anything else, the process of testing and analyzing the different mechanisms involved in interference coupling leads to a much improved understanding of these mechanisms. This leads to the ability to assess any given wiring assembly and any given interface circuit in the light of the underlying physics. Given this ability, it is possible to identify groups of techniques which have the same objective. These are:

- Common-mode rejection
- Differential-mode damping
- Common-mode damping
- Shielding

The design of the interface circuitry and system shielding is critically dependent on an understanding of cable-coupling mechanisms. Every circuit described in Chapter 8 includes a definition of the complete signal link; there are no loose ends where the circuit terminates at a set of plug pins or socket inserts. This allows the design of the interface circuitry to be related to the coupling characteristics of the cable. For example, the performance characteristics of 'grounded' and 'floating' configurations are compared.

Packing density on printed circuit boards (pcb) precludes the option of carrying out a detailed analysis of every signal link on a board. Even so, anyone who understands the method of modeling will also understand the interference coupling mechanisms, and will avoid most of the obvious errors.

It would be good practice to fit a buffer circuit at each pcb interface to decouple the signals on the interconnecting cables from the multiple branches and stubs that must exist on the board. A variety of such buffer circuits is described in Chapter 8. Any interference experienced would then be due to internal coupling. A well-designed board would not experience such coupling; and this could be confirmed the first time the board is functionally checked.

Since there are books which provide excellent advice on the detailed design of pcbs, there is no point in duplicating their content. One book which stands out by virtue of the wealth of detailed design information it provides is 'EMC Design Techniques for Electronic Engineers' [1.10]. Another book which is well worth purchasing is 'Introduction to Electromagnetic Compatibility' [1.11], since it provides detailed information on the theory underlying electromagnetic field propagation and relates this to practical design problems. The chapters on shielding and electrostatic discharge also provide valuable information.

1.8 System design

1.8.1 Guidelines

The set of guidelines listed below is based on deductions derived from the analysis of electromagnetic coupling. If agreed by members of the design team at the outset of a project, it will avoid the pitfalls suffered by many previous electronic systems.

1. The best use of the structure is as a conducting shield. It should not be used as a convenient return path for any of the signals or power supplies in the system. Any current caused to flow in the structure will be a source of interference.
2. Assign a 'return' conductor to every 'send' conductor and keep the two conductors as close together as possible along the entire route from source to load.
3. Encourage the use of ground loops, since these enhance the shielding properties of the system.

4. Design the interface circuitry to enhance the balance between the current in the 'send' conductor and the current in the 'return' conductor. Ideally, the net current flow through any cable section should be zero.
5. Treat every signal link as a transmission line and use resistors similar in value to the characteristic impedance in the interface circuitry; as far as is practicable.
6. Implement common-mode damping on critical signal links.
7. Test and analyze critical signal links.

1.8.2 Top-down approach

The process described in the following pages starts off by defining the basic building blocks of all circuit models and then develops the models to a point where they can reliably simulate the performance of actual signal links. These signal links can then be regarded as basic building blocks in the design of complete systems.

The modular approach to system design is already well established, in the use of block diagrams. A block diagram of the complete system is defined at the beginning of a project with each block identifying a unit of equipment, or even a set of related equipment units, and with lines between the blocks used to identify the signal and power lines. Each block can then be treated as a separate entity, and its function represented by a separate block diagram. At the lowest level the functional block is defined in terms of a circuit diagram such as a logic gate, a J-K flip flop, a band-pass filter, or a buffer circuit.

By incorporating the models of the signal links into the process, it becomes possible to invoke the top-down approach to the analysis of the EMC of the system. Section 9.2 describes the relationships between the different types of diagram.

1.8.3 Formal EMC requirements

Any proper outline of the formal EMC requirements would take many chapters to describe, and is way beyond the scope of this book. However, it is still necessary to establish some sort of relationship between these requirements and the actual performance of the system.

Susceptibility requirements can be analyzed by defining the threat environment in terms of a graph relating the maximum amplitude of the external field strength to frequency. Using the circuit model of the link-under-review the response of the differential-mode current to the external field can be compared directly with the characteristics of the actual signal being transmitted. section 9.4 provides an example.

Emission requirements can be analyzed by simulating the signal carried by the link-under-review and applying this signal to the differential-mode input of the model. The frequency response of current in the common-mode mode loop can then be determined. In the worst-case situation, this is the current delivered to the environment. The frequency response of the common-mode current can be compared directly with the maximum acceptable limits defined by the formal requirements.

For both susceptibility and emission, this analysis is based on the assumption that the shielding effectiveness of the structure is zero. Section 8.7 gives some guidance on methods of estimating the shielding effectiveness.

Lumped parameter models

A review of the method used to calculate circuit component values for three-phase power lines leads to the identification of the first step; the derivation of a formula relating the capacitance of an isolated conductor to its length and radius. This formula can be regarded as a basic building block from which the capacitive parameters of multi-conductor assemblies can be derived.

In computer terminology, a low-level object or operation from which higher-level, more complex objects or operations can be constructed is termed a *primitive*. So it seems reasonable to use the term *primitive capacitance* to identify the basic relationship.

The first equation quoted in section 2.1 can be found in textbooks on electromagnetic theory [2.1]. Although the derivation involves some complex integration, the end result is a formula which defines the primitive capacitance, $Cp_{i,j}$. This relates the voltage on conductor i to the charge on conductor j.

Primitive inductance, $Lp_{i,j}$, can be regarded as the twin of primitive capacitance. In this case, textbooks can be found which outline the derivation process [2.2]. However, these earlier derivations do not include the contribution made by internal linkages. Nor do they cater for the fact that in any circuit analysis the current must flow in a complete loop. The derivation provided by section 2.2 includes these considerations.

In these earlier textbooks, the name given to the derived parameter is 'partial inductance'. However, in the analysis of Chapter 3, an array of elemental conductors is used to represent a composite conductor. In this treatment, the *partial parameters* associated with each composite conductor are derived from a collection of *primitive parameters*. With any computation process, it is necessary to distinguish between the two types. So they have been given different names.

Section 2.3 identifies the duality between primitive inductance and primitive capacitance. The square root of the product of these two parameters is the propagation delay; the time it takes for a transient pulse to propagate along the length of a conductor. *Primitive parameters* define the properties of conductors when they are acting as antennae.

Section 2.4 describes how primitive parameters can be combined to derive the capacitance between a pair of parallel conductors, and the inductance of a loop formed when the terminals at the far end are connected. *Loop parameters* are those which can be measured with electronic test equipment.

With EMC analysis, an important requirement is to maintain visibility of the voltage developed along each conductor as well as the voltages appearing between conductor terminals. Section 2.5 derives component values for a circuit model which maintains a one-to-one correlation between each conductor and its inductive and capacitive properties. *Circuit parameters* are those which are used in circuit diagrams.

The resistance of the conductors is also an important factor in determining the performance of any cable. Since this parameter can vary due to skin effect, an equation is derived which relates conductor resistance to frequency.

Each conductor of a two-conductor cable can be represented as a T-network, where each horizontal branch contains an inductor and a resistor and the vertical branch is formed by a single capacitor. The circuit model of the conductor pair forms a bridge network. Duality between inductors and capacitors means that the bridge is balanced. Voltage across each capacitor is balanced by a voltage across its related inductor, whatever the frequency. Section 2.6 identifies the relationship and indicates that this property can be exploited.

Section 2.7 describes a systematic process which derives relationships between the geometry of a three-conductor cable and the component values of the circuit model which simulates its performance. This process starts with a set of three primitive equations relating currents and voltages of the assembly when it is acting as an antenna. Since signals in an electrical system are carried by loop currents, it is possible to derive two loop equations.

At this point, it is possible to create a circuit model which creates a similar pair of circuit equations. Correlating the parameters of the circuit equations with those of the loop equations allows a relationship to be established with the loop parameters. Since the loop parameters have been derived from primitives, it is possible to define the circuit components in terms of the primitive formulae.

As well as enabling component values to be calculated, the process identifies the essential difference between electromagnetic field theory and circuit theory. The former theory recognizes the fact that voltage of a conductor is affected by the current in every conductor. The latter theory assumes that there is a one-to-one correlation between the current in a circuit branch and the voltage across that branch. This results in a dramatic simplification of the mathematics, and is probably the main reason why circuit models are so useful.

A desirable characteristic of any signal link is that the effect of other signals in the system is minimal. Section 2.8 uses the formulae derived in the previous section to define the most significant feature of the interconnecting cable. The signal and return conductors should be as close together as possible. This ensures maximum magnetic coupling and maximum electric coupling between these two conductors, and minimizes the coupling between this conductor pair and other conductors in the vicinity. In turn, this minimizes any common-mode current created by the signal link and reduces the susceptibility of the link to external interference.

Section 2.9 identifies the most significant parameter associated with EMC; the *Transfer Admittance*. This is the ratio of the current in the victim loop to the voltage in the culprit loop. It is shown that this parameter defines both the conducted susceptibility and the conducted emission.

The general circuit model of the three-conductor assembly can be applied to simulate the coupling between the differential-mode signal carried by a co-axial cable and a common-mode signal in the loop formed by structure and screen. Section 2.10 shows that for an ideal co-axial cable the only coupling parameter is the resistance of the screen.

However, the screens of most co-axial cables are braided. At higher frequencies a number of capacitive and inductive effects due to the screen's construction come into play and dominate the effect that is caused by screen resistance. The effect can be simulated by representing the screen as a T-network, with very small values assigned to the reactive components. These values can be derived from manufacturer's data on transfer impedance, or from tests on a cable sample.

Analyzing the coupling between two conductors over a ground plane is a problem faced by most designers at one time or another. The method of images can be used to derive formulae for the components of the three-conductor model. The formulation is described in section 2.11.

A basic assumption inherent in the textbook theory of circuits is that resistors, inductors, capacitors, voltage sources, and current sources are 'lumped parameters'; that is, they are discrete devices which link two nodes of a network. Another assumption is that action and reaction are instantaneous everywhere in the network. Since conventional circuit theory does not cater for the fact that it takes time for current to propagate along a conductor, its use is limited to low-frequency applications, where the length of the cable is less than one-tenth the wavelength of the signal.

This limitation applies to the models derived in this chapter and in Chapter 3. Subsequent chapters describe how this restriction can be overcome.

2.1 Primitive capacitance

Consider a short length, l, of a single conductor of radius $r_{1,1}$ and assume that there is a line charge along that length. If the charge density is ρ, then the intensity of the electric field at a point $P(r,z)$ can be calculated. Figure 2.1.1 illustrates the relationships.

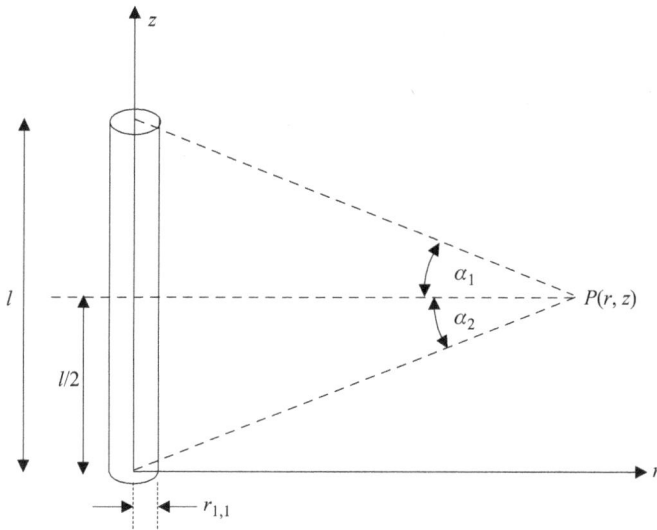

Figure 2.1.1   Electric field at a point.

The radial component of the electric field strength at the point P is [2.1]:

$$E_r = \frac{\rho}{4 \cdot \pi \cdot \varepsilon} \cdot \frac{1}{r} \cdot [\sin(\alpha_1) + \sin(\alpha_2)] \quad \text{(V/m)} \tag{2.1.1}$$

where ε is the permittivity.

Since the point P lies on the plane intersecting the mid-point of the conductor, α_1 is equal to α_2. Re-writing (2.1.1) in terms of radial and axial components gives:

$$E_r = \frac{\rho}{2 \cdot \pi \cdot \varepsilon} \cdot \frac{l/2}{r \cdot \sqrt{(l/2)^2 + r^2}} \tag{2.1.2}$$

At all points on the surface defined by the equation $z = \frac{l}{2}$ the E_z component of the electric field is zero. If unit charge were to follow a path along this surface from infinity to the radius $r_{1,1}$ then no axial force would be involved; all the electric forces would be radial. The work done on conductor 1 due to the charge on conductor 1 would be:

$$Vp_{1,1} = \frac{\rho}{2 \cdot \pi \cdot \varepsilon} \cdot \int\limits_{r_{1,1}}^{\infty} \frac{l/2}{r \cdot \sqrt{(l/2)^2 + r^2}} \cdot dr \quad \text{(V)} \tag{2.1.3}$$

Performing the integration:

$$Vp_{1,1} = \frac{\rho}{2 \cdot \pi \cdot \varepsilon} \cdot \ln \left[\frac{l/2 + \sqrt{(l/2)^2 + r_{1,1}^2}}{r_{1,1}} \right] \tag{2.1.4}$$

In situations where $r_{1,1} \ll l$:

$$Vp_{1,1} = \frac{\rho}{2 \cdot \pi \cdot \varepsilon} \cdot \ln \left(\frac{l}{r_{1,1}} \right) \tag{2.1.5}$$

The capacitance of the conductor is the ratio of charge to voltage. That is:

$$Cp_{1,1} = \frac{\rho \cdot l}{Vp_{1,1}} \quad \text{(F)} \tag{2.1.6}$$

Substituting for the voltage:

$$Cp_{1,1} = \frac{2 \cdot \pi \cdot \varepsilon \cdot l}{\ln \left(\frac{l}{r_{1,1}} \right)} \tag{2.1.7}$$

If a second conductor were to be placed parallel to conductor 1, as shown on Figure 2.1.2, then the energy level at the axis of conductor 2 due to the charge on conductor 1 would be:

$$Vp_{2,1} = \frac{\rho}{2 \cdot \pi \cdot \varepsilon} \cdot \ln \left(\frac{l}{r_{2,1}} \right) \tag{2.1.8}$$

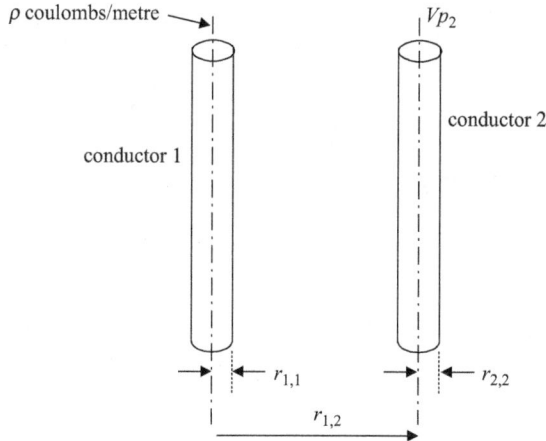

Figure 2.1.2   Voltage on conductor 2 due to charge on conductor 1.

and the primitive capacitance would be:

$$Cp_{2,1} = \frac{2 \cdot \pi \cdot \varepsilon \cdot l}{\ln\left(\dfrac{l}{r_{2,1}}\right)} \tag{2.1.9}$$

The general formula for primitive capacitance would be:

$$Cp_{i,j} = \frac{2 \cdot \pi \cdot \varepsilon \cdot l}{\ln\left(\dfrac{l}{r_{i,j}}\right)} \tag{2.1.10}$$

where i and j are integers which identify the conductors. If $j = i$ then the value is the primitive capacitance of conductor i due to the charge it carries.

This formulation depends on the assumption that the length is much greater than any radius and that the charge is uniformly distributed along the length. It has the practical advantage that it is simple. It works well in practice, in that it can be used as a building block in the construction of all circuit models. Moreover, any inaccuracies in the derivation are completely masked by the uncertainty in the value of the relative permittivity.

Capacitance values are a function of length, radii, and permittivity. Length and radii can be determined simply by measuring the physical parameters. Permittivity is a combination of two parameters:

$$\varepsilon = \varepsilon_o \cdot \varepsilon_r \quad (\text{F/m}) \tag{2.1.11}$$

where ε_o is defined as the permittivity of free space, 8.854 pF/m, and ε_r is the relative permittivity.

In classical theory, the value of the relative permittivity is different for each type of material. When dealing with EMC problems, a variety of insulating materials are present in the vicinity of the conductor. Rather than taking the individual relative permittivity of each material and using data on the cross section of the assembly to calculate the overall effect, an approximate 'overall' or 'effective' value can be used. In this book ε_r is defined as the effective value of the relative permittivity. Its value can be determined by tests such as those described in sections 7.4 and 7.5.

2.2 Primitive inductance

The derivation of primitive inductance starts off with basically the same configuration as with primitive capacitance. Figure 2.2.1 illustrates a short section of conductor of radius $r_{1,1}$ with a constant current Ip flowing in the z-direction.

The distribution of magnetic field will be the same as that due to a filamentary current along the axis of the conductor. The magnetic field strength H at point $P(r,z)$ is [2.3]:

$$H = \frac{Ip}{4 \cdot \pi \cdot r} \cdot [\sin(\alpha_1) + \sin(\alpha_2)] \quad (\text{T}) \tag{2.2.1}$$

Converting this equation to one which uses the co-ordinates z and r gives:

$$H = \frac{Ip}{4 \cdot \pi \cdot r} \cdot \left[\frac{l-z}{\sqrt{(l-z)^2 + r^2}} + \frac{z}{\sqrt{z^2 + r^2}} \right] \tag{2.2.2}$$

To determine the inductance of this section of conductor, it is necessary to calculate the value of the total flux ϕ which passes through a rectangular strip extending from the surface of the conductor to a very large distance. This means that the flux density will need to be integrated over the defined area. Since:

$$B = \mu \cdot H \quad (\text{Wb/m}^2) \tag{2.2.3}$$

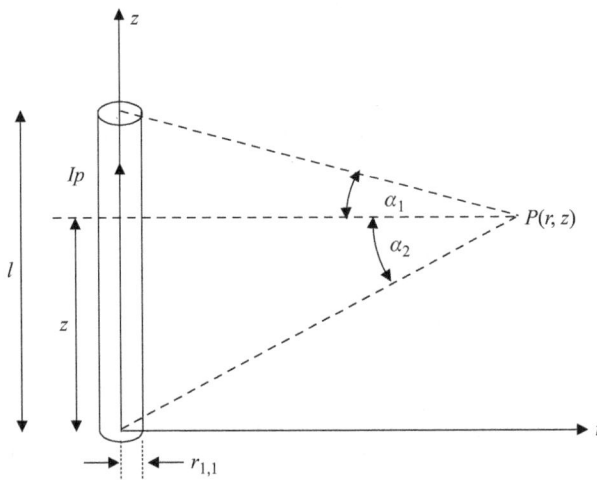

Figure 2.2.1   Magnetic field at a point.

and

$$\phi = \int B \cdot ds \quad (\text{Wb}) \tag{2.2.4}$$

where s is the surface. Then:

$$\phi = \frac{\mu \cdot Ip}{4 \cdot \pi} \cdot \int\limits_{r_{1,1}}^{\infty} \int\limits_{0}^{l} \frac{1}{r} \cdot \left(\frac{l-z}{\sqrt{(l-z)^2 + r^2}} + \frac{z}{\sqrt{z^2 + r^2}} \right) \cdot dz \cdot dr$$

$$= \frac{\mu \cdot Ip}{2 \cdot \pi} \cdot \int\limits_{r_{1,1}}^{\infty} \frac{\sqrt{l^2 + r^2} - r}{r} \cdot dr$$

$$= \frac{\mu \cdot Ip \cdot l}{2 \cdot \pi} \cdot \left[\ln\left(\frac{l + \sqrt{l^2 + r_{1,1}^2}}{r_{1,1}} \right) + \left(\frac{r_{1,1} - \sqrt{l^2 + r_{1,1}^2}}{l} \right) \right]$$

The above derivation can be found in at least one textbook [2.2]. Inductance is defined as the ratio of magnetic flux to the current creating that flux. Since the flux ϕ is external to the conductor:

$$Lexternal = \frac{\phi}{Ip} \quad (\text{H}) \tag{2.2.5}$$

This leads to:

$$Lexternal = \frac{\mu \cdot l}{2 \cdot \pi} \cdot \left[\ln\left(\frac{l + \sqrt{l^2 + r_{1,1}^2}}{r_{1,1}} \right) + \left(\frac{r_{1,1} - \sqrt{l^2 + r_{1,1}^2}}{l} \right) \right] \tag{2.2.6}$$

There are also internal flux linkages.

$$Linternal = \frac{\mu \cdot l}{2 \cdot \pi} \cdot \frac{1}{4} \quad (\text{H}) \tag{2.2.7}$$

Hence, the total inductance is the sum of *Lexternal* and *Linternal*:

$$Lp_{1,1} = \frac{\mu \cdot l}{2 \cdot \pi} \cdot \left[\ln\left(\frac{l + \sqrt{l^2 + r_{1,1}^2}}{r_{1,1}} \right) + \left(\frac{r_{1,1} - \sqrt{l^2 + r_{1,1}^2}}{l} \right) + \frac{1}{4} \right] \tag{2.2.8}$$

If it is assumed that $l \gg r_{1,1}$, then:

$$Lp_{1,1} = \frac{\mu \cdot l}{2 \cdot \pi} \cdot \left[\ln\left(\frac{2 \cdot l}{r_{1,1}} \right) - 1 + \frac{1}{4} \right]$$

$$= \frac{\mu \cdot l}{2 \cdot \pi} \cdot \left[\ln\left(\frac{l}{r_{1,1}} \right) + \ln 2 - 1 + \frac{1}{4} \right]$$

$$= \frac{\mu \cdot l}{2 \cdot \pi} \cdot \left[\ln\left(\frac{l}{r_{1,1}} \right) - 0.057 \right]$$

To a first approximation:

$$Lp_{1,1} = \frac{\mu \cdot l}{2 \cdot \pi} \cdot \ln\left(\frac{l}{r_{1,1}}\right) \tag{2.2.9}$$

Similar reasoning applies to a two-conductor assembly, as shown on Figure 2.2.2. It is assumed that there is a current Ip flowing in conductor 1. From (2.2.6), the inductance due to the flux linking conductor 2 due to current in conductor 1 is:

$$Lmutual = \frac{\mu \cdot l}{2 \cdot \pi} \cdot \left[\ln\left(\frac{l + \sqrt{l^2 + r_{2,1}{}^2}}{r_{2,1}}\right) + \left(\frac{r_{2,1} - \sqrt{l^2 + r_{2,1}{}^2}}{l}\right) \right] \tag{2.2.10}$$

When the pair of conductors is analyzed as part of a complete loop, it must be assumed that the return path for the current Ip is along conductor 2. Magnetic flux due to current Ip in conductor 1 will link the region between infinity and the outer surface of conductor 2. It will also counteract the internal linkages of the return current Ip in conductor 2.

Hence, it is necessary to add the component $Linternal$ to the value of the mutual inductance.

This leads to:

$$L_{2,1} = Lmutual + Linternal$$

$$= \frac{\mu \cdot l}{2 \cdot \pi} \cdot \left[\ln\left(\frac{l + \sqrt{l^2 + r_{2,1}{}^2}}{r_{2,1}}\right) + \left(\frac{r_{2,1} - \sqrt{l^2 + r_{2,1}{}^2}}{l}\right) + \frac{1}{4} \right] \tag{2.2.11}$$

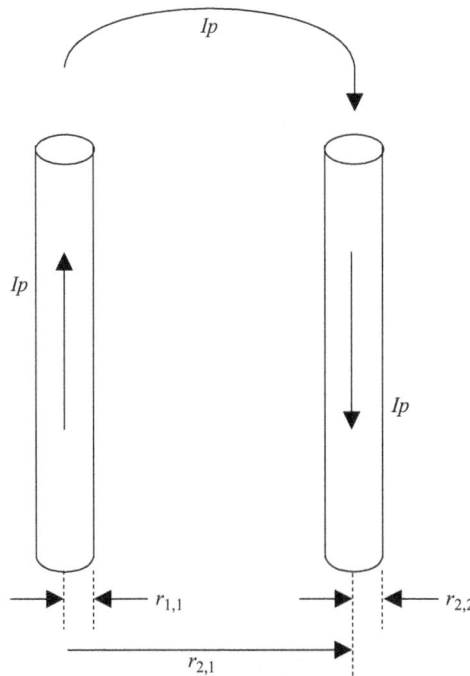

Figure 2.2.2   Current in a two-conductor assembly.

If $l \gg r_{2,1}$ then, to a first approximation:

$$Lp_{2,1} = \frac{\mu \cdot l}{2 \cdot \pi} \cdot \ln\left(\frac{l}{r_{2,1}}\right) \qquad (2.2.12)$$

Since $r_{1,2} = r_{2,1}$ then $Lp_{1,2} = Lp_{2,1}$.

The general formula for primitive inductance can be defined as:

$$Lp_{i,j} = \frac{\mu_o \cdot \mu_r \cdot l}{2 \cdot \pi} \cdot \ln\left(\frac{l}{r_{i,j}}\right) \quad \text{(H)} \qquad (2.2.13)$$

where i and j are integers which identify the conductors.

As with the derivation of primitive capacitance, this formulation depends on the assumption that the length is much greater than any of the radii. It is also assumed that the current does not vary along the length. The fact that the formulation of (2.2.13) lacks precision is compensated by the fact that it is simple. From an engineering point of view, it is accurate enough for its intended purpose; to act as a building block for circuit models. The tests described in sections 7.4 and 7.5 demonstrate that the value of the measured inductance was the same as that predicted by measurements of the geometry of the assembly-under-test.

It is also worth noting that any inaccuracies in the formulae for primitive inductance (and primitive capacitance) disappear when the loop parameters and circuit parameters are formulated. Section 2.4 explains why.

The permeability μ is a combination of two parameters:

$$\mu = \mu_o \cdot \mu_r \quad \text{(H/m)} \qquad (2.2.14)$$

where $\mu_o = 4 \cdot \pi \cdot 10^{-7}$ H/m. The parameter μ_r is a pure number, similar in concept to ε_r. It can be defined here as the effective value of the relative permeability of the loop-under-review.

In this book, it is assumed that the value of μ_r is unity because in the majority of applications there is no magnetic material in the cable assembly. In those situations where magnetic material is present, the value of μ_r can be established by referring to data on the properties of materials, or by carrying out tests of the assembly-under-review.

Such tests would involve measurements of the frequency response of the current in a short-circuited loop. A circuit model of the assembly can be created by using data on the physical construction. This will produce a similar curve. The initial slope of both frequency response curves will be -20 dB per decade. By modifying the value of μ_r in the model, both curves can be made to coincide over this range. The value of the relative permeability of the loop-under-test will be that of the model.

The test described in section 7.5 illustrates the process used to measure ε_r, using an open-circuit line. If both ends of the line are short-circuited, the same process can be used to measure μ_r.

2.3 Duality of L and C

In general terms, the primitive components can be defined as:

$$Cp_{i,j} = \frac{2 \cdot \pi \cdot \varepsilon_o \cdot \varepsilon_o \cdot l}{\ln\left(\frac{l}{r_{i,j}}\right)} \quad \text{(F)} \tag{2.3.1}$$

and

$$Lp_{i,j} = \frac{\mu_o \cdot \mu_r \cdot l}{2 \cdot \pi} \cdot \ln\left(\frac{l}{r_{i,j}}\right) \quad \text{(H)} \tag{2.3.2}$$

where i and j are integers which identify the conductors. For example, $Lp_{i,j}$ gives the value of the primitive inductance of conductor i due to current in conductor j. If $j = i$ then the value is the primitive inductance of conductor i due to the current it carries.

Since $r_{i,j} = r_{j,i}$, then $Lp_{i,j} = Lp_{j,i}$ and $Cp_{i,j} = Cp_{j,i}$.

Multiplying gives:

$$Lp_{i,j} \cdot Cp_{i,j} = \mu_o \cdot \mu_r \cdot \varepsilon_o \cdot \varepsilon_r \cdot l^2 \tag{2.3.3}$$

From electromagnetic theory:

$$\sqrt{\mu_o \cdot \mu_r \cdot \varepsilon_o \cdot \varepsilon_r} = \frac{1}{v} \tag{2.3.4}$$

and

$$\sqrt{\mu_o \cdot \varepsilon_o} = \frac{1}{c} \tag{2.3.5}$$

where c is the velocity of light in vacuum and v is the velocity of propagation of the electromagnetic field. Also

$$v = \lambda \cdot f \quad \text{(m/s)} \tag{2.3.6}$$

where λ is the wavelength and f is the frequency. At the quarter-wave frequency, fq, of a monopole antenna, the current amplitude will peak. If the wavelength at this frequency is λq. then the length of the cable will be:

$$l = \frac{\lambda q}{4} \quad \text{(m)} \tag{2.3.7}$$

Taking the square root of (2.3.3) and using (2.3.4) to substitute for $\sqrt{\mu_o \cdot \mu_r \cdot \varepsilon_o \cdot \varepsilon_r}$ gives:

$$\sqrt{Lp_{i,j} \cdot Cp_{i,j}} = \frac{l}{v} \quad \text{(s)} \tag{2.3.8}$$

The ratio l/v is the propagation delay; the length of time it takes a transient pulse to propagate along the length of the conductor assembly. It can also be defined as the time constant T, as used in transient analysis. Section 6.3 derives its relationship to the inductance, capacitance, and characteristic impedance of the line.

Using (2.3.6) to substitute for v and (2.3.7) to substitute for l provides a relationship between the circuit components and the first resonant frequency:

$$\sqrt{Lp_{i,j} \cdot Cp_{i,j}} = \frac{1}{4 \cdot fq} \tag{2.3.9}$$

This means that, if the values of $Cp_{i,j}$ and the frequency of quarter-wave resonance, fq, are known, then the value of $Lp_{i,j}$ can be calculated.

Rearranging (2.3.7) gives:

$$\lambda q = 4 \cdot l$$

Using (2.3.6) to substitute for λq:

$$\frac{v}{fq} = 4 \cdot l$$

This leads to:

$$v = 4 \cdot l \cdot fq \tag{2.3.10}$$

From (2.3.4) and (2.3.5)

$$\frac{\sqrt{\mu_r \cdot \varepsilon_r}}{c} = \frac{1}{v}$$

This leads to:

$$\mu_r \cdot \varepsilon_r = \left(\frac{c}{v}\right)^2 \tag{2.3.11}$$

Since the conductors of most cables do not use magnetic material, it can usually be assumed that $\mu_r = 1$.

Equations (2.3.10) and (2.3.11) are extremely useful in establishing the value of the propagation velocity and the relative permittivity when the frequency of quarter-wave resonance is known. The worksheet of Figure 7.5.8 makes use of this relationship to demonstrate that the antenna-mode current propagates at a higher velocity than the differential-mode current.

Knowledge of the relative permittivity of the cable allows the values of the associated capacitors to be calculated.

2.4 Loop parameters

Any cable can act either as an antenna or as a transmission line. Figure 2.4.1 illustrates the situation where the conductors are acting as an antenna. Currents Ip_1 and Ip_2 are assumed to flow in the same direction; in this case, from left to right.

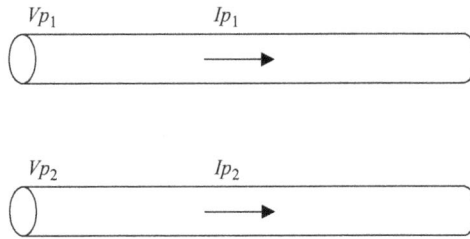

Figure 2.4.1 Primitive parameters.

It is possible to deal with inductive and capacitive effects separately. Voltages Vp_1 and Vp_2 can initially be defined as the energy levels due to magnetic effects on conductors 1 and 2. If it is assumed that currents and voltages are sinusoidal functions of time, then the relationship between them can be defined. Restricting consideration to magnetic effects, the primitive equations for two conductors are:

$$Vp_1 = j \cdot \omega \cdot Lp_{1,1} \cdot Ip_1 + j \cdot \omega \cdot Lp_{1,2} \cdot Ip_2$$
$$Vp_2 = j \cdot \omega \cdot Lp_{2,1} \cdot Ip_1 + j \cdot \omega \cdot Lp_{2,2} \cdot Ip_2 \qquad (2.4.1)$$

Primitive equations define the behavior of the conductor assembly as an antenna. When analyzing the behavior of the assembly as part of a circuit, it is necessary to deal with loop currents and loop voltages, as illustrated by Figure 2.4.2.

The relationship between voltages is:

$$Vl = Vp_1 - Vp_2 \qquad (2.4.2)$$

The relationship between currents is:

$$Ip_1 = Il = -Ip_2 \qquad (2.4.3)$$

Using (2.4.3) to substitute loop currents for primitive currents in (2.4.1) and then invoking (2.4.2) leads to:

$$Vl = j \cdot \omega \cdot (Lp_{1,1} - Lp_{1,2} - Lp_{2,1} + Lp_{2,2}) \cdot Il \qquad (2.4.4)$$

This gives the value of the loop inductance of the conductor pair:

$$Ll = Lp_{1,1} - 2 \cdot Lp_{1,2} + Lp_{2,2} \qquad (2.4.5)$$

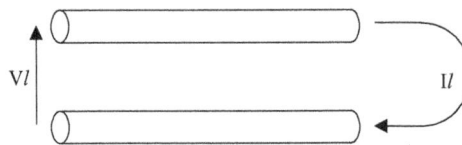

Figure 2.4.2 Loop parameters.

This relationship takes into account the fact that $Lp_{1,2} = Lp_{2,1}$. Using (2.3.2) to substitute for the primitive inductors:

$$Ll = \frac{\mu_0 \cdot \mu_r \cdot l}{2 \cdot \pi} \cdot \ln\left(\frac{r_{1,2} \cdot r_{1,2}}{r_{1,1} \cdot r_{2,2}}\right) \tag{2.4.6}$$

The step between (2.4.5) and (2.4.6) is highly significant, in that the length parameter disappears from the logarithmic term. There is no longer any inaccuracy due to variation in the ratio between length and separation distance at any particular cross section of the assembly-under-review. It does assume that the cross section is uniform along the length l and that $l > 10 \cdot r$.

If $l \leq 10 \cdot r$, then propagation time needs to be considered and it becomes necessary to invoke the concepts of electromagnetic field theory. Even then, however, any inaccuracy in the predicted value of inductance does not negate the fact that loop inductance exists. The inaccuracy can be catered for by carrying out tests on the assembly-under-review.

Loop capacitance can be derived the same way. The primitive equations are:

$$\begin{aligned}
Vp_1 &= \frac{1}{j \cdot \omega \cdot Cp_{1,1}} \cdot Ip_1 + \frac{1}{j \cdot \omega \cdot Cp_{1,2}} \cdot Ip_2 \\
Vp_2 &= \frac{1}{j \cdot \omega \cdot Cp_{2,1}} \cdot Ip_1 + \frac{1}{j \cdot \omega \cdot Cp_{2,2}} \cdot Ip_2
\end{aligned} \tag{2.4.7}$$

Using (2.4.3) to substitute loop currents for primitive currents in (2.4.7) and then invoking (2.4.2) leads to:

$$Vl = \frac{1}{j \cdot \omega} \cdot \left(\frac{1}{Cp_{1,1}} - \frac{1}{Cp_{1,2}} - \frac{1}{Cp_{2,1}} + \frac{1}{Cp_{2,2}}\right) \cdot Il \tag{2.4.8}$$

Hence:

$$\frac{1}{Cl} = \frac{1}{Cp_{1,1}} - \frac{1}{Cp_{1,2}} - \frac{1}{Cp_{2,1}} + \frac{1}{Cp_{2,2}} \tag{2.4.9}$$

Using equation (2.3.1) to substitute for the primitive capacitors:

$$\frac{1}{Cl} = \frac{1}{2 \cdot \pi \cdot \varepsilon_o \cdot \varepsilon_r \cdot l} \cdot \left[\ln\frac{r_{1,2} \cdot r_{1,2}}{r_{1,1} \cdot r_{2,2}}\right]$$

So:

$$Cl = \frac{2 \cdot \pi \cdot \varepsilon_o \cdot \varepsilon_r \cdot l}{\ln\left(\dfrac{r_{1,2} \cdot r_{1,2}}{r_{1,1} \cdot r_{2,2}}\right)} \tag{2.4.10}$$

Equations (2.4.6) and (2.4.10) define the loop inductance and loop capacitance of a conductor pair. These are the values which could be measured with an LCR meter.

This derivation makes the assumption that all the current flowing in one conductor returns via the other. The nature of electromagnetic coupling ensures that this never happens in practice. Nevertheless, (2.4.6) and (2.4.10) can be regarded as providing a good approximation to the properties of an isolated twin-conductor cable. Section 5.2 develops these relationships to deal with the properties of such a cable as an antenna.

2.5 Circuit parameters

With EMC analysis, an important requirement is to maintain visibility of the voltage developed along each conductor, as well as the voltages appearing between conductor terminals. For the two-conductor configuration of Figure 2.4.1, there is little difficulty in meeting this requirement. It is assumed that all the current flowing in conductor 1 returns via conductor 2.

2.5.1 Inductance

The primitive equations for inductance define the voltage developed along each conductor in terms of the primitive currents. Reproducing (2.4.1):

$$Vp_1 = j \cdot \omega \cdot Lp_{1,1} \cdot Ip_1 + j \cdot \omega \cdot Lp_{1,2} \cdot Ip_2$$
$$Vp_2 = j \cdot \omega \cdot Lp_{2,1} \cdot Ip_1 + j \cdot \omega \cdot Lp_{2,2} \cdot Ip_2$$

$$(2.4.1)$$

As far as inductance is concerned, the circuit model is shown in Figure 2.5.1.

The circuit equations for inductive coupling, derived from Figure 2.5.1 are:

$$Vc_1 = j\omega Lc_1 \cdot Ic_1$$
$$Vc_2 = j\omega Lc_2 \cdot Ic_1$$

$$(2.5.1)$$

Equations (2.4.1) and (2.5.1) were derived using different criteria. Equation (2.4.1) was derived from electromagnetic theory, while (2.5.1) came from circuit theory. Different assumptions were involved in their derivation. The equations can be correlated by defining the relationship between voltages and currents:

$$Vp_1 = Vc_1$$
$$-Vp_2 = Vc_2$$

$$(2.5.2)$$

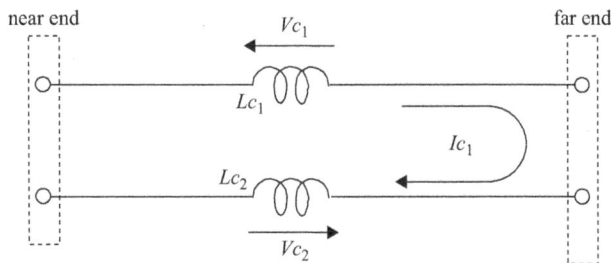

Figure 2.5.1 Circuit model for magnetic coupling.

and

$$Ip_1 = Ic_1$$
$$Ip_2 = -Ic_1 \tag{2.5.3}$$

Making these substitutions allows the circuit parameters to be derived from the primitives:

$$Lc_1 = Lp_{1,1} - Lp_{1,2}$$
$$Lc_2 = Lp_{2,2} - Lp_{1,2} \tag{2.5.4}$$

Using (2.3.2) to relate primitive inductors to physical parameters gives formulae for the circuit inductors for a conductor pair:

$$Lc_1 = \frac{\mu_o \cdot \mu_r \cdot l}{2 \cdot \pi} \cdot \ln\left(\frac{r_{1,2}}{r_{1,1}}\right)$$
$$Lc_2 = \frac{\mu_o \cdot \mu_r \cdot l}{2 \cdot \pi} \cdot \ln\left(\frac{r_{1,2}}{r_{2,2}}\right) \tag{2.5.5}$$

2.5.2 Capacitance

Figure 2.5.2 illustrates the capacitive parameters associated with each conductor.

The primitive equations for capacitance define the voltage developed between each conductor and a theoretical surface at zero voltage in terms of the current flowing into the environment or out of the environment. Reproducing (2.4.7):

$$Vp_1 = \frac{1}{j \cdot \omega \cdot Cp_{1,1}} \cdot Ip_1 + \frac{1}{j \cdot \omega \cdot Cp_{1,2}} \cdot Ip_2$$
$$Vp_2 = \frac{1}{j \cdot \omega \cdot Cp_{2,1}} \cdot Ip_1 + \frac{1}{j \cdot \omega \cdot Cp_{2,2}} \cdot Ip_2 \tag{2.4.7}$$

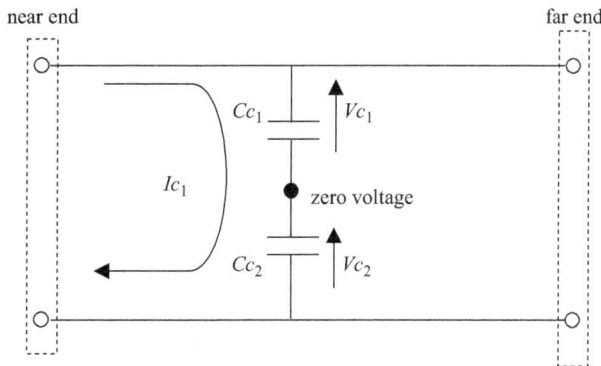

Figure 2.5.2 Circuit model for capacitive coupling.

The circuit equations for capacitive coupling derived from Figure 2.5.2 are:

$$Vc_1 = \frac{1}{j \cdot \omega \cdot Cc_1} \cdot Ic_1$$

$$Vc_2 = \frac{1}{j \cdot \omega \cdot Cc_2} \cdot Ic_1$$

(2.5.6)

Invoking (2.5.2) and (2.5.3) to substitute circuit parameters in the primitive equations leads to:

$$Vc_1 = Vp_1 = \frac{1}{j \cdot \omega} \cdot \left[\frac{1}{Cp_{1,1}} - \frac{1}{Cp_{1,2}} \right] \cdot Ic_1$$

$$Vc_2 = -Vp_2 = \frac{1}{j \cdot \omega} \cdot \left[\frac{1}{Cp_{2,2}} - \frac{1}{Cp_{2,1}} \right] \cdot Ic_1$$

(2.5.7)

Correlating (2.5.6) and (2.5.7) allows circuit capacitors to be defined in terms of primitive capacitors:

$$\frac{1}{Cc_1} = \frac{1}{Cp_{1,1}} - \frac{1}{Cp_{1,2}}$$

$$\frac{1}{Cc_2} = \frac{1}{Cp_{2,2}} - \frac{1}{Cp_{1,2}}$$

(2.5.8)

Using (2.3.1) to relate primitive capacitors to length and radii gives formulae for the circuit capacitors for a conductor pair:

$$Cc_1 = \frac{2 \cdot \pi \cdot \varepsilon_o \cdot \varepsilon_r \cdot l}{\ln \frac{r_{1,2}}{r_{1,1}}}$$

$$Cc_2 = \frac{2 \cdot \pi \cdot \varepsilon_o \cdot \varepsilon_r \cdot l}{\ln \frac{r_{1,2}}{r_{2,2}}}$$

(2.5.9)

2.5.3 Maintaining duality

From (2.5.5) and (2.5.9):

$$Lc_i \cdot Cc_i = \mu_o \cdot \mu_r \cdot \varepsilon_o \cdot \varepsilon_r \cdot l^2$$

(2.5.10)

where the subscript i defines the conductor.

Section 2.6 exploits this duality to identify further simplifications in the derivation of circuit models. This critical relationship between the inductive and capacitive properties of conductors is retained by avoiding the temptation to remove the zero-volt node. (see figures 2.5.2, 2.7.6 and 5.2.5)

2.5.4 Resistance

The resistance of each conductor is a function of its cross-sectional area and its length. If conductor i is circular in section, the steady-state resistance is:

$$Rss_i = \frac{\rho \cdot l}{\pi \cdot (r_{i,i})^2} \text{ ohm} \qquad (2.5.11)$$

where ρ is the resistivity of the conducting material. Skin effect makes resistance a function of frequency. Textbooks on electromagnetic theory derive the formula [2.4]:

$$Rskin_i = \frac{l}{2 \cdot r_{i,i}} \cdot \sqrt{\frac{\mu \cdot f}{\pi \cdot \sigma}} \text{ ohm} \qquad (2.5.12)$$

where σ is the conductivity. Conductivity and resistivity are related:

$$\rho = \frac{1}{\sigma} \text{ ohm m} \qquad (2.5.13)$$

At high frequencies, resistance increases as the square root of the frequency; that is, at 10 dB per decade. The crossover point occurs when $Rskin$ is equal to Rss. From (2.5.11), (2.5.12) and (2.5.13):

$$\frac{l}{2 \cdot r_{i,i}} \cdot \sqrt{\frac{\mu \cdot \rho \cdot Fx}{\pi}} = \frac{\rho \cdot l}{\pi \cdot r_{i,i}^2}$$

where Fx is the crossover frequency. This leads to the formula:

$$Fx = \frac{4 \cdot \rho}{\mu \cdot \pi} \cdot \frac{1}{r_{i,i}^2} \qquad (2.5.14)$$

Hence, the general formula for conductor resistance is:

$$Rc_i = Rss_i \cdot \sqrt{1 + \frac{f}{Fx_i}} \qquad (2.5.15)$$

The graph of Figure 2.5.3 illustrates the variation of resistance with frequency. The curves for Rc and $Rskin$ were derived from (2.5.15) and (2.5.12), respectively. These give the relationships for a 15-m length of 1-mm diameter copper conductor. In this case the crossover frequency is 69 kHz. This leads to a third representation of a twin-conductor cable; Figure 2.5.4.

Unlike inductance and capacitance, resistance is independent of the current in other conductors. This means that:

$$Rc_i = Rp_{i,i} \qquad (2.5.16)$$

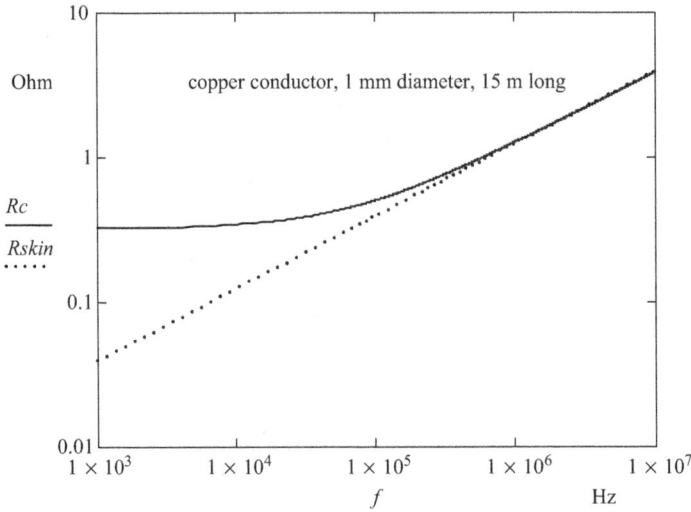

Figure 2.5.3 Relationship between resistance and frequency.

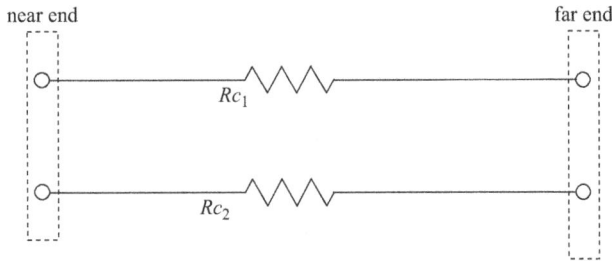

Figure 2.5.4 Circuit model for resistive coupling.

2.5.5 Basic assumption

A basic assumption made in the derivation of the above formulae for inductance and capacitance of the conductors was that the current in the return conductor is equal in value but opposite in direction to that in the send conductor. This is never the case for a twin conductor; some current radiates out into the environment. However, when this is taken into account, the formulae for conductor inductance and capacitance remain unaltered. Section 5.2 on the "Virtual Conductor" provides the more comprehensive derivation.

2.6 Twin-conductor model

Since the phenomenon causing interference is an electromagnetic field, there is no point in trying to separate out the electric field effects and the magnetic field effects. Resistive effects also have a significant role in the coupling mechanism. Any form of analysis of interference must take into account the combined effects of resistance, inductance, and capacitance. Since

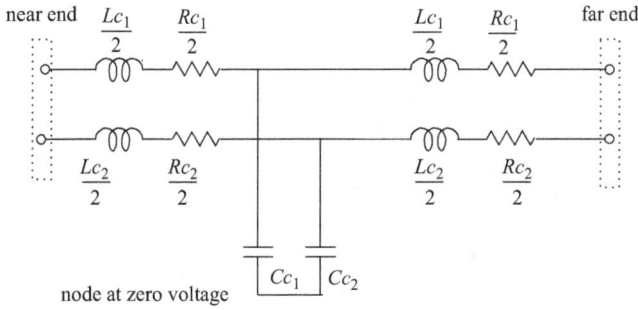

Figure 2.6.1 Circuit model of conductor pair.

every conductor possesses these properties, it is certainly not valid to assume that a zero-impedance path for current exists anywhere in the system.

Since the simplest configuration capable of efficiently carrying an electronic signal from one location to another is the twin-core cable, the best place to start is to analyze the properties of such a cable. Figure 2.6.1 departs from the conventional approach by recognizing the fact that both conductors have the same set of properties.

The most significant property of this model is the duality of the inductors and capacitors. If a voltage source is connected to the near end and the terminals at the far end are short-circuited, then, as far as reactive parameters are concerned, the configuration behaves as a bridge network.

From (2.5.10):

$$Lc_1 \cdot Cc_1 = Lc_2 \cdot Cc_2$$

Hence:

$$\frac{Lc_1}{Lc_2} = \frac{Cc_2}{Cc_1} \tag{2.6.1}$$

From the circuit model of Figure 2.6.2:

$$V_1 = \frac{I_1 - I_2}{j \cdot \omega \cdot C_1} \tag{2.6.2}$$

$$V_2 = \frac{I_1 - I_2}{j \cdot \omega \cdot C_2} \tag{2.6.3}$$

$$V_3 = j \cdot \omega \cdot \frac{Lc_1}{2} \cdot I_2 \tag{2.6.4}$$

$$V_4 = j \cdot \omega \cdot \frac{Lc_2}{2} \cdot I_2 \tag{2.6.5}$$

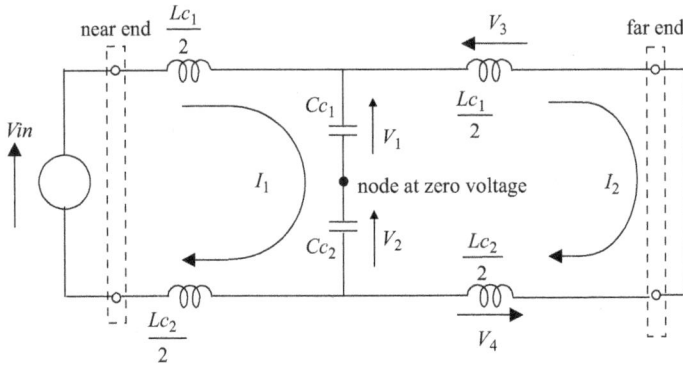

Figure 2.6.2   Bridge network.

Hence:

$$\frac{V_1}{V_2} = \frac{Cc_2}{Cc_1} \tag{2.6.6}$$

and

$$\frac{V_3}{V_4} = \frac{Lc_1}{Lc_2} \tag{2.6.7}$$

From (2.6.1):

$$\frac{V_1}{V_2} = \frac{Cc_1}{Cc_2} = \frac{Lc_2}{Lc_1} = \frac{V_3}{V_4} \tag{2.6.8}$$

Since the voltage at the junction of the capacitors is zero, the voltage at the junction of the inductors is also zero. Hence the bridge circuit can be redrawn, as shown in Figure 2.6.3.

This form of the circuit model indicates that it is valid to represent the reactive components of each conductor as a single impedance; a parallel combination of inductance and capacitance in series with an inductance. Such a representation is extremely useful in simplifying the analysis of the transfer admittance in section 2.9 and in deriving parameter values for the virtual conductor in section 5.2.

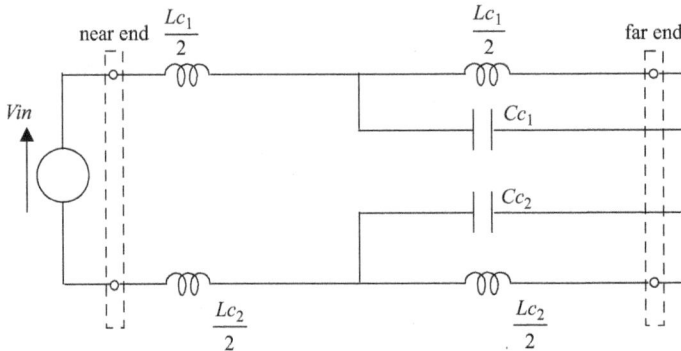

Figure 2.6.3   Equivalent circuit.

2.7 Three-conductor model

The process established in the previous sections can be simplified by using Z-parameters, where 'Z' can represent the impedance presented by any individual component, be it a capacitance, an inductance, or a resistance. Primitive impedances can be identified as Zp, loop impedances by Zl, and circuit impedances by Zc. Figure 2.7.1 illustrates a three-conductor assembly.

The primitive equations for three conductors are:

$$\begin{aligned}
Vp_1 &= Zp_{1,1} \cdot Ip_1 + Zp_{1,2} \cdot Ip_2 + Zp_{1,3} \cdot Ip_3 \\
Vp_2 &= Zp_{2,1} \cdot Ip_1 + Zp_{2,2} \cdot Ip_2 + Zp_{2,3} \cdot Ip_3 \\
Vp_3 &= Zp_{3,1} \cdot Ip_1 + Zp_{3,2} \cdot Ip_2 + Zp_{3,3} \cdot Ip_3
\end{aligned} \tag{2.7.1}$$

Figure 2.7.2 illustrates a configuration in which the cable assembly is terminated by short-circuits at both ends, with voltage sources Vl_1 and Vl_2 inserted in the loops at the near end.

Comparing Figures (2.7.1) and (2.7.2) allows the relationships between primitive parameters and loop parameters to be defined. Voltages are related by:

$$\begin{aligned}
Vl_1 &= Vp_1 - Vp_2 \\
Vl_2 &= Vp_2 - Vp_3
\end{aligned} \tag{2.7.2}$$

Currents are related by:

$$\begin{aligned}
Ip_1 &= Il_1 \\
Ip_2 &= Il_2 - Il_1 \\
Ip_3 &= -Il_2
\end{aligned} \tag{2.7.3}$$

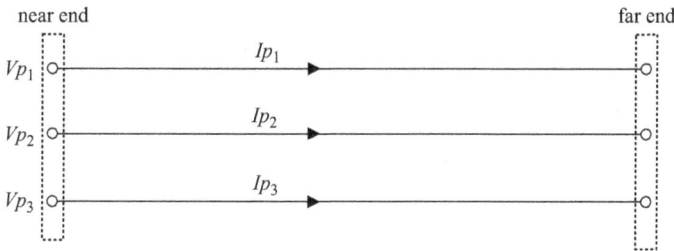

Figure 2.7.1   Three-conductor assembly.

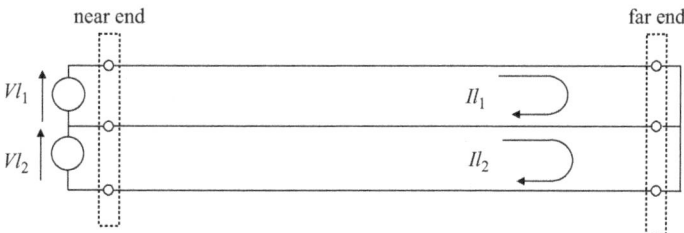

Figure 2.7.2   Loop voltages and currents.

Substituting loop currents for primitive currents in (2.7.1) gives:

$$Vp_1 = Zp_{1,1} \cdot Il_1 + Zp_{1,2} \cdot (Il_2 - Il_1) - Zp_{1,3} \cdot Il_2$$
$$Vp_2 = Zp_{2,1} \cdot Il_1 + Zp_{2,2} \cdot (Il_2 - Il_1) - Zp_{2,3} \cdot Il_2 \qquad (2.7.4)$$
$$Vp_3 = Zp_{3,1} \cdot Il_1 + Zp_{3,2} \cdot (Il_2 - Il_1) - Zp_{3,3} \cdot Il_2$$

Subtracting adjacent rows of (2.7.4) gives:

$$Vp_1 - Vp_2 = (Zp_{1,1} - Zp_{1,2} - Zp_{2,1} + Zp_{2,2}) \cdot Il_1 + (Zp_{1,2} - Zp_{1,3} - Zp_{2,2} + Zp_{2,3}) \cdot Il_2$$
$$Vp_2 - Vp_3 = (Zp_{2,1} - Zp_{2,2} - Zp_{3,1} + Zp_{3,2}) \cdot Il_1 + (Zp_{2,2} - Zp_{2,3} - Zp_{3,2} + Zp_{3,3}) \cdot Il_2$$
$$(2.7.5)$$

The loop equations for three conductors can be defined as:

$$Vl_1 = Zl_{1,1} \cdot Il_1 + Zl_{1,2} \cdot Il_2$$
$$Vl_2 = Zl_{2,1} \cdot Il_1 + Zl_{2,2} \cdot Il_2 \qquad (2.7.6)$$

where:

$$Zl_{1,1} = Zp_{1,1} - Zp_{1,2} - Zp_{2,1} + Zp_{2,2}$$
$$Zl_{1,2} = Zp_{1,2} - Zp_{1,3} - Zp_{2,2} + Zp_{2,3}$$
$$Zl_{2,1} = Zp_{2,1} - Zp_{2,2} - Zp_{3,1} + Zp_{3,2} \qquad (2.7.7)$$
$$Zl_{2,2} = Zp_{2,2} - Zp_{2,3} - Zp_{3,2} + Zp_{3,3}$$

Since $Zp_{i,j} = Zp_{j,i}$, then:

$$Zl_{1,2} = Zl_{2,1} \qquad (2.7.8)$$

The circuit model which can simulate the behavior of the loop equations is shown in Figure 2.7.3. The circuit equations for three conductors are:

$$Vc_1 = (Zc_1 + Zc_2) \cdot Ic_1 - Zc_2 \cdot Ic_2$$
$$Vc_2 = -Zc_2 \cdot Ic_1 + (Zc_2 + Zc_3) \cdot Ic_2 \qquad (2.7.9)$$

A comparison between (2.7.6) and (2.7.9) shows clearly that there is a one-to-one correlation between loop equations and circuit equations Hence, loop impedances can be related to circuit impedances:

$$Zl_{1,1} = Zc_1 + Zc_2$$
$$Zl_{1,2} = -Zc_2$$
$$Zl_{2,2} = Zc_2 + Zc_3$$

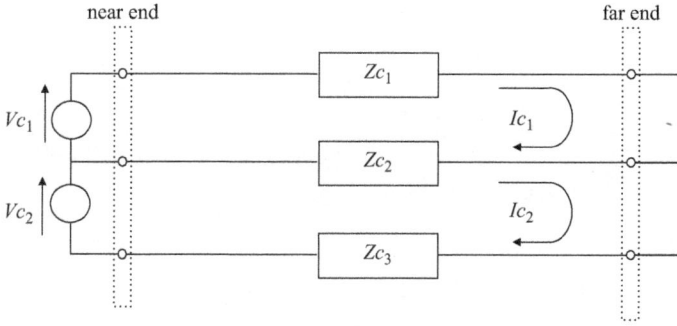

Figure 2.7.3   Circuit impedances for three-conductor assembly.

Defining circuit impedances in terms of loop impedances gives:

$$Zc_1 = Zl_{1,1} + Zl_{1,2}$$
$$Zc_2 = -Zl_{1,2}$$
$$Zc_3 = Zl_{2,2} + Zl_{1,2}$$

(2.7.10)

Using (2.7.7) to substitute primitive impedances for loop impedances gives the circuit impedances for three conductors in terms of the primitive impedances:

$$Zc_1 = Zp_{1,1} - Zp_{2,1} - Zp_{1,3} + Zp_{2,3}$$
$$Zc_2 = Zp_{2,2} - Zp_{1,2} - Zp_{2,3} + Zp_{1,3}$$
$$Zc_3 = Zp_{3,3} - Zp_{3,1} - Zp_{2,3} + Zp_{2,1}$$

(2.7.11)

If the radial dimensions of the conductor assembly are defined as shown in Figure 2.7.4, then circuit components can be defined in terms of spatial parameters.

If the impedances are assumed to be purely inductive, that is $Zp_{i,j} = j \cdot \omega \cdot Lp_{i,j}$, then (2.3.2) can be used to define the circuit inductors for three conductors:

$$Lc_1 = \frac{\mu_o \cdot \mu_r \cdot l}{2 \cdot \pi} \cdot \ln \frac{r_{1,2} \cdot r_{1,3}}{r_{1,1} \cdot r_{2,3}}$$

$$Lc_2 = \frac{\mu_o \cdot \mu_r \cdot l}{2 \cdot \pi} \cdot \ln \frac{r_{1,2} \cdot r_{2,3}}{r_{2,2} \cdot r_{1,3}}$$

$$Lc_3 = \frac{\mu_o \cdot \mu_r \cdot l}{2 \cdot \pi} \cdot \ln \frac{r_{1,3} \cdot r_{2,3}}{r_{3,3} \cdot r_{1,2}}$$

(2.7.12)

where l is the length of the assembly.

This establishes the inductance values for the assembly depicted by Figure 2.7.4.

If Figure 2.7.3 were to be redrawn as shown on Figure 2.7.5, then the same process could be used to derive capacitor values. Equation (2.7.11) applies to both figures.

This time, however, the impedances are assumed to be purely capacitive. The impedance parameters would then be related to capacitors using:

$$Zp_{i,j} = \frac{1}{j \cdot \omega \cdot Cp_{i,j}}$$

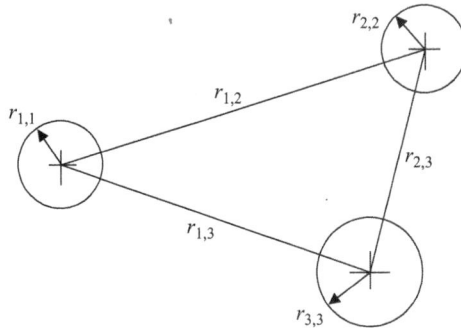

Figure 2.7.4   Cross section of conductor assembly.

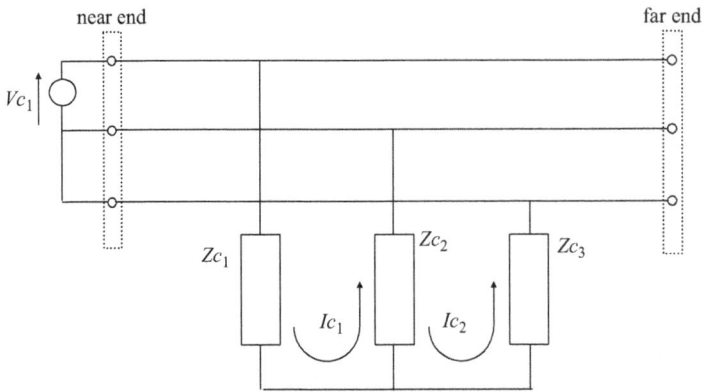

Figure 2.7.5   Circuit model of capacitive coupling.

Making this substitution in (2.7.11) and then invoking (2.3.1) leads to formulae for the circuit capacitors for three conductors:

$$Cc_1 = \frac{2 \cdot \pi \cdot \varepsilon_o \cdot \varepsilon_r \cdot l}{\ln \dfrac{r_{1,2} \cdot r_{1,3}}{r_{1,1} \cdot r_{2,3}}}$$

$$Cc_1 = \frac{2 \cdot \pi \cdot \varepsilon_o \cdot \varepsilon_r \cdot l}{\ln \dfrac{r_{1,2} \cdot r_{2,3}}{r_{2,2} \cdot r_{1,3}}} \qquad (2.7.13)$$

$$Cc_1 = \frac{2 \cdot \pi \cdot \varepsilon_o \cdot \varepsilon_r \cdot l}{\ln \dfrac{r_{1,3} \cdot r_{2,3}}{r_{3,3} \cdot r_{1,2}}}$$

The relationship between resistors and conductors is self-evident. Equations (2.5.11), (2.5.14), and (2.5.15) can be used to estimate the resistance at any frequency. This leads to the triple-T model for the three-conductor assembly; Figure 2.7.6.

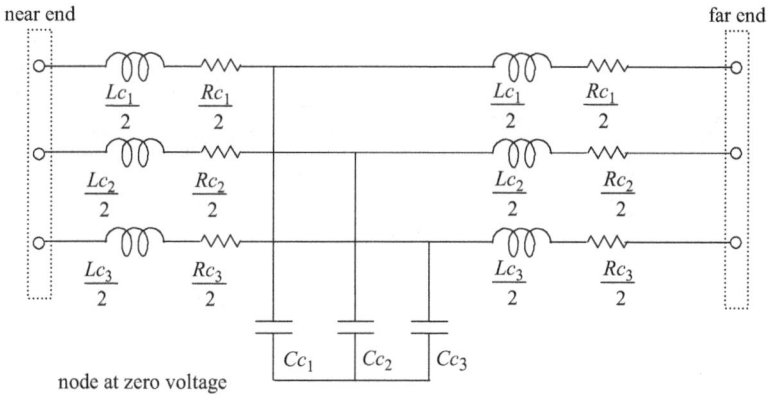

Figure 2.7.6 Triple-T circuit model of three-conductor assembly.

Equations (2.7.12) and (2.7.13) can be correlated with the formulae developed for phase inductance [1.7] and phase capacitance [1.8] of three-phase power lines. In fact, the derivation above is simply a formalization of the process described in that basic textbook.

2.8 Optimum coupling

The formulae derived in section 2.7 provide useful guidance as to how to reduce interference during the design process. If the cross-sectional view of the assembly of Figure 2.7.4 is compared with the circuit model in the light of (2.7.12) and (2.7.13), then some fundamental relationships can be established.

First and foremost, mesh analysis, rather than nodal analysis, has been used to establish the formulae. There are two interdependent loops. In Figure 2.8.1, conductors 1 and 2 are assigned to carry the signal from the near end to the far end. This signal link can be defined as the differential-mode loop. In this loop, all the current flowing down the signal conductor is assumed to return via the return conductor. Figure 2.8.2 illustrates this.

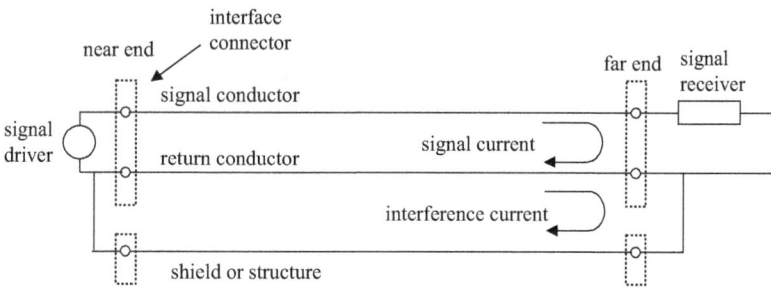

Figure 2.8.1 Defining the function of the conductors.

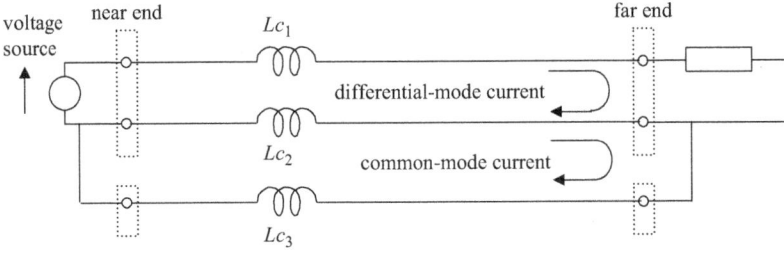

Figure 2.8.2   Defining the loop currents.

Inevitably, the differential-mode current in the return conductor develops a voltage across its inductance. This creates a current flow in the loop formed by the return conductor and the structure – the common-mode loop, as shown on Figure 2.8.2. Common-mode current in the inductance of the structure creates an unwanted voltage between any two points on that conductor.

Similarly, any voltage developed along the structure by an external source will create a current in the common-mode loop. In turn, this will induce an unwanted signal in the differential-mode loop.

In either case, any current in the common-mode loop is undesirable, since it constitutes interference. Whether the level of interference is acceptable or not is a matter of system design.

Low frequency simulation of this configuration is modeled by the circuit of Figure 2.8.2. As far as EMC is concerned, the objective would be to maximize the differential-mode current and minimize the common-mode current.

The physical design should ensure that the impedance of the differential-mode loop is as low as possible. That is, the inductors Lc_1 and Lc_2 should be as small as possible. Conversely, the value of Lc_3 should be as high as possible.

If the three conductors are circular in cross section as illustrated in Figure 2.8.3, then the inductor values would be as defined by (2.7.12). These are re-written below:

$$Lc_1 = \frac{\mu_o \cdot \mu_r \cdot l}{2 \cdot \pi} \cdot \left[\ln\frac{r_{1,2}}{r_{1,1}} + \ln\frac{r_{1,3}}{r_{2,3}} \right]$$

$$Lc_2 = \frac{\mu_o \cdot \mu_r \cdot l}{2 \cdot \pi} \cdot \left[\ln\frac{r_{1,2}}{r_{2,2}} + \ln\frac{r_{2,3}}{r_{1,3}} \right] \qquad (2.8.1)$$

$$Lc_3 = \frac{\mu_o \cdot \mu_r \cdot l}{2 \cdot \pi} \cdot \left[\ln\frac{r_{1,3}}{r_{1,2}} + \ln\frac{r_{2,3}}{r_{3,3}} \right]$$

Given this set of equations, it can be seen that reducing r_{12} would reduce the value of Lc_1 and Lc_2, while increasing the value of Lc_3. That is, altering just one physical dimension will affect the value of all three inductors. Reducing the spacing between the signal and return conductors will have a significant effect in improving EMC.

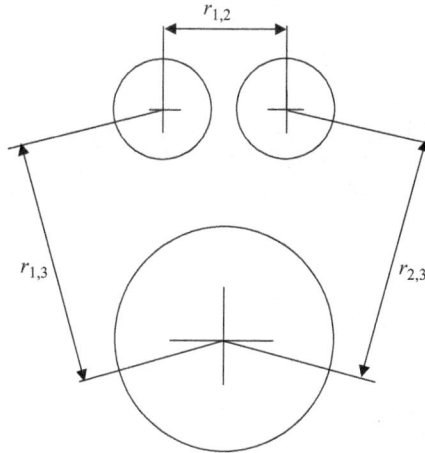

Figure 2.8.3 Cross section of conductor assembly.

Further improvement can be achieved by locating conductors 1 and 2 an equal distance away from conductor 3.

$$\text{If} \qquad r_{1,3} = r_{2,3}$$

$$\text{then :} \qquad \ln\frac{r_{1,3}}{r_{2,3}} = \ln\frac{r_{2,3}}{r_{1,3}} = 0$$

That is, if the conductors carrying the differential-mode current are equidistant from the structure, then these logarithmic factors can be deleted from (2.8.1). This means that Lc_1 and Lc_2 can be reduced further.

If, in addition, the radii of the signal and return conductors were the same, then:

$$\frac{r_{1,2}}{r_{1,1}} = \frac{r_{1,2}}{r_{2,2}}$$

Giving the relationship $Lc_1 = Lc_2$.

This is a desirable characteristic in configurations where balanced drivers and balanced receiver are utilized.

When capacitive coupling is brought into the picture, the circuit model changes to that shown by Figure 2.8.4.

If Lc_1 and Lc_2 are reduced in value, then Cc_1 and Cc_2 are increased. If the value of Lc_3 increases, then the value of Cc_3 reduces. It can be seen from Figure 2.8.4 that these changes to capacitance values will act in the same way as inductance value changes; more current will flow in the signal loop and less will flow in the common-mode loop.

This means that, when the separation between signal and return conductors is reduced, all the reactive components will change in a way that enhances EMC.

Hence, the basic requirement for optimum coupling is that the signal and return conductors be held as closely together as possible along the entire route from signal driver to signal receiver.

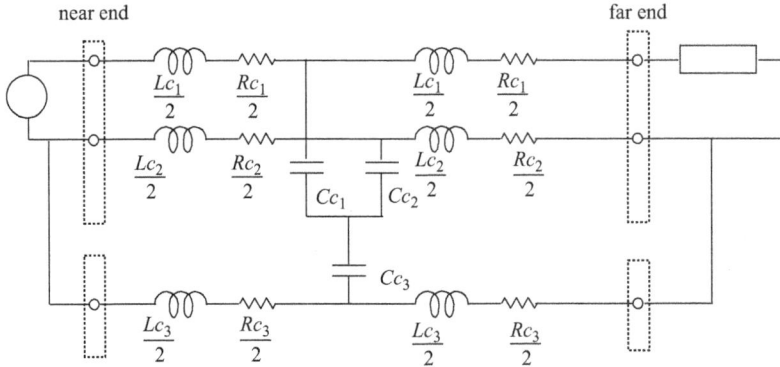

Figure 2.8.4   Adding capacitors and resistors to model.

2.9 Transfer admittance

It was shown in section 2.6 that the reactive components of the model will form a bridge circuit. The same natural balance exists in the three-conductor model of Figure 2.9.1. If the terminations at the far end are short-circuited, then the voltage at the junction of the three inductors will be the same as the voltage at the junction of the three capacitors.

If the voltages at these two nodes are at the same potential, it is valid to join them together. Hence it is valid to redraw the model to give the diagram of Figure 2.9.2.

At any particular frequency, this model can be further simplified to that shown in Figure 2.9.3, where $Z1$, $Z2$, and $Z3$ represent the impedances associated with each conductor.

The circuit equations for Figure 2.9.3 are:

$$V_1 = (Z1 + Z2) \cdot I_1 - Z2 \cdot I_2 \qquad (2.9.1)$$

$$0 = -Z2 \cdot I_1 + (Z2 + Z3) \cdot I_2 \qquad (2.9.2)$$

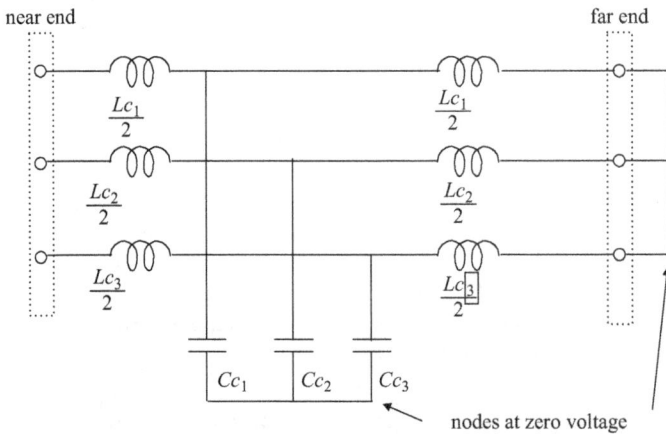

Figure 2.9.1   Voltage balance in a three-conductor assembly.

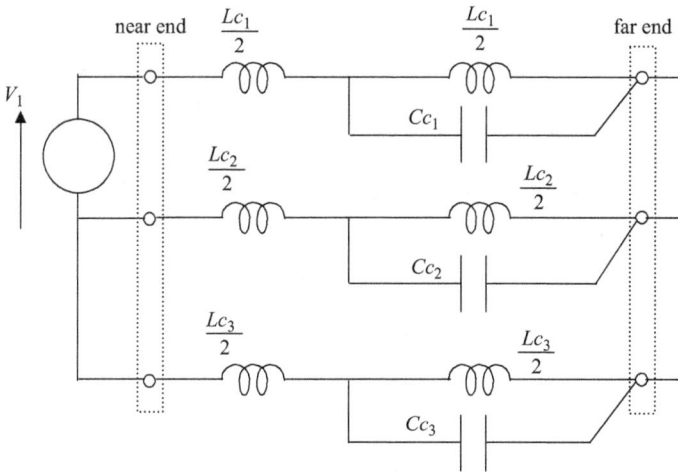

Figure 2.9.2   Equivalent circuit of configuration of Figure 2.9.1.

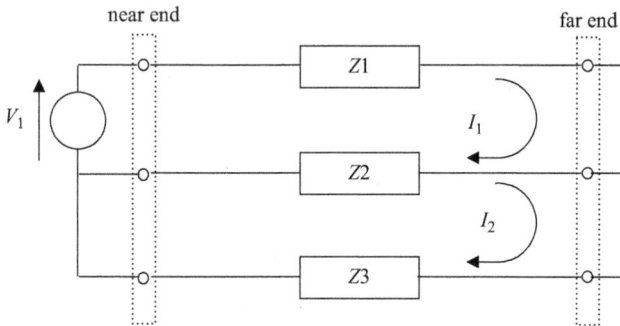

Figure 2.9.3   Simplified model.

from (2.9.2):

$$I_1 = \frac{Z2 + Z3}{Z2} \cdot I_2 \qquad (2.9.3)$$

substituting for I_1 in (2.9.1):

$$V_1 = \frac{(Z1 + Z2) \cdot (Z2 + Z3)}{Z2} \cdot I_2 - Z2 \cdot I_2$$

this expands to:

$$V_1 = \frac{Z1 \cdot Z2 + Z1 \cdot Z3 + Z2 \cdot Z2 + Z2 \cdot Z3 - Z2 \cdot Z2}{Z2} \cdot I_2$$

giving:

$$\frac{I_2}{V_1} = \frac{Z2}{Z1 \cdot Z2 + Z1 \cdot Z3 + Z2 \cdot Z3} \qquad (2.9.4)$$

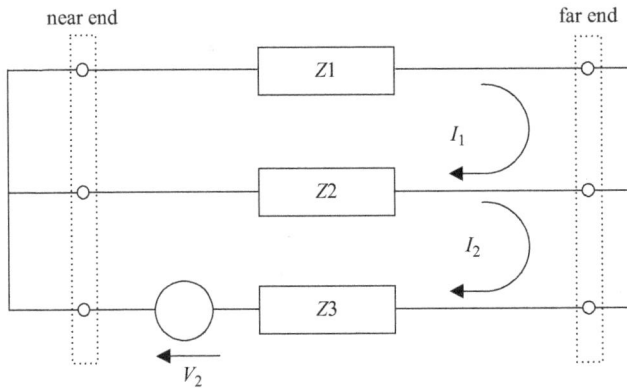

Figure 2.9.4 Voltage source in loop 2.

If the voltage source is placed in the second loop, as shown in Figure 2.9.4, then the same process leads to the ratio:

$$\frac{I_1}{V_2} = \frac{Z2}{Z1 \cdot Z2 + Z1 \cdot Z3 + Z2 \cdot Z3} \tag{2.9.5}$$

This identifies the fact that there is a duality in any circuit network. Comparing Figures (2.9.3) and (2.9.4)

$$YT = \frac{I_1}{V_2} = \frac{I_2}{V_1} \tag{2.9.6}$$

Since the ratio of current to voltage is termed admittance, and since the ratio is between parameters in different loops, then the parameters of (2.9.4) and (2.9.5) are described as *transfer admittance, YT*.

The transfer admittance provides a direct measure of the ratio of unwanted signal in the victim loop to the source voltage in the culprit loop. It is the basic parameter which enables the interference characteristic of any assembly to be defined.

Figure 2.9.3 is effectively a circuit model of the conducted emission test carried out by engineers at an EMC Test House. It allows the ratio of common-mode current to injection voltage to be simulated over a range of frequencies. Figure 2.9.4 simulates the setup used in a conducted susceptibility test. It defines the ratio of unwanted signal current to a source voltage in the common-mode loop.

This reasoning leads to a definition of transfer admittance:

Transfer admittance is the ratio of the current in the victim loop to the source voltage in the culprit loop when there are no other voltage sources.

Most significantly, transfer admittance has the same value for conducted emission as for conducted susceptibility. This means that a conducted susceptibility test on a cable assembly can be used to predict the response of a conducted emission test, and vice versa. It also identifies a useful check which can be made on the integrity of any program used to analyze any circuit model:

> Interchange the locations of source voltage and monitored current and re-run the program. If the results from both runs are not identical, then there is an error somewhere.

In practice, there are many factors which tend to compromise this analysis of transfer admittance. Most notable is the fact that that the test methods and test equipment used with the two types of test are different. Even so, it is fair to say that a cable configuration which exhibits high susceptibility at a particular frequency will also create a high level of emission at that frequency.

2.10 Co-axial coupling

The special property of a co-axial cable is that the separation between the axis of the inner conductor and any point on the outer conductor is the radius of the outer conductor. The fact that the inner and outer conductors share the same axis also means that the separation between the inner conductor and the external conductor is exactly the same as the separation between the outer conductor and external conductor. Figure 2.10.1 shows a typical cross section.

Hence:

$$r_{1,2} = r_{2,1} = r_{2,2} \qquad\qquad (2.10.1)$$

and

$$r_{1,3} = r_{3,1} = r_{2,3} = r_{3,2} \qquad\qquad (2.10.2)$$

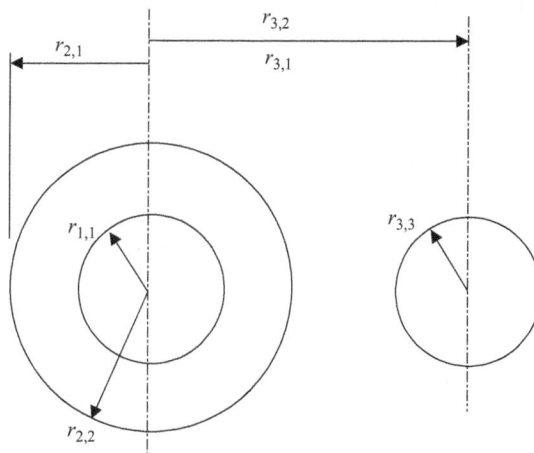

Figure 2.10.1   Co-axial cable with external conductor.

The circuit model for a three-conductor cable is shown in Figure 2.7.6 and the formula for inductance values is given by (2.7.12). When the relationships of (2.10.1) and (2.10.2) are incorporated, the values of the inductors become:

$$
\begin{aligned}
Lc_1 &= \frac{\mu_o \cdot \mu_r \cdot l}{2 \cdot \pi} \cdot \ln \frac{r_{2,2}}{r_{1,1}} \\
Lc_2 &= 0 \\
Lc_3 &= \frac{\mu_o \cdot \mu_r \cdot l}{2 \cdot \pi} \cdot \ln \frac{r_{3,2} \cdot r_{2,3}}{r_{3,3} \cdot r_{2,2}}
\end{aligned}
\tag{2.10.3}
$$

The value of Lc_1 is the familiar formula for the loop inductance of co-axial cable. The value of Lc_2 turns out to be zero, while the value of Lc_3 is the loop inductance of conductors 2 and 3 acting as a conductor pair. Since Lc_2 is zero, this inductance disappears from the circuit model.

Values for the capacitors of the circuit model are related to the inductors by (2.3.3). This means that the theoretical value of Cc_2 is infinite. In effect, it becomes a short-circuit. When these modifications are carried out on the model of Figure 2.7.6, it changes to the circuit model of co-axial coupling shown by Figure 2.10.2.

The circuit model of Figure 2.10.2 is that of an ideal co-axial cable, where the screen is constructed of solid conducting material. For this ideal case, the only component at the interface between common-mode loop and the differential-mode loop is the resistance of the screen. This means that the use of co-axial cable for the signal link will ensure that the value of the transfer admittance is extremely small. Put another way, the common-mode rejection will be extremely high.

However, it is usually impractical to use such a cable. In the vast majority of assemblies, the outer shield of the screened cables is made from thin wires inter-wound to form a flexible braid. Gaps in the braid allow penetration of external field and emission of internal field. At high frequencies, the shielding effectiveness of the screen deteriorates.

A more realistic model for a braided co-axial cable is shown in Figure 2.10.3. The magnetic field coupling between inner and outer loops is simulated by Lc_2, while Cc_2 represents the effect of the electric field coupling. Effectively, the model reverts to that shown in Figure 2.7.6. There is a significant difference. The values of the reactive

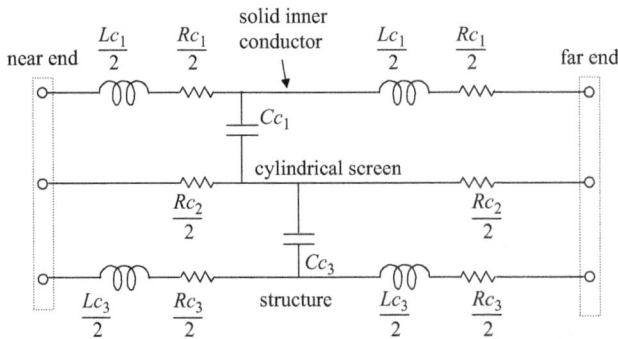

Figure 2.10.2 Circuit model of ideal co-axial coupling.

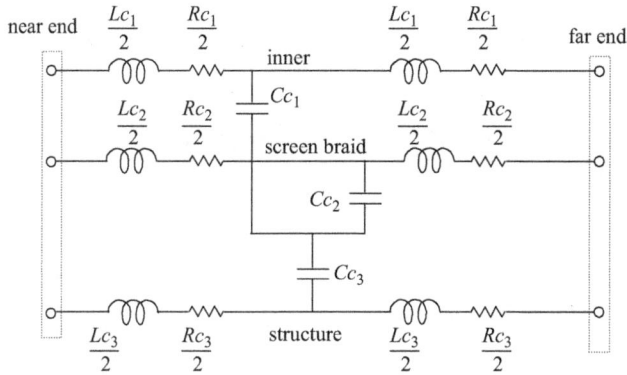

Figure 2.10.3   Circuit model of braided co-axial cable.

parameters associated with the screen conductor are much less than those of the separate return conductor derived in section 2.7.

A rule-of-thumb estimate for the value of these parameters would be to assume:

$$Lc_2 = \frac{Lc_1}{10} \quad \text{and} \quad Cc_2 = \frac{Cc_1}{10}$$

The best way of establishing more accurate values is to carry out tests on a representative assembly. Such tests usually result in a parameter described as the transfer impedance, and are defined in terms of a frequency response relating impedance to frequency. The relationship between transfer admittance and transfer impedance indicates that $Z2$ in Figure 2.9.3 can be defined as the transfer impedance of the assembly-under-review.

Some manufacturers provide data on the transfer impedance in the form of a frequency response curve for a 1 m length. It should not be too difficult to create a model which replicates this curve and to derive values for the resistance, inductance, and capacitance.

In any event, it can be said that co-axial cable will provide much better EMC than the twin-conductor cable, since the transfer impedance is much less. It may be that a shielded twisted pair could provide even better immunity to interference. However, it is more difficult to derive parameter values for such a configuration. This topic is dealt with in detail in the next chapter.

2.11 The ground plane

The concept of the conducting plane can be found in every textbook on electromagnetic theory. It exploits the fact that, on a non-conducting plane surface midway between two charged conductors, the electric field is orthogonal to that surface. If this surface was conducting, then the electric field distribution would not change. Figure 2.11.1 illustrates a cross section of two conductors over a ground plane.

Conversely, a flat conducting surface can be represented by a non-conducting surface midway between the two conductors and a second pair of identical conductors. That is, the configuration of Figure 2.11.1 can be replaced by that of Figure 2.11.2.

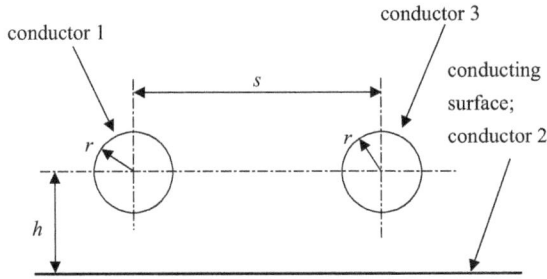

Figure 2.11.1 Two conductors over a ground plane.

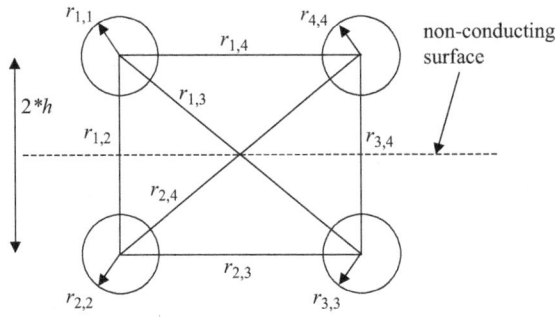

Figure 2.11.2 Simulating the effect of the plane.

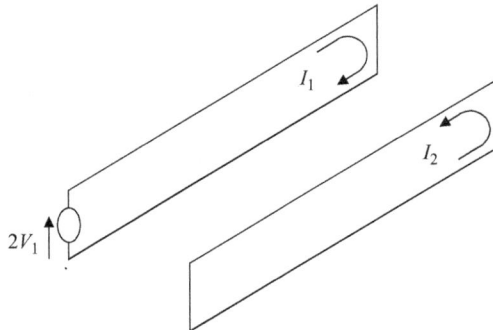

Figure 2.11.3 Transformer coupling.

If these four conductors are connected as two separate circuit loops, then the picture changes to that shown by Figure 2.11.3. Here, the coupling is between two isolated loops. As far as magnetic coupling is concerned, this is essentially a transformer; a transformer where the primary and secondary have one turn each and there is no magnetic material.

The circuit model for this configuration is the familiar transformer model of Figure 2.11.4. Here, Lc_1 represents the leakage inductance of the primary, Lc_2 represents the mutual inductance, and Lc_3 represents the leakage inductance of the secondary.

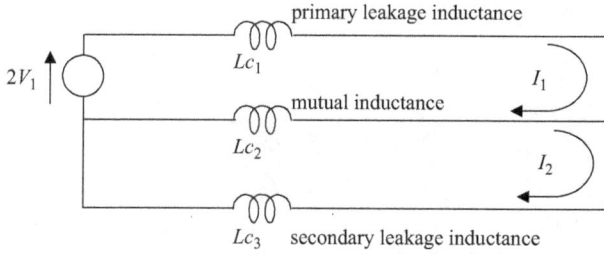

Figure 2.11.4   Circuit model of transformer coupling.

Following the procedure outlined in section 2.7, the three inductors can be defined for a length l of the assembly:

$$Lc_1 = \frac{\mu_o \cdot \mu_r \cdot l}{2 \cdot \pi} \cdot \ln \frac{r_{1,2} \cdot r_{2,1} \cdot r_{1,4} \cdot r_{2,3}}{r_{1,1} \cdot r_{2,2} \cdot r_{1,3} \cdot r_{2,4}}$$

$$Lc_2 = \frac{\mu_o \cdot \mu_r \cdot l}{2 \cdot \pi} \cdot \ln \frac{r_{1,3} \cdot r_{2,4}}{r_{1,4} \cdot r_{2,3}} \tag{2.11.1}$$

$$Lc_3 = \frac{\mu_o \cdot \mu_r \cdot l}{2 \cdot \pi} \cdot \ln \frac{r_{3,2} \cdot r_{4,1} \cdot r_{3,4} \cdot r_{4,3}}{r_{3,1} \cdot r_{4,2} \cdot r_{3,3} \cdot r_{4,4}}$$

The next step is to use the transformer coupling model to derive a model which simulates the coupling between the conductors of Figure 2.11.1. The relationships between the radial parameters of Figure 2.11.2 and the dimensions of Figure 2.11.1 are:

$$r_{1,2} = r_{2,1} = r_{3,4} = r_{4,3} = 2 \cdot h$$

$$r_{1,3} = r_{3,1} = r_{2,4} = r_{4,2} = \sqrt{s^2 + 4 \cdot h^2}$$

$$r_{1,4} = r_{4,1} = r_{2,3} = r_{3,2} = s \tag{2.11.2}$$

$$r_{1,1} = r_{2,2} = r_{3,3} = r_{4,4} = r$$

The voltage source is defined as $2 \cdot V_1$ in Figure 2.11.3 because the voltage between any conductor and its image is twice that which would exist between conductor and ground plane. Since the formulae of (2.11.1) have been derived for a voltage source of $2 \cdot V_1$, the inductor values are twice that associated with conductors over a plane.

Dividing each of the inductive parameters of (2.11.1) by two and invoking the relationships of (2.11.2) leads to:

$$Ld_1 = \frac{Lc_1}{2} = \frac{\mu_o \cdot \mu_r \cdot l}{2 \cdot \pi} \cdot \ln \frac{2 \cdot h \cdot s}{r \cdot \sqrt{s^2 + 4 \cdot h^2}}$$

$$Ld_2 = \frac{Lc_2}{2} = \frac{\mu_o \cdot \mu_r \cdot l}{2 \cdot \pi} \cdot \ln \frac{\sqrt{s^2 + 4 \cdot h^2}}{s} \tag{2.11.3}$$

$$Ld_3 = \frac{Lc_3}{2} = \frac{\mu_o \cdot \mu_r \cdot l}{2 \cdot \pi} \cdot \ln \frac{2 \cdot h \cdot s}{r \cdot \sqrt{s^2 + 4 \cdot h^2}}$$

Duality between inductance and capacitance can be used to calculate values for the associated capacitors:

$$Cd_i = \frac{\mu_0 \cdot \mu_r \cdot \varepsilon_0 \cdot \varepsilon_r \cdot l^2}{Ld_i} \tag{2.11.4}$$

If the spatial dimensions of the setup are known, then a one to one correlation can be established between the conductors of Figure 2.11.1 and the components of the triple-T circuit model of Figure 2.7.6. Inductance Ld_1 and capacitance Cd_1 can be assigned to conductor 1, while inductance Ld_3 and capacitance Cd_3 can be assigned to conductor 3.

The most significant feature of (2.11.3) is that a value can be assigned to Ld_2. This is the inductance of the ground plane. Similarly, the capacitance of the ground plane is Cd_2.

The existence of Lc_2 means that any transient current in the plane will create an end-to-end voltage along the surface. If a voltage exists along the surface, then that surface cannot possibly be equipotential. To assume that the conducting plane is an equipotential surface is to ignore the lessons learnt from electromagnetic theory.

The relationship between separation distance and inductance values can be determined by inspection of (2.11.3). As the separation between conductors 1 and 2 reduces, the values of Ld_1 and Ld_3 reduce while the value of Ld_2 increases. When the spacing is at a minimum, the assembly behaves like a transformer. Conversely, as the spacing increases, the coupling between the two loops decreases, and the value of Ld_2 reduces.

The reasoning applied to inductor values can also be applied to capacitor values. As the spacing between the conductors increases, capacitive coupling reduces.

This means that, if the ground plane is acting as a return conductor for signals in conductors 1 and 3, (as with a printed circuit board) then interference coupling can be reduced by separating conductors 1 and 3 as much as possible and by reducing the separation between these conductors and the ground plane. The reasoning applies to both inductive effects and capacitive effects. The action of the ground plane in printed circuit boards is analyzed further in sections 8.2 and 9.3.

If the ground plane is used to represent the properties of the structure when it is acting as a shield, then it is desirable for conductors 1 and 3 to be as close together as possible. This will increase the coupling between the send and return conductors and enhance the balance between the currents in those conductors, while reducing the amplitude of common-mode current in the ground conductor.

CHAPTER 3

Other cross sections

It is possible to develop the process to determine the properties of conductor assemblies of virtually any cross section. The starting point is a technique devised by researchers at Culham to predict induced voltages in aircraft cables [1.9].

In this technique, the assembly-under-review is represented by an array of parallel conductors. It is assumed that the conductors at each end are short-circuited. So the end-to-end voltage of each conductor is the same. Since the voltage along the length of one conductor of this array is determined by the currents in all the conductors, then a set of primitive equations can be defined. Solving this set of equations allows the current in each conductor to be calculated. When the currents are known, it is possible to calculate the magnetic potential of any point in the vicinity. This allows the magnetic field pattern in the region to be determined.

The composite conductor can be defined as a set of elemental conductors, aligned in parallel, which enables the distribution of currents or voltages in the actual conductor to be simulated.

An elemental conductor can be defined as a conductor which represents a small segment of the surface of a composite conductor.

In the method described here, the primitive equations are set up and the currents in the elemental conductors are calculated, but the focus remains on the behavior of those currents.

Section 3.1 illustrates the process by representing the cross section of a circular tube as an array of elemental conductors. The current in each element is calculated. Adding all these currents together gives a figure for the total current in the section. The ratio of the voltage to the total current gives a figure for the partial inductance of the composite conductor.

A circular section was chosen for this first illustration because it was easy to check the accuracy of the result.

Since the process necessarily involves the use of a computer program, Mathcad software was used to illustrate the details of the computation. Since mathematical notation is used throughout, the program is much easier to follow than one written in, say, the JAVA language. A few special features of Mathcad are described in Appendix A1.

Section 3.2 develops the method to deal with the differential-mode current in two conductors to derive a value for the loop inductance. Duality between inductance and capacitance is then used to determine the value of the loop capacitance. This value is the same as that obtained when the 'method of images' is invoked. Such a check effectively validates the technique.

Section 3.3 develops the technique one stage further; to enable a circuit model to be created for a three-conductor assembly. The example chosen is that of a screened pair cable.

The program defines the geometry of the elemental conductors of the three composite conductors, creates an array of primitive inductors, and converts this to an array of loop inductors. If it is assumed that a sinusoidal voltage of one volt is applied between the screen and the inner pair, the loop currents in the elemental conductors can be calculated.

Then the primitive current in every elemental conductor is calculated. That is, current flow from left to right is deemed to be positive current. Negative current flows from right to left. Summing the currents in the appropriate elements allows the current in each composite to be calculated.

Knowledge of currents and impedance values allows a three-by-three matrix of partial voltages to be determined. From this it is possible to calculate the values for a three-by-three array of *partial inductors*. Then the loop inductance values for the composite conductors are calculated. This is a two-by-two matrix.

Using this information, the value of each of the three inductors of the circuit model can be calculated. Invoking the duality between inductance and capacitance gives the value of each of the capacitors. This allows a circuit model to be defined for a 10 meter length of twin-core, screened cable.

The Mathcad worksheet which carries out all the necessary computations is reproduced. This can be hand copied into a new worksheet by the reader. Alternatively, the worksheet file can be downloaded from the website at www.designemc.info. MATLAB® files which replicate the calculations in every worksheet are also available for download. Modifying this program, virtually any cross section can be modeled. The only changes would be a re-definition of the geometry of the cross section and of the length of cable.

Intermediate results from the calculation are used to create a bubble chart which defines the current distribution in the cable. This is more meaningful to a circuit designer than a diagram of the magnetic field distribution. It also illustrates the fact that skin effect is not uniform. Differential-mode current concentrates on adjacent surfaces.

Figure 3.3.5 illustrates the fact that common-mode current tends to flow on surfaces of the signal-carrying conductors that are as far apart as possible. This is because the common-mode current in the wire pair is flowing in the same direction. Since this current returns via the outer shield, it concentrates on that part of the shield which is closest to the current concentration in the wire pair.

Current in any conductor causes a magnetic field which has both internal and external linkages with that conductor, and the formula derived for primitive inductance takes account of both. By representing the surface of the conductor as an array of elemental conductors, this treatment enables the effect of internal linkages to be included, even though they are concentrated on that surface.

3.1 Single composite conductor

To derive values corresponding to the primitive inductance and capacitance of any non-circular conductor, the first step is to represent the cross section as an array of parallel conductors. The process is best illustrated by using it to represent a circular-section conductor. This allows the resultant values to be confirmed by comparing them with known values.

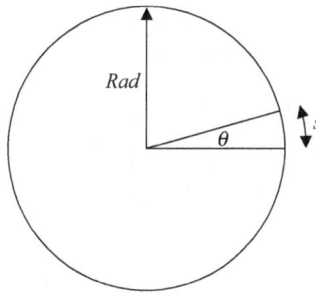

Figure 3.1.1   Circular section conductor.

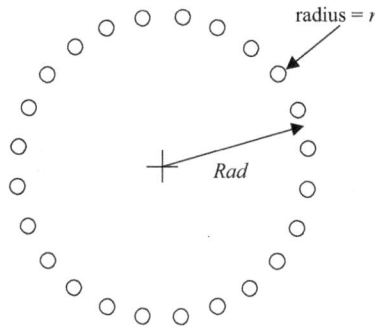

Figure 3.1.2   Array of elemental conductors.

Figure 3.1.1 illustrates a section of the conductor to be analyzed, while Figure 3.1.2 shows the method of simulation. Each elemental conductor represents a small segment, s, of the surface. Hence:

$$2 \cdot \pi \cdot r = s = \theta \cdot Rad$$

giving a general formula for the radius of each elemental conductor:

$$r = \frac{\theta \cdot Rad}{2 \cdot \pi} \tag{3.1.1}$$

In this particular case, the conductor being simulated is circular. So there are n elemental conductors equally spaced round the periphery, and this leads to the relationship:

$$n \cdot \theta = 2 \cdot \pi \tag{3.1.2}$$

Substituting for $2 \cdot \pi$ in (3.1.1):

$$r = \frac{Rad}{n} \tag{3.1.3}$$

Since these conductors each have a defined radius, they cannot be termed 'filamentary'.

If it is assumed that a current Ip_i is flowing in elemental conductor i, then a set of primitive equations can be set up. The voltage on each conductor will be a function of the current in every other conductor, as well as the current it carries itself. In the case of the three-conductor assembly of section 2.7 there were three equations. For an assembly of n conductors, there are n equations:

$$Vp_i = \sum_{j=1}^{n} Zp_{i,j} \cdot Ip_j \tag{3.1.4}$$

Defining the primitive equations for elemental conductors in terms of vector algebra:

$$\mathbf{Vp} = \mathbf{Zp} \cdot \mathbf{Ip} \tag{3.1.5}$$

where $\mathbf{Vp}$ and $\mathbf{Ip}$ are vectors of n elements, and $\mathbf{Zp}$ is a square matrix of n^2 elements.

If the impedances are assumed to be inductive and the terminals at each end are shorted together, then the voltage along the length of each elemental conductor will be the same.

Impedance and inductance are related by:

$$\mathbf{Zp} = j \cdot \omega \cdot \mathbf{Lp}$$

Whatever value is chosen for ω, it will remain the same throughout this computation process. If it is chosen to be unity, then:

$$\mathbf{Zp} = |\mathbf{Lp}| \quad \text{where} \quad \omega = 1$$

Since the voltage along the length of each elemental conductor is a fixed value, it is possible to define every component of the $\mathbf{Vp}$ vector. This being so, there is enough information to calculate the value of every element of the $\mathbf{Ip}$ vector. With personal computers, the task is completed in a fraction of a second. The relevant function in Mathcad is:

$$\mathbf{Ip} = \text{lsolve}(\mathbf{Zp}, \mathbf{Vp}) \tag{3.1.6}$$

The total current in the composite conductor is the sum of the currents in the elemental conductors. That is:

$$Iq = \sum_{i=1}^{n} Ip_i \tag{3.1.7}$$

The impedance of the composite can be defined as:

$$Zq = \frac{Vq}{Iq} \tag{3.1.8}$$

If the voltage along the length is assumed to be unity, then:

$$Zq = \frac{1}{|Iq|} \quad \text{where} \quad Vq = 1$$

If the value of ω is set at unity:

$$Lq = \frac{1}{|Iq|} \quad \text{where} \quad Vq = 1 \quad \text{and} \quad \omega = 1 \tag{3.1.9}$$

The parameter *Lq* can be correlated with the primitive inductance *Lp* defined by (2.2.9) in that it defines the inductance of a composite conductor when that conductor is acting as an antenna. To avoid confusion between the two terms, it is useful to describe *Lq* as a *partial inductance*.

It follows that *Vq*, *Iq*, and *Zq* can be described as *partial voltage, partial current,* and *partial impedance.*

The set of equations developed above is amenable to the creation of a computer program to calculate the partial inductance of virtually any cross section. Two stages are involved

- defining the co-ordinates of the elemental conductors,
- calculating the values of the electrical parameters.

Figure 3.1.3 is a copy of a Mathcad worksheet which carries out the computations for the first stage. In this case, the process is quite simple. For a more complex cross section, the

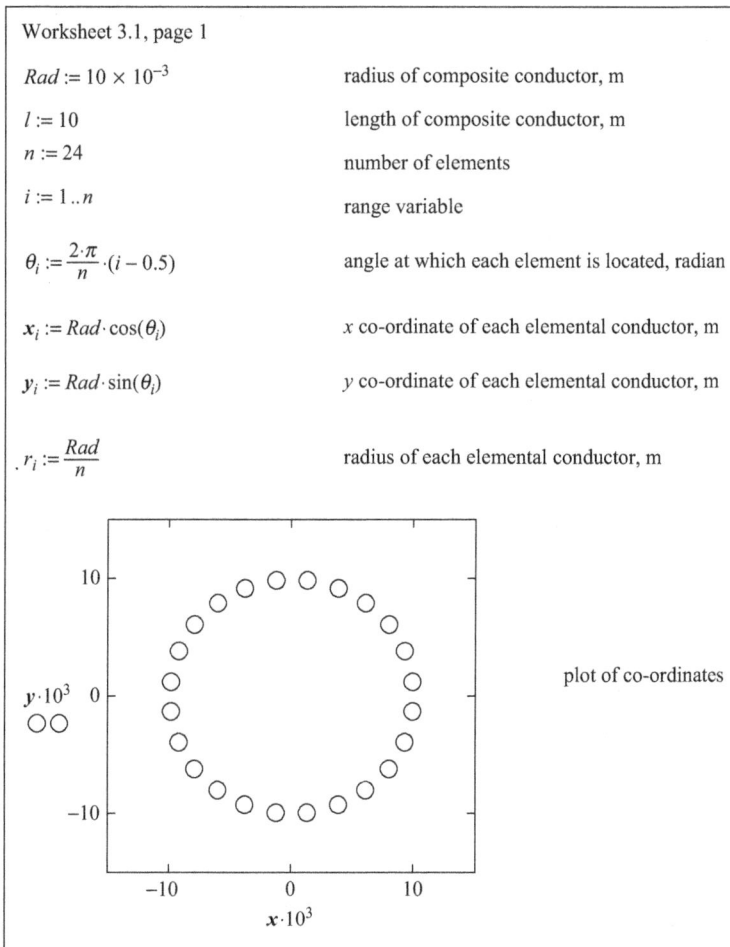

Worksheet 3.1, page 1

$Rad := 10 \times 10^{-3}$ radius of composite conductor, m

$l := 10$ length of composite conductor, m

$n := 24$ number of elements

$i := 1 .. n$ range variable

$\theta_i := \dfrac{2 \cdot \pi}{n} \cdot (i - 0.5)$ angle at which each element is located, radian

$x_i := Rad \cdot \cos(\theta_i)$ x co-ordinate of each elemental conductor, m

$y_i := Rad \cdot \sin(\theta_i)$ y co-ordinate of each elemental conductor, m

$. r_i := \dfrac{Rad}{n}$ radius of each elemental conductor, m

plot of co-ordinates

Figure 3.1.3   Defining the physical parameters of the composite conductor.

definition of the co-ordinates would probably take the form of a three-column table of values.

Figure 3.1.4 invokes equations (3.1.4) to (3.1.9) to determine the value of Lq, the partial inductance of the composite conductor. It then carries out a check to confirm the resultant value is the same as the primitive inductance of a conductor with the same radius as the composite. This check provides a level of confidence that the process is valid.

Finally, the value of the *partial capacitance* is calculated, using the relationship between inductance and capacitance defined by (2.3.3).

Worksheet 3.1, page 2

$\mu_o := 4 \cdot \pi \cdot 10^{-7}$ H/m $\qquad\qquad\qquad \mu_r := 1$

$K := \dfrac{\mu_o \cdot \mu_r \cdot l}{2 \cdot \pi} = 2 \times 10^{-6}$ H $\qquad\qquad$ K is a constant for this computation

$Zp := \Bigg|$ for $i \in 1 .. n$ $\qquad\qquad\qquad$ Zp = array of $n \times n$ inductance values, H

$\qquad\qquad$ for $j \in 1 .. n$

$\qquad\qquad\qquad\Big| h \leftarrow x_j - x_i$

$\qquad\qquad\qquad v \leftarrow y_j - y_i$

$\qquad\qquad\qquad rad \leftarrow \sqrt{h^2 + v^2}$ $\qquad\qquad$ rad = separation between pair of elements

$\qquad\qquad\qquad rad \leftarrow r_i$ if $rad = 0$ $\qquad\qquad\qquad$ = radius of element if separation = 0

$\qquad\qquad\qquad Lp_{i,j} \leftarrow K \cdot \ln\left(\dfrac{l}{rad}\right)$ $\qquad\qquad$ see (2.3.2)

$\qquad\Big| Lp$

$Vp_i := 1$ $\qquad\qquad\qquad\qquad\qquad\qquad$ input voltage, V

$Ip := \text{lsolve}(Zp, Vp)$ $\qquad\qquad\qquad\qquad$ output current, A

$Iq := \displaystyle\sum_{i=1}^{n} Ip_i = 7.238 \times 10^4$ $\qquad\qquad$ sum of currents in elements

$\omega := 1$ radian/s $\qquad\qquad Lq := \dfrac{Vp_1 \cdot \omega}{Iq}$ $\qquad\qquad\qquad Lq = 1.382 \times 10^{-5}$ H

Check:- $\qquad\qquad Lp = 2 \cdot 10^{-7} \cdot l \cdot \ln\left(\dfrac{l}{Rad}\right)$ $\qquad\qquad Lp = 1.382 \times 10^{-5}$ H

$\varepsilon_o := 8.854 \cdot 10^{-12}$ F/m $\qquad\qquad \varepsilon_r := 1$

$Cq := \dfrac{\varepsilon_o \cdot \varepsilon_r \cdot \mu_o \cdot \mu_r \cdot l^2}{Lq}$ $\qquad\qquad\qquad Cq = 8.053 \times 10^{-11}$ F

Figure 3.1.4 Partial inductance and capacitance of single composite conductor.

3.2 The composite pair

The concept of an array of elemental conductors can be extended to the modeling of a pair of conductors of any cross section. The following illustration assumes circular sections.

Figure 3.2.1 depicts the cross section of a pair of composite conductors, while Figure 3.2.2 illustrates how the primitive currents can be related to the loop currents. Only one loop can have a voltage applied; that between the composite pair. All the other loop voltages must be zero.

Following the approach of section 2.7, the first step is to create the primitive equations for the composite pair. Using vector notation:

$$\mathbf{Vp} = \mathbf{Zp} \cdot \mathbf{Ip} \qquad (3.2.1)$$

If it is assumed that there are $n1$ elemental conductors in composite 1 and $n2$ elemental conductors in composite 2, then the voltage source will be in loop $n1$. The total number of conductors is N, where:

$$N = n1 + n2 \qquad (3.2.2)$$

Inspection of (2.7.7) reveals a clear correlation between primitive impedances and loop impedances. The general formula for loop impedance can be defined as:

$$Zl_{i,j} = Zp_{i,j} - Zp_{i,j+1} - Zp_{i+1,j} + Zp_{i+1,j+1} \qquad (3.2.3)$$

Using this relationship, it is possible to create a matrix of loop impedances $\mathbf{Zl}$. The loop equations can then be set up. Again in vector notation:

$$\mathbf{Vl} = \mathbf{Zl} \cdot \mathbf{Il} \qquad (3.2.4)$$

It is clear from Figure 3.2.2 that the number of elements in the $\mathbf{Vl}$ and $\mathbf{Il}$ vectors is $N - 1$.

Voltages in every loop can be defined:

$$\begin{aligned} Vl_i &= 1 \;\; if \;\; i = n1 \\ Vl_i &= 0 \;\; if \;\; i \neq n1 \end{aligned} \qquad (3.2.5)$$

Since all the voltages and impedances have been defined for the loop equations, it is possible to determine the loop currents. The relevant function in Mathcad is:

$$\mathbf{Il} = \mathrm{lsolve}(\mathbf{Zl}, \mathbf{Vl}) \qquad (3.2.6)$$

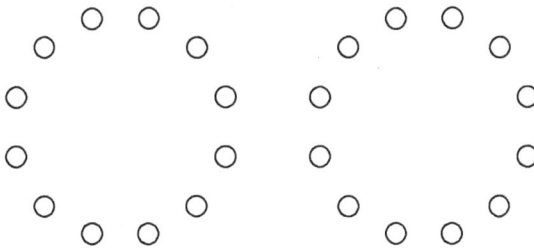

Figure 3.2.1   Cross section of composite pair.

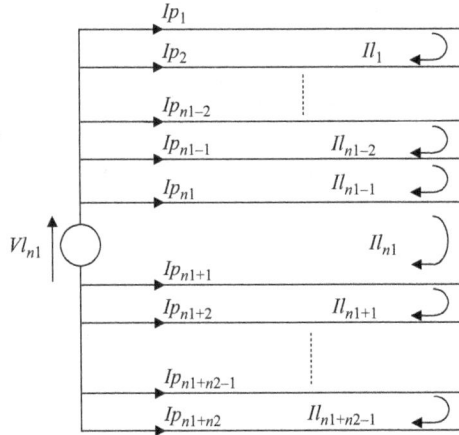

Figure 3.2.2   Primitive currents and loop currents.

Once the loop currents have been determined, it is a simple matter to calculate the values of the primitive currents. The relationship can be derived from Figure 3.2.2.

$$
\begin{aligned}
Ip_1 &= Il_1 \\
Ip_i &= Il_i \quad \text{for} \quad i = 2 \text{ to } N - 1 \\
Ip_N &= -Il_{N-1}
\end{aligned}
\tag{3.2.7}
$$

It now becomes possible to calculate every voltage component in every elemental conductor.

$$
v_{i,j} = Zp_{i,j} \cdot Ip_j
\tag{3.2.8}
$$

If the variables are set out in tabular form, the result would be an array of N by N values. Since there are only two composite conductors, the objective is to reduce this to an array of 2 by 2 values. This can be done by dividing the array into four sections, each representing the contribution made by the current in a composite conductor. Figure 3.2.3 provides a picture of the resultant sub-matrices. Since there are $n1$ elemental conductors in composite 1, the average value of the contributions of current in these conductors to the voltage experienced by composite 1 is:

$$
vq_{1,1} = \frac{1}{n1} \cdot \sum_{i=1}^{n1} \sum_{j=1}^{n1} v_{i,j}
\tag{3.2.9}
$$

Similar reasoning applies to the other three sections of Figure 3.2.3, giving:

$$
vq_{1,2} = \frac{1}{n1} \cdot \sum_{i=1}^{n1} \sum_{j=n1+1}^{n1+n2} v_{i,j}
\tag{3.2.10}
$$

$$
vq_{2,1} = \frac{1}{n2} \cdot \sum_{i=n1+1}^{n1+n2} \sum_{j=1}^{n1} v_{i,j}
\tag{3.2.11}
$$

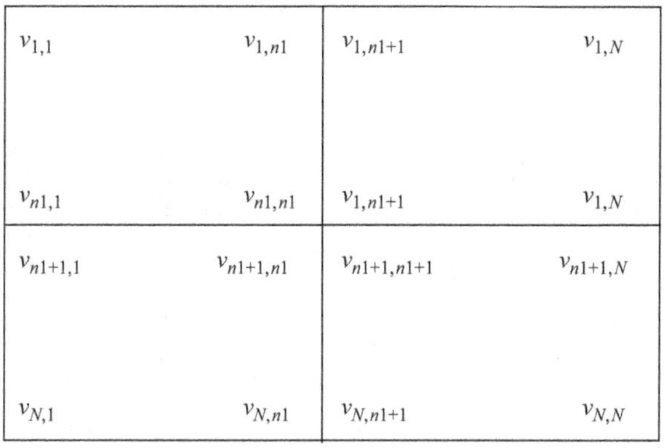

$v_{1,1}$	$v_{1,n1}$	$v_{1,n1+1}$	$v_{1,N}$
$v_{n1,1}$	$v_{n1,n1}$	$v_{1,n1+1}$	$v_{1,N}$
$v_{n1+1,1}$	$v_{n1+1,n1}$	$v_{n1+1,n1+1}$	$v_{n1+1,N}$
$v_{N,1}$	$v_{N,n1}$	$v_{N,n1+1}$	$v_{N,N}$

Figure 3.2.3   Dividing the voltage matrix into four sub-matrices.

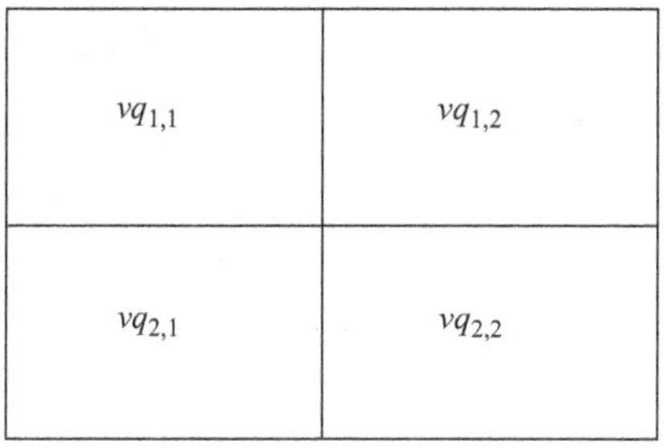

$vq_{1,1}$	$vq_{1,2}$
$vq_{2,1}$	$vq_{2,2}$

Figure 3.2.4   Voltage components derived from Figure 3.2.3.

$$vq_{2,2} = \frac{1}{n2} \cdot \sum_{i=n1+1}^{n1+n2} \sum_{j=n1+1}^{n1+n2} v_{i,j} \tag{3.2.12}$$

Calculating the average value of the current in each conductor is much simpler:

$$Iq_1 = \sum_{i=1}^{n1} Ip_i \tag{3.2.13}$$

$$Iq_2 = \sum_{i=n1+1}^{n1+n2} Ip_i \tag{3.2.14}$$

The four voltage components can be set out in the form of the small array of Figure 3.2.4, where:

$vq_{1,1}$ is the voltage in composite 1 due to Iq_1
$vq_{1,2}$ is the voltage in composite 1 due to Iq_2
$vq_{2,1}$ is the voltage in composite 2 due to Iq_1
$vq_{2,2}$ is the voltage in composite 2 due to Iq_2

Partial impedance values can be derived by using the relationship:

$$Zq_{h,k} = \frac{vq_{h,k}}{Iq_k} \qquad (3.2.15)$$

The integers h and k are used to identify the two composite conductors.

The number of equations has now been reduced from N to 2:

$$Vq_1 = Zq_{1,1} \cdot Iq_1 + Zq_{1,2} \cdot Iq_2$$
$$Vq_2 = Zq_{2,1} \cdot Iq_1 + Zq_{2,2} \cdot Iq_2 \qquad (3.2.16)$$

Comparing this equation with (2.4.1) or (2.4.7) indicates that the parameters Vq, Iq, and Zq can be treated as primitives. As indicated in section 3.1, a clear distinction between the properties of elemental conductors and those of composite conductors can be achieved by invoking the term 'partial' to describe the parameters involved. Hence (3.2.16) can be described as the *partial equations* for the composite pair.

If the impedances $Zq_{i,j}$ are assumed to be due to inductive effects, then the relationship of (2.5.4) can be used to determine the circuit inductance associated with each conductor.

$$Zc_1 = j \cdot \omega \cdot Lc_1 = j \cdot \omega \cdot (Lq_{1,1} - Lq_{1,2})$$
$$Zc_2 = j \cdot \omega \cdot Lc_2 = j \cdot \omega \cdot (Lq_{2,2} - Lq_{2,1})$$

Hence:

$$Lc_1 = Lq_{1,1} - Lq_{1,2}$$
$$Lc_2 = Lq_{2,2} - Lq_{2,1} \qquad (3.2.17)$$

In summary, the process above has established a relationship between the primitive parameters of the elemental conductors and the inductances of a circuit model of the composite pair. Since the primitive parameters of the elemental conductors can be related to the parameters of radius and length, it becomes possible to derive the inductive components from knowledge of the structure of the assembly. Circuit capacitances can be derived by invoking the duality between inductance and capacitance described in section 2.3.

A three-page worksheet derived from this set of equations is illustrated by Figures 3.2.5 to 3.2.7. It is assumed that the diameter of the conductors is 2 mm, that the centers are spaced 4 mm apart, and that the length of the cable is 1 m.

Page 1 of the worksheet derives the co-ordinates of the elemental conductors, Figure 3.2.5.

Page 2 calculates the value of the current in each elemental conductor, Figure 3.2.6.

Page 3 processes this data to derive component values for the circuit model, Figure 3.2.7.

It is useful to print out the intermediate results in the computation. If there are any errors in the program, then the results would become implausible. For example, the sum of the currents in the **Iq** vector should always be zero.

The **Vq** vector indicates that the input voltage is split equally between the two conductors. This is intuitively correct since the two conductors are defined to be identical. Vq_2 is negative because Iq_2 is negative. Figure 3.2.8 is a circuit model which effectively summarizes the results of the computations.

It is possible to check this result by comparing it with that derived from a textbook on electromagnetic theory [3.1]. Figure 3.2.9 is a copy of the final three steps in the

Worksheet 3.2, page 1

$l := 1$ length of assembly, m

$X1 := 0$ $X2 := 4 \cdot 10^{-3}$ co-ordinates of center of composite conductors 1 and 2, m

$Y1 := 0$ $Y2 := 0$

$R1 := 1 \cdot 10^{-3}$ $R2 := 1 \cdot 10^{-3}$ radius of composite conductors 1 and 2, m.

$n_1 := 12$ $n_2 := 12$ number of elemental conductors in composite 1 and 2

$N := n_1 + n_2$ total number of elemental conductors

defining elemental conductors for composite 1:

$i := 1..n_1$ range variable

$\theta_i := \dfrac{2 \cdot \pi}{n_1} \cdot (i - 0.5)$

$x_i := X1 + R1 \cdot \cos(\theta_i)$ $y_i := Y1 + R1 \cdot \sin(\theta_i)$ $r_i := \dfrac{R1}{n_1}$

defining elemental conductors for composite 2:

$i := n_1 + 1..n$ extending the range variable

$\theta_i := \dfrac{2 \cdot \pi}{n_2} \cdot (i - n_1 - 0.5)$

$x_i := X2 + R2 \cdot \cos(\theta_i)$ $y_i := Y2 + R2 \cdot \sin(\theta_i)$ $r_i := \dfrac{R2}{n_2}$

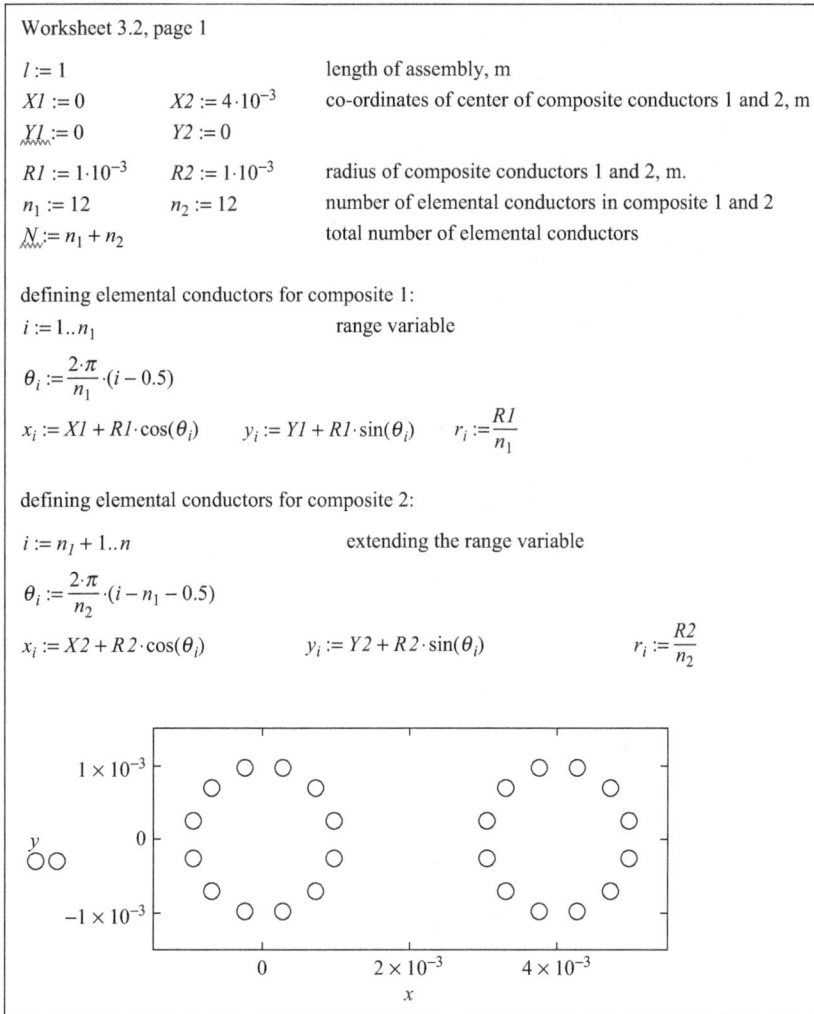

Figure 3.2.5   Deriving the co-ordinates of the elemental conductors.

Mathcad worksheet. The fact that the results agree to three decimal places provides a high degree of confidence in the method. The fact that the value of the loop capacitance is the last parameter to be computed indicates that all the preceding results are also correct.

However, it is worth bearing in mind that the relative permittivity and the relative permeability are both assumed to be unity. It is best to confirm the component values using electrical tests. Alternatively, estimated values for these parameters can be defined at the start of the program.

A noteworthy feature of this configuration is the distribution of current in the conductors, illustrated by the bubble plot of Figure 3.2.10. The diameter of each circle is proportional to the amplitude of the current in that conductor. Current in the left-hand conductor is assumed

Worksheet 3.2, page 2

$$\mu_o := 4 \cdot \pi \cdot 10^{-7} \text{ H/m} \qquad\qquad \mu_r := 1 \qquad K := \frac{\mu_o \cdot \mu_r \cdot l}{2 \cdot \pi} = 2 \times 10^{-7} \text{ H}$$

$$Zp := \begin{vmatrix} \text{for } i \in 1 .. N \\ \quad \begin{vmatrix} \text{for } j \in 1 .. N \\ \quad \begin{vmatrix} h \leftarrow x_j - x_i \\ v \leftarrow y_j - y_i \\ rad \leftarrow \sqrt{h^2 + v^2} \\ rad \leftarrow r_i \text{ if } rad = 0 \\ Lp_{i,j} \leftarrow K \cdot \ln\left(\dfrac{l}{rad}\right) \end{vmatrix} \end{vmatrix} \\ Lp \end{vmatrix} \qquad\qquad \text{see Figure 3.1.4}$$

$$Zloop := \begin{vmatrix} \text{for } i \in 1 .. N - 1 \\ \quad \begin{vmatrix} \text{for } j \in 1 .. N - 1 \\ \quad Lloop_{i,j} \leftarrow Zp_{i,j} - Zp_{i,j+1} - Zp_{i+1,j} + Zp_{i+1,j+1} \end{vmatrix} \\ Lloop \end{vmatrix} \qquad\qquad \text{see (3.2.3)}$$

$$Vloop := \begin{vmatrix} \text{for } i \in 1 .. N - 1 \\ \quad \begin{vmatrix} V_i \leftarrow 0 \\ V_i \leftarrow 1 \text{ if } i = n_1 \end{vmatrix} \\ V \end{vmatrix} \qquad\qquad \text{see (3.2.5)}$$

$$Iloop = \text{lsolve}(Zloop, Vloop) \qquad\qquad \text{see (3.2.6)}$$

$$Ip := \begin{vmatrix} I_1 \leftarrow Iloop_1 \\ \text{for } i \in 2 .. N - 1 \\ \quad I_i \leftarrow Iloop_i - Iloop_{i-1} \\ I_N \leftarrow -Iloop_{N-1} \\ I \end{vmatrix} \qquad\qquad \text{see (3.2.7)}$$

Figure 3.2.6 Calculating the currents in the elemental conductors.

to be flowing down *into* the page, while current in the right-hand conductor is assumed to flow up *out of* the page.

This illustration shows that the current is concentrated in the two facing surfaces. This asymmetrical distribution means that conductor resistance will probably increase more rapidly with frequency than the increase predicted in section 2.5. However, tests indicate that this aspect of the response can be catered for by assuming a slightly higher value for the steady-state resistance than that given by (2.5.11) and by retaining the use of (2.5.15).

Worksheet 3.2, page 3

$$Start := \begin{pmatrix} 1 \\ n_1 + 1 \end{pmatrix} \qquad End := \begin{pmatrix} n_1 \\ n_1 + n_2 \end{pmatrix} \qquad \text{see Figure 3.2.3}$$

$$h := 1..2 \qquad k := 1..2 \qquad\qquad\qquad \text{pointers to sub-matrices}$$

$$vq_{h,k} := \begin{vmatrix} v \leftarrow 0 \\ \text{for } i \in Start_h..End_h \\ \quad \text{for } j \in Start_k..End_k \\ \quad\quad v \leftarrow v + Zp_{i,j} \cdot Ip_j \\ \dfrac{v}{n_h} \end{vmatrix}$$

see (3.2.8) and (3.2.12)

$$vq = \begin{pmatrix} 2.622 & -2.122 \\ 2.122 & -2.622 \end{pmatrix} \qquad \text{V}$$

$$Vq_h := \begin{vmatrix} v \leftarrow 0 \\ \text{for } k \in 1..2 \\ \quad v \leftarrow v + vq_{h,k} \\ v \end{vmatrix}$$

check voltages along conductors

$$Vq = \begin{pmatrix} 0.5 \\ -0.5 \end{pmatrix} \qquad \text{V}$$

$$Iq_h := \begin{vmatrix} I \leftarrow 0 \\ \text{for } i \in Start_h..End_h \\ \quad I \leftarrow I + Ip_i \\ I \end{vmatrix}$$

see (3.2.13) and (3.2.14)

$$Iq = \begin{pmatrix} 1.898 \times 10^6 \\ -1.898 \times 10^6 \end{pmatrix} \qquad \text{A}$$

$$Lq_{h,k} := \dfrac{vq_{h,k}}{Iq_k} \qquad \text{see (3.2.15)} \qquad Lq = \begin{pmatrix} 1.382 \times 10^{-6} & 1.118 \times 10^{-6} \\ 1.118 \times 10^{-6} & 1.382 \times 10^{-6} \end{pmatrix}$$

$$Lc_1 := Lq_{1,1} - Lq_{1,2}$$
$$Lc_2 := Lq_{2,2} - Lq_{2,1} \qquad \text{see (2.5.4)} \qquad \dfrac{Lc}{2} = \begin{pmatrix} 1.317 \times 10^{-7} \\ 1.317 \times 10^{-7} \end{pmatrix} \qquad \text{H}$$

$$\varepsilon_o := 8.854 \times 10^{-12} \qquad \varepsilon_r := 1$$

$$Cc := \dfrac{\mu_o \cdot \mu_r \cdot \varepsilon_o \cdot \varepsilon_r \cdot l^2}{Lc} \qquad \text{see (2.3.3)} \qquad Cc = \begin{pmatrix} 4.224 \times 10^{-11} \\ 4.224 \times 10^{-11} \end{pmatrix} \qquad \text{F}$$

Figure 3.2.7 Calculating circuit components for composite conductors.

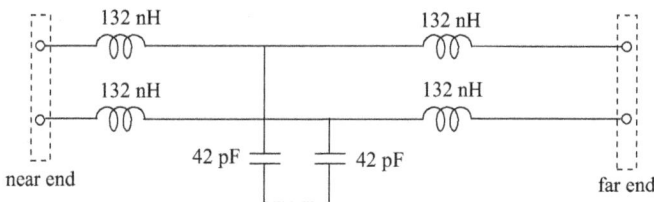

Figure 3.2.8 Circuit model of the composite pair.

Worksheet 3.2, page 4

$$Cloop := \frac{Cc_1 \cdot Cc_2}{Cc_1 + Cc_2}$$ $Cloop = 2.112 \times 10^{-11}$

$$b := \frac{X2}{2}$$ $r := R1$ Electromagnetic Concepts and Applications
GG Skitek & SV Marshall. Pages 208 to 209.
Capacitance between two cylindrical conductors

$$Ctheory := \frac{\pi \cdot \varepsilon_o \cdot \varepsilon_r \cdot l}{\ln\left(\frac{b + \sqrt{b^2 - r^2}}{r}\right)}$$

$Ctheory = 2.112 \times 10^{-11}$

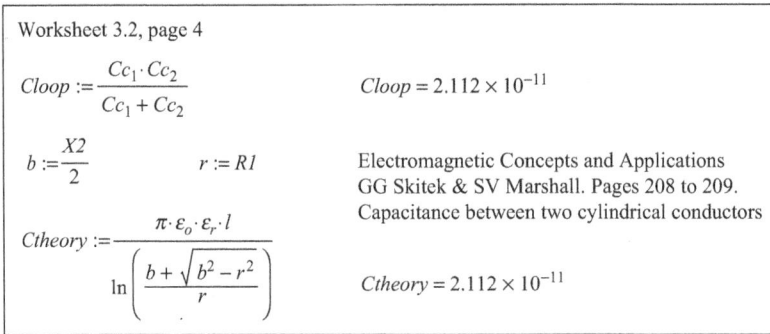

Figure 3.2.9   Comparison of results with textbook theory.

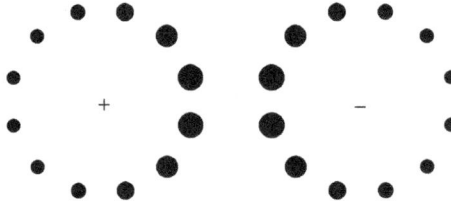

Figure 3.2.10   Bubble plot for composite pair, showing distribution of current.

If it were assumed that the two conductors were isolated at the far end, then the plot would show the distribution of charge on the conductors, where the left-hand conductor is positively charged and the right-hand conductor is assumed to hold negative charge.

3.3 The screened pair

To properly utilize composite conductors for EMC analysis, the number of such conductors in the model needs to be three. So it is necessary to develop the twin-conductor model one more stage to derive component values for an assembly comprising the three composite conductors. Since the screened pair is a widely used configuration, that is the one simulated here. The particular cable chosen for simulation has conductors of 0.8 mm diameter spaced 1.2 mm apart, enclosed by a screen of 3 mm diameter. It is assumed that the length of the cable is 10 m.

Figure 3.3.1 is a copy of the first page of a four-page worksheet. It illustrates the calculations used to define the co-ordinates of the conductors. This is only a small development of the process illustrated by the worksheet of Figure 3.2.5. Three conductors are defined, rather than two. Here, the screen is defined as composite conductor 1, while the signal and return conductors are identified as conductors 2 and 3.

To analyze the common-mode current flow, it is assumed that all three conductors are shorted together at the far end, that the signal and return conductors are shorted together at the near end, and that a sinusoidal voltage source of 1 V is applied between screen and signal conductors at the near end. The task is to determine the three inductive parameters, and use them to calculate the capacitor values.

Worksheet 3.3, page 1

$l := 10$

$X1 := 0$ $\qquad$ $X2 := 0.6 \cdot 10^{-3}$ $\qquad$ $X3 := -0.6 \cdot 10^{-3}$

$Y1 := 0$ $\qquad$ $Y2 := 0$ $\qquad$ $Y3 := 0$

$R1 := 1.5 \cdot 10^{-3}$ $\qquad$ $R2 := 0.4 \cdot 10^{-3}$ $\qquad$ $R3 := 0.4 \cdot 10^{-3}$

$n_1 := 30$ $\qquad$ $n_2 := 12$ $\qquad$ $n_3 := 12$

$N := n_1 + n_2 + n_3$

$i := 1..n1$ $\qquad\qquad\qquad$ $\theta_i := \dfrac{2 \cdot \pi}{n_1} \cdot (i - 0.5)$

$x_i := X1 + R1 \cdot \cos(\theta_i)$ $\qquad$ $y_i := Y1 + R1 \cdot \sin(\theta_i)$ $\qquad$ $r_i := \dfrac{R1}{n_1}$

$i := n_1 + 1..n_1 + n_2$ $\qquad\qquad$ $\theta_i := \dfrac{2 \cdot \pi}{n_2} \cdot (i - n_1 - 0.5)$

$x_i := X2 + R2 \cdot \cos(\theta_i)$ $\qquad$ $y_i := Y2 + R2 \cdot \sin(\theta_i)$ $\qquad$ $r_i := \dfrac{R2}{n_2}$

$i := n_1 + n_2 + 1..N$ $\qquad\qquad$ $\theta_i := \dfrac{2 \cdot \pi}{n_2} \cdot (i - n_1 - n_2 - 0.5)$

$x_i := X3 + R3 \cdot \cos(\theta_i)$ $\qquad$ $y_i := Y3 + R3 \cdot \sin(\theta_i)$ $\qquad$ $r_i := \dfrac{R3}{n_3}$

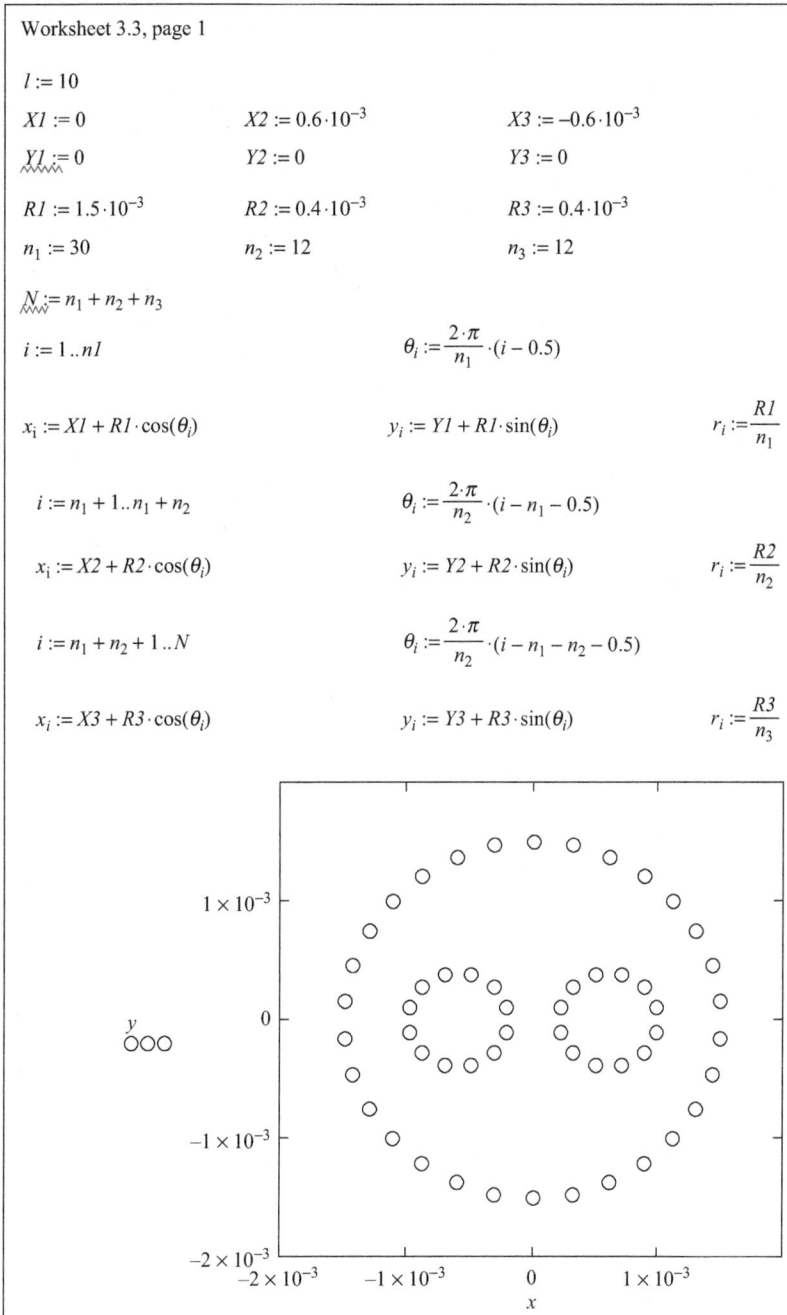

Figure 3.3.1 Definition of elemental conductors.

Worksheet 3.3, page 3   (worksheet 3.3 page 2 is identical to Figure 3.2.6)

$$n := \begin{pmatrix} 30 \\ 12 \\ 12 \end{pmatrix}$$

number of elemental conductors
in each composite.

$$Start := \begin{pmatrix} 1 \\ n_1 + 1 \\ n_1 + n_2 + 1 \end{pmatrix} \qquad End := \begin{pmatrix} n_1 \\ n_1 + n_2 \\ N \end{pmatrix}$$

pointers to nine sub-matrices.

$$h := 1..3 \qquad k := 1..3$$

range variables

$$vq_{h,k} := \begin{vmatrix} v \leftarrow 0 \\ \text{for } i \in Start_h .. End_h \\ \quad \text{for } j \in Start_k .. End_k \\ \qquad v \leftarrow v + Zp_{i,j} \cdot Ip_j \\ \dfrac{v}{n_h} \end{vmatrix}$$

voltage components of sub-matrices

$$vq = \begin{pmatrix} 12.667 & -6.334 & -6.334 \\ 12.694 & -7.284 & -6.41 \\ 12.694 & -6.41 & -7.284 \end{pmatrix}$$

$$Vq_h := \begin{vmatrix} v \leftarrow 0 \\ \text{for } k \in 1..3 \\ \quad v \leftarrow v + vq_{h,k} \\ v \end{vmatrix}$$

voltage along composite conductors:-

$$Vq = \begin{pmatrix} -5.874 \times 10^{-8} \\ -1 \\ -1 \end{pmatrix}$$

$$Iq_h := \begin{vmatrix} I \leftarrow 0 \\ \text{for } i \in Start_h .. End_h \\ \quad I \leftarrow I + Ip_i \\ I \end{vmatrix}$$

current in composite conductors:-

$$Iq = \begin{pmatrix} 7.193 \times 10^5 \\ -3.597 \times 10^5 \\ -3.597 \times 10^5 \end{pmatrix}$$

$$Itotal := Iq_1 + Iq_2 + Iq_3 = 0$$

sum of currents in conductors

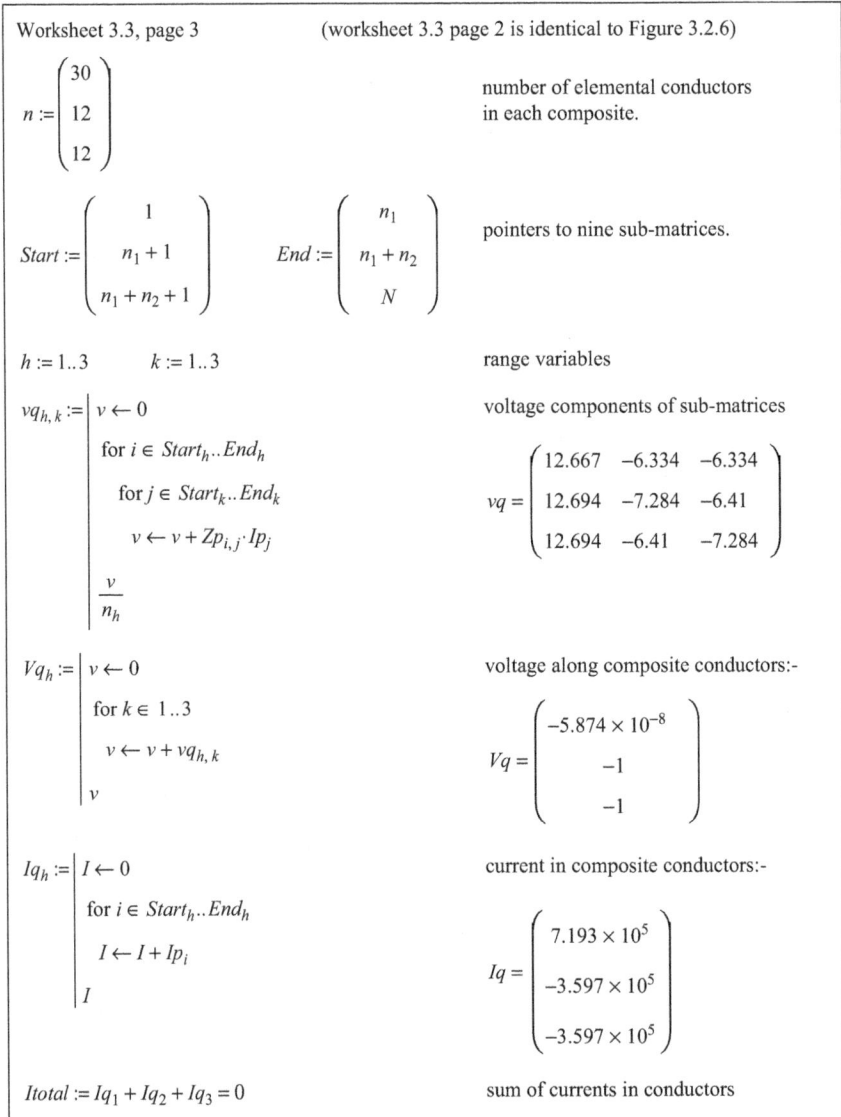

Figure 3.3.2 Computing values for voltages and currents in composite conductors.

Calculating the current in each elemental conductor is carried out by the second page of the worksheet. Since this page is identical to that of Figure 3.2.6, there is no need to replicate it.

Once the currents in all the elemental conductors have been defined, the next stage is to compute the values of the currents and voltages in the composite conductors, and this set of subroutines is illustrated in Figure 3.3.2. This process is very similar to that of Figure 3.2.7,

Worksheet 3.3, page 4

$$Lq_{h,k} := \frac{vq_{h,k}}{Iq_k}$$

deriving primitive inductance values from current and voltage data; see (3.2.15)

$$Lq = \begin{pmatrix} 1.761 \times 10^{-5} & 1.761 \times 10^{-5} & 1.761 \times 10^{-5} \\ 1.765 \times 10^{-5} & 2.025 \times 10^{-5} & 1.782 \times 10^{-5} \\ 1.765 \times 10^{-5} & 1.782 \times 10^{-5} & 2.025 \times 10^{-5} \end{pmatrix}$$

deriving loop inductance values for three-conductor assembly; see (3.2.3)

$$L\_loop := \begin{vmatrix} \text{for } h \in 1..2 \\ \quad \text{for } k \in 1..2 \\ \qquad L_{h,k} \leftarrow Lq_{h,k} - Lq_{h,k+1} - Lq_{h+1,k} + Lq_{h+1,k+1} \\ \quad L \end{vmatrix}$$

$$L\_loop := \begin{pmatrix} 2.606 \times 10^{-6} & -2.432 \times 10^{-6} \\ -2.432 \times 10^{-6} & 4.864 \times 10^{-6} \end{pmatrix}$$

$$Lc_1 := L\_loop_{1,1} + L\_loop_{1,2}$$

$$Lc_2 := -L\_loop_{1,2}$$

$$Lc_3 := L\_loop_{2,2} + L\_loop_{2,1}$$

deriving circuit inductors for three-conductor assembly; see (2.7.10)

$$\frac{Lc}{2} = \begin{pmatrix} 8.711 \times 10^{-8} \\ 1.216 \times 10^{-6} \\ 1.216 \times 10^{-6} \end{pmatrix} \text{ H}$$

$$\varepsilon_o := 8.854 \times 10^{-12} \text{ F/m}$$

$$\varepsilon_r := 1$$

$$Cc := \frac{\mu_o \cdot \mu_r \cdot \varepsilon_o \cdot \varepsilon_r \cdot l^2}{Lc}$$

deriving capacitor values; see (2.3.3)

$$Cc = \begin{pmatrix} 6.386 \times 10^{-9} \\ 4.575 \times 10^{-10} \\ 4.575 \times 10^{-10} \end{pmatrix} \text{ F}$$

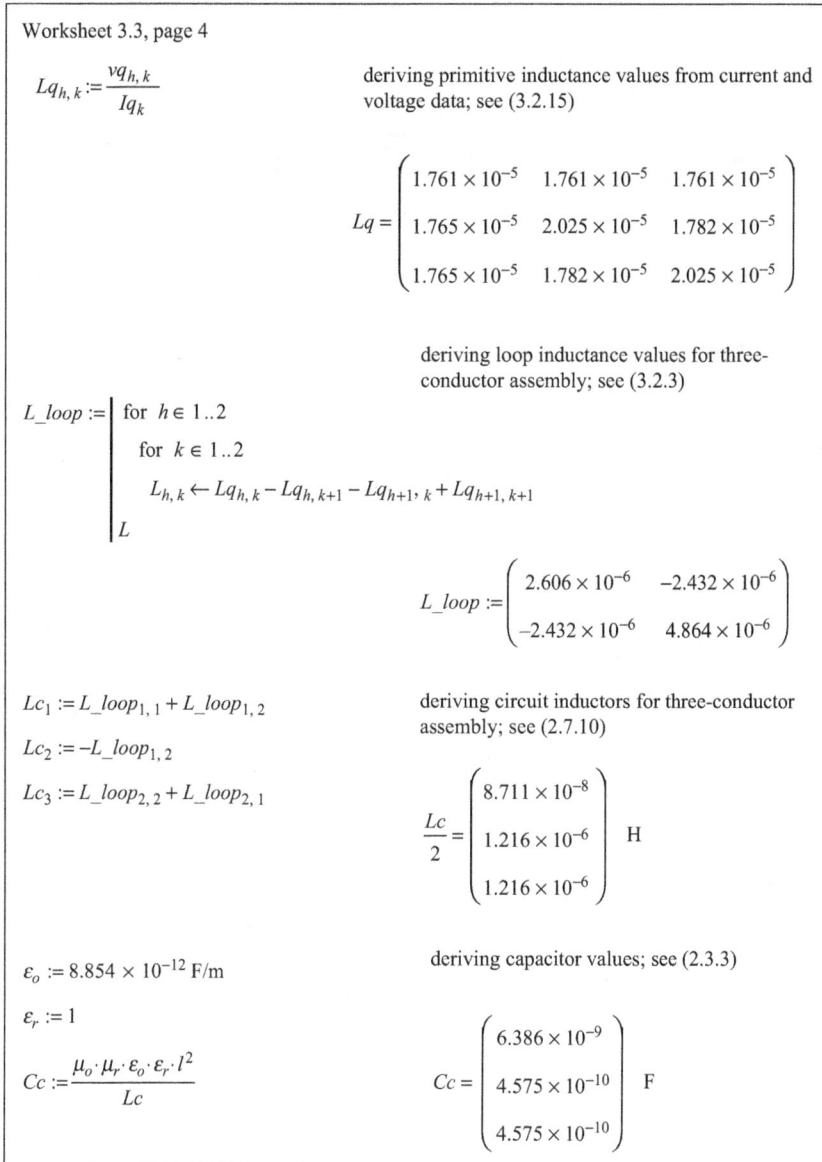

Figure 3.3.3   Calculating values of circuit components.

the main difference being that the number of composites has increased from two to three. Two sets of intermediate results are worthy of note; the vectors for partial voltage Vq and partial current Iq.

The value of Vq_1 illustrates the fact that the sum of the voltages induced in the screen is effectively zero. This is essentially the same as a statement that all internally generated electromagnetic fields are contained within the confines of the screen.

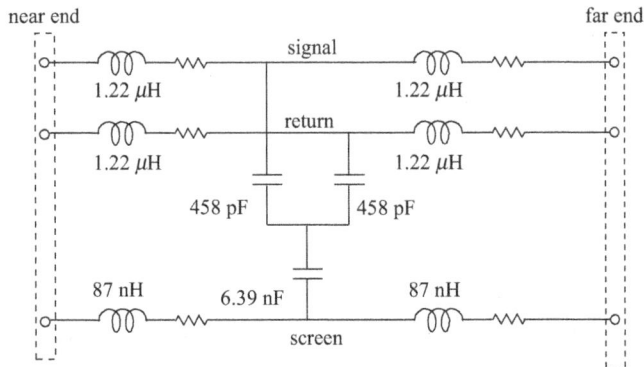

Figure 3.3.4   Representative circuit model of the screened pair example.

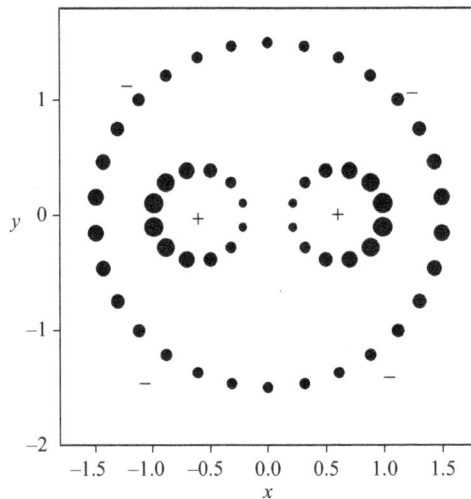

Figure 3.3.5   Bubble plot for screened pair, showing distribution of current.

The net effect of currents in the three conductors is to balance out the voltages induced in the screen. As far as external circuits are concerned, the surface of the screen is at zero voltage. The screen behaves in exactly the same way as the outer conductor of a co-axial cable.

The voltages on conductors 2 and 3 are both 1 V in amplitude, since these conductors are shorted together and the voltage source is set at 1 V. The sign is negative because the currents Iq_1 and Iq_2 are both negative.

An important check to make is to add all the partial currents in the composite conductors together. If the sum of the currents is not zero, then there can be no confidence in any subsequent calculations. It is to be expected that the returning current is equally split between composites 2 and 3, so this is also a useful check.

Figure 3.3.3 illustrates the final set of calculations in the worksheet. The partial inductance values are derived from current and voltage data, and these values are used to create the matrix of loop inductors. For this particular cross section, the loop matrix L_loop is symmetrical, so there is no problem in proceeding to the final two stages; the determination of circuit inductance and circuit capacitance values. Since half-values of inductors are used in the circuit model, then these are the values printed out. This leads to the model of Figure 3.3.4.

The only parameters left to define are the resistors, and this can be done by using general purpose test equipment to measure conductor resistance, or by invoking (2.5.11).

If the cross section of the cable assembly had been asymmetrical, the loop matrix would also have been asymmetrical, leading to a problem in determining a circuit model. Due to the nature of circuit theory, the impedance matrices derived from mesh equations are always symmetrical. It would not have been possible to achieve a one-to-one correlation between an asymmetrical matrix of loop impedances and a symmetrical matrix of circuit impedances. This does not preclude the creation of circuit models, but it does mean that the models for susceptibility and emission would be different.

A useful check on the plausibility of the computations is to produce a bubble plot of the distribution of current in the assembly. This is essentially the same as the cross section illustrated in Figure 3.3.1, but with the radius at each location proportional to the amplitude of the current rather than the actual radius.

Figure 3.3.5 illustrates the resultant plot. It is assumed that the current in the outer screen is flowing out of the page, while the current flow in the inner conductors is into the page. Since the signal and return conductors are shorted together at each end, the picture that emerges is one showing the distribution of common-mode current. In contrast to Figure 3.2.10, the current is concentrated on the outer surfaces of the wire pair.

CHAPTER 4

Transmission line models

It is often required to extend the model to frequencies well beyond the one-tenth wavelength limit of the lumped component model. Such a task can be carried out by invoking the concepts of transmission line theory. This involves the need for circuit components to be defined in terms of Ω/m, F/m, H/m, and S/m.

Using this approach, transmission line theory derives a pair of hybrid equations relating current and voltage at the sending end of the line to the current and voltage at the far end.

Although at least one book on electromagnetic theory derives these equations [4.1], some books do not. So a derivation is provided by Appendix C.

In section 4.1, it is postulated that a T-network circuit model can be used to replicate the relationships of the hybrid equations. A pair of loop equations is derived from this model. It is assumed that a one-to-one relationship exists between the currents and voltages of the two sets of equations. This allows formulae to be developed which define the impedances of the T-network model. These impedances can be described as distributed parameters since they are based on the assumption that R, L, and C are distributed along the length of the cable.

Since this model represents a defined length of cable, the impedances of the branches can be defined in terms of the lumped parameters; that is, in terms of resistors, capacitors, inductors, and frequency, as used in textbooks on circuit theory. A one-to-one correlation is thus established between the components of the distributed parameter model and the lumped parameter model. The distributed parameter model is defined by Figure 4.1.2 and the relationships are defined by (4.2.1)–(4.2.4). The key feature of this set of relationships is that it eliminates the need for the circuit designer to use 'per unit length' parameters.

There is still a fundamental limitation. The formulation assumes that action and reaction are instantaneous between adjacent conductors. An electromagnetic wave takes a finite time to cross the gap between these conductors. Section 4.1 provides a way of estimating the maximum usable frequency of the model.

Section 4.2 develops this model to represent a three-conductor line, by representing each conductor as a T-network. Although the impedance values of each branch are not simple functions of inductance, capacitance, and resistance, the distributed parameter model is still amenable to analysis using circuit theory. (This means that it should be possible to develop SPICE software to implement this approach.)

In section 4.3, a Mathcad program is used to simulate the coupling between two conductors over a ground plane. This is done for three situations; where the terminations at the far end are open-circuit, short-circuit, and loaded by the characteristic impedance. The resultant graphs are extremely informative in comparing the frequency response characteristics of the grounded configuration and the floating configuration.

Section 4.4 shows how the circuit model can be used in conjunction with conducted susceptibility tests.

4.1 Single-T model

The twin-T model of Figure 2.6.1 can be simplified to the T-network of Figure 4.1.1.

Component values are related using:

$$Rc = Rc_1 + Rc_2 \tag{4.1.1}$$

$$Lc = Lc_1 + Lc_2 \tag{4.1.2}$$

$$Cc = \frac{Cc_1 \cdot Cc_2}{Cc_1 + Cc_2} \tag{4.1.3}$$

where Lc_1 and Lc_2 are defined by (2.5.5) and where Cc_1 and Cc_2 are as defined by (2.5.8).

This simplification to the model does not mean that the reference terminals at the near and far ends are at the same voltage. It merely reduces the number of parameters that the mathematics has to handle.

A two-conductor transmission line can be simulated by connecting a large number of such models in series. However, there is a better way. The model can be constructed using distributed parameters, as shown in Figure 4.1.2.

Here, Vs and Is are the voltage and current at the sending end, while Vr and Ir are the voltage and current at the receiving end.

Appendix C shows how a pair of hybrid equations relating these four parameters can be derived from fundamental concepts:

$$Vs = Vr \cdot \cosh(\gamma \cdot l) + Zo \cdot Ir \cdot \sinh(\gamma \cdot l) \tag{4.1.4}$$

$$Is = \frac{Vr}{Zo} \cdot \sinh(\gamma \cdot l) + Ir \cdot \cosh(\gamma \cdot l) \tag{4.1.5}$$

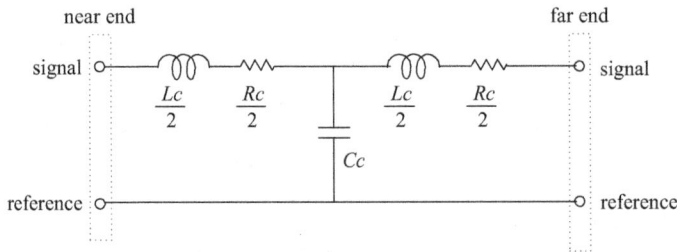

Figure 4.1.1   Single-T model, using lumped parameters.

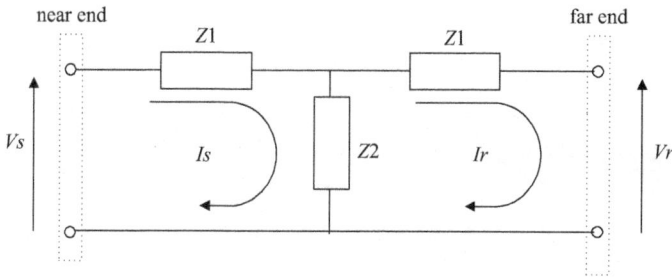

Figure 4.1.2 Single-T model, using distributed parameters.

The circuitry at each interface can be anything from a short-circuit to an open-circuit; it can be resistive, capacitive, inductive, or any combination of all three. It can also be another transmission line. Appendix C is well worth reading, since it recognises the fact that the equations relate to conducting loops. It does not assume the existence of a zero-volt surface.

The propagation constant γ is a function of distributed parameters; resistance per meter, inductance per meter, conductance per meter, and capacitance per meter:

$$\gamma = \sqrt{\left(\frac{Rc}{l} + \frac{j \cdot \omega \cdot Lc}{l}\right) \cdot \left(\frac{Gc}{l} + \frac{j \cdot \omega \cdot Cc}{l}\right)}$$

Simplifying this gives:

$$\gamma = \frac{1}{l} \cdot \sqrt{(Rc + j \cdot \omega \cdot Lc) \cdot (Gc + j \cdot \omega \cdot Cc)} \tag{4.1.6}$$

The characteristic impedance Zo is also a function of distributed parameters:

$$Zo = \sqrt{\frac{Rc + j \cdot \omega \cdot Lc}{l} \cdot \frac{l}{Gc + j \cdot \omega \cdot Cc}}$$

This can also be simplified:

$$Zo = \sqrt{\frac{Rc + j \cdot \omega \cdot Lc}{Gc + j \cdot \omega \cdot Cc}} \tag{4.1.7}$$

The parameter Gc is assumed to represent the conductance of the insulation. That is, the inverse of the insulation resistance. Since most cables use high quality insulation, it can be assumed that $Gc = 0$. However, this can only be an initial assumption.

Circuit theory gives the relationships between currents and voltages of Figure 4.1.2 as:

$$Vs = (Z1 + Z2) \cdot Is - Z2 \cdot Ir \tag{4.1.8}$$

$$Vr = Z2 \cdot Is - (Z1 + Z2) \cdot Ir \tag{4.1.9}$$

Rearranging (4.1.9) gives:

$$Is = \frac{Vr}{Z2} + \left(1 + \frac{Z1}{Z2}\right) \cdot Ir \qquad (4.1.10)$$

Substituting for Is in (4.1.8) gives:

$$Vs = (Z1 + Z2) \cdot \left[\frac{Vr}{Z2} + \left(1 + \frac{Z1}{Z2}\right) \cdot Ir\right] - Z2 \cdot Ir$$

$$= \left(1 + \frac{Z1}{Z2}\right) \cdot Vr + \left[Z1 + Z2 + \frac{Z1^2}{Z2} + Z1 - Z2\right] \cdot Ir$$

Simplifying further:

$$Vs = \left(1 + \frac{Z1}{Z2}\right) \cdot Vr + Z1 \cdot \left(2 + \frac{Z1}{Z2}\right) \cdot Ir \qquad (4.1.11)$$

Equations (4.1.11) and (4.1.10) form a pair of equations which can be correlated with (4.1.4) and (4.1.5).

Comparing (4.1.4) with (4.1.11) gives:

$$1 + \frac{Z1}{Z2} = \cosh(\gamma \cdot l) \qquad (4.1.12)$$

and

$$Z1 \cdot \left(2 + \frac{Z1}{Z2}\right) = Zo \cdot \sinh(\gamma \cdot l) \qquad (4.1.13)$$

Comparing (4.1.10) with (4.1.5) gives:

$$\frac{1}{Z2} = \frac{1}{Zo} \cdot \sinh(\gamma \cdot l) \qquad (4.1.14)$$

and also replicates (4.1.12):

$$1 + \frac{Z1}{Z2} = \cosh(\gamma \cdot l)$$

To make subsequent equations less unwieldy, let:

$$\theta = \gamma \cdot l \qquad (4.1.15)$$

The parameter θ can be described as a phase variable since it effectively determines the phase shift between input and output signals of a transmission line and since it varies with frequency. From (4.1.12) and (4.1.15):

$$\frac{Z1}{Z2} = \cosh \theta - 1$$

Substituting for $\dfrac{Z1}{Z2}$ in (4.1.13) gives:

$$Z1 \cdot (1 + \cosh \theta) = Zo \cdot \sinh \theta$$

This leads to:

$$Z1 = \frac{\sinh \theta}{1 + \cosh \theta} \cdot Zo$$

Invoking the half-angle formula:

$$Z1 = \frac{2 \cdot \sinh \dfrac{\theta}{2} \cdot \cosh \dfrac{\theta}{2}}{\cosh^2 \left(\dfrac{\theta}{2}\right) + \sinh^2 \left(\dfrac{\theta}{2}\right) + \cosh^2 \left(\dfrac{\theta}{2}\right) - \sinh^2 \left(\dfrac{\theta}{2}\right)} \cdot Zo$$

This can be simplified to:

$$Z1 = Zo \cdot \tanh \left(\frac{\theta}{2}\right) \tag{4.1.16}$$

Rearranging (4.1.14):

$$Z2 = Zo \cdot \operatorname{cosech} \theta \tag{4.1.17}$$

Equations (4.1.16) and (4.1.17) relate the Z-parameters of Figure 4.1.2 to the characteristic impedance and the phase variable of the transmission line.

Finally, from (4.1.6) and (4.1.15):

$$\theta = \sqrt{(Rc + j \cdot \omega \cdot Lc) \cdot (Gc + j \cdot \omega \cdot Cc)} \tag{4.1.18}$$

Using (4.1.7) and (4.1.18), Zo and θ can be calculated. These can then be used in (4.1.16) and (4.1.17) to obtain values for the components of Figure 4.1.2. This means that the lumped parameter model of Figure 4.1.1 can easily be transformed into one which used distributed parameters. The transformation would be:

$$\frac{1}{2} \cdot (Rc + j \cdot \omega \cdot Lc) \to Zo \cdot \tanh \left(\frac{\theta}{2}\right)$$
$$\frac{1}{(Gc + j \cdot \omega \cdot Cc)} \to Zo \cdot \operatorname{cosech} \theta \tag{4.1.19}$$

Invoking this transformation allows the circuit model of Figure 4.1.2 to simulate the response of a transmission line over the same range of frequencies at which the hybrid equations are valid.

The characteristic impedance Zo is independent of the length of the line. Given knowledge of the values of Rc, Lc, Gc, and Cc, the phase variable θ is also independent of the line length. The variable l does not appear in (4.1.18). Although it is necessary to invoke the concept of 'distributed parameters' to derive the hybrid equations, it is not necessary to use the concept when calculating impedance values for the model of Figure 4.1.2.

An electromagnetic wave takes a finite time to cross the gap between the conductors of any cableform. Just as SPICE modelling is limited by the electrical length of the assembly, modeling using distributed parameters is limited by the radial separation of the adjacent conductors. It can be reasoned that, if *rmax* is the maximum separation between the centers of two conductors in the cross section of the assembly-under-review, then the maximum frequency at which the distributed parameter model is usable is *fmax*, where:

$$fmax = \frac{vmin}{10 \cdot rmax} \tag{4.1.20}$$

and *vmin* is the propagation velocity of the electromagnetic wave in the insulating medium with the highest relative permittivity in the section-under-review.

That is, if the velocity of propagation is 200 m/µs and the maximum spacing between conductors is 10 mm, then the maximum frequency at which this modelling is usable would be about 2 GHz.

When distributed parameters are used, there is no theoretical limit to the length of the cable which can be modeled. It is assumed that the cross section of the cable is constant and that the lengths of the conductors are identical.

4.2 Triple-T model

The next step is to create a distributed parameter model for the three-conductor transmission line. Figure 4.2.1 illustrates the starting point; the lumped parameter model developed in section 2.7.

Representing each conductor by the T-network of section 4.1 leads to Figure 4.2.2. The relationships between components of the two models are:

$$Zo_i = \sqrt{\frac{Rc_i + j \cdot \omega \cdot Lc_i}{Gc_i + j \cdot \omega \cdot Cc_i}} \tag{4.2.1}$$

$$\theta_i = \sqrt{(Rc_i + j \cdot \omega \cdot Lc_i) \cdot (Gc_i + j \cdot \omega \cdot Cc_i)} \tag{4.2.2}$$

$$Z_{1,i} = Zo_i \cdot \tanh\left(\frac{\theta_i}{2}\right) \tag{4.2.3}$$

$$Z_{2,i} = Zo \cdot \text{cosech}(\theta_i) \tag{4.2.4}$$

These four equations can be collected together to compute the values of a pair of Z-parameters from a set of R, L, C, and G values. Indices are used to retain correlation between the parameters and the conductors.

Given the definitions above, the result of the computation would be an array of two rows by three columns. The first row would hold values correlating with horizontal branches of Figure 4.2.2. The second row would hold values for the vertical branches. Each column would be associated with a conductor. The conductance parameter G_i is included in the formulae to allows losses in the insulation to be simulated. It is fair to make an initial assumption that all conductance values are zero.

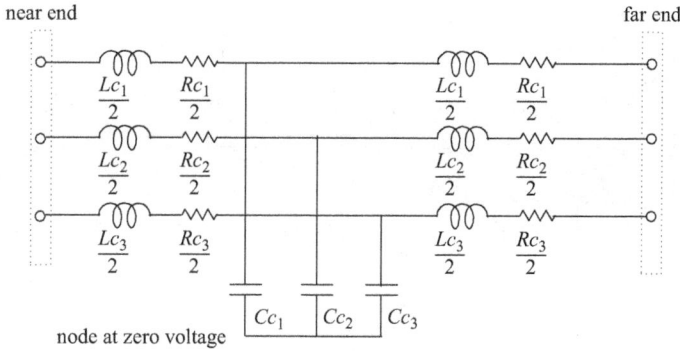

Figure 4.2.1   Triple-T network of lumped parameters.

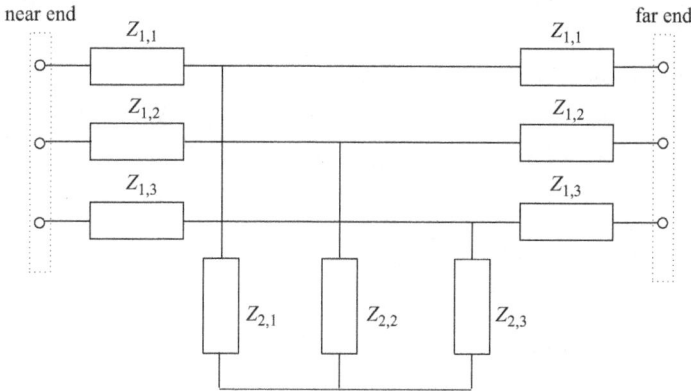

Figure 4.2.2   Triple-T network of distributed parameters.

One feature of the Triple-T model is that it only includes components representing the cable assembly.

It is necessary for the components at the cable terminations to be included before any simulation is possible. This results in the full model of Figure 4.2.3.

Impedances within the unit at the near end are identified as Zn_i, while those within the equipment at the far end are designated Zf_i.

Four separate loops can be identified. Currents in each loop have been defined, as well as all possible voltage sources. It is assumed that conductors 1 and 2 carry the differential-mode signal current. So voltage sources V_1 and V_3 are located within the equipment units. Voltage sources V_2 and V_4 represent possible interference sources in the common-mode loop.

The four loop equations for the circuit model of Figure 4.2.3 are:

$$
\begin{aligned}
V_1 &= Z11 \cdot I_1 + Z12 \cdot I_2 + Z13 \cdot I_3 + Z14 \cdot I_4 \\
V_2 &= Z12 \cdot I_1 + Z22 \cdot I_2 + Z23 \cdot I_3 + Z24 \cdot I_4 \\
V_3 &= Z13 \cdot I_1 + Z23 \cdot I_2 + Z33 \cdot I_3 + Z34 \cdot I_4 \\
V_4 &= Z14 \cdot I_1 + Z24 \cdot I_2 + Z34 \cdot I_3 + Z44 \cdot I_4
\end{aligned}
\tag{4.2.5}
$$

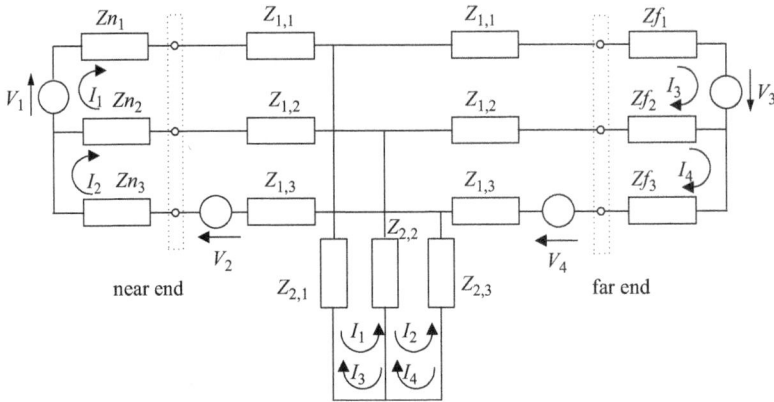

Figure 4.2.3 Full circuit model, using lumped and distributed parameters.

where the loop impedances are:

$$Z11 = Zn_1 + Z_{1,1} + Z_{2,1} + Z_{2,2} + Z_{1,2} + Zn_2$$
$$Z12 = -(Z_{1,2} + Z_{2,2} + Zn_2)$$
$$Z13 = -(Z_{2,1} + Z_{2,2})$$
$$Z14 = Z_{2,2}$$
$$Z22 = Zn_2 + Z_{1,2} + Z_{2,2} + Z_{2,3} + Z_{1,3} + Zn_3$$
$$Z23 = Z_{2,2}$$
$$Z24 = -(Z_{2,2} + Z_{2,3})$$
$$Z33 = Zf_2 + Z_{1,2} + Z_{2,2} + Z_{2,1} + Z_{1,1} + Zf_1$$
$$Z34 = -(Z_{1,2} + Z_{2,2} + Zf_2)$$
$$Z44 = Zf_3 + Z_{1,3} + Z_{2,3} + Z_{2,2} + Z_{1,2} + Zf_2$$

$$(4.2.6)$$

The loop impedances are derived by inspection of Figure 4.2.3. For example, $Z11$ is the sum of the impedances in the loop carrying the current I_1, while $Z34$ is the sum of the impedances which share currents I_3 and I_4. The negative sign indicates that these two currents flow in opposite directions.

Equations (4.2.1)–(4.2.6) provide a traceable relationship between conventional circuit components and the array of loop impedances. If all the component values are defined, and all the source voltages are defined, then vector algebra can be used to derive all the currents.

It then becomes a routine matter to calculate the voltage between any two points or the current in any component, at any particular frequency.

Since there is now a one-to-one correlation between the branches of the lumped parameter and distributed parameter networks, it becomes possible to define the circuit model in terms of the lumped parameters, but analyze it using the loop impedances of (4.2.6). This leads to the general circuit model of Figure 4.2.4. It is entirely possible to replace the circuitry at the near and far ends with components representing the interface circuits of the equipment-under-review.

In this model, the values for the components of the triple-T network can be derived from geometrical data of the cable and structure, while values for the components of the interface circuitry at the near end and far end of the cable are defined by the designer.

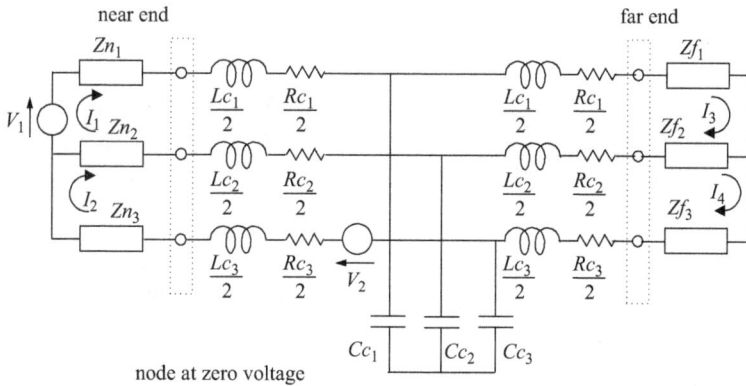

Figure 4.2.4 General circuit model of three-conductor signal link.

It is also possible to assign values to all the components of the model by carrying out electrical tests on an actual assembly. Chapter 7 describes how this can be done.

4.3 Cross-coupling

Enough analytical tools now exist to allow an initial assessment to be made of a wide range of intra-system interference problems. Perhaps the simplest illustration of the process lies in the analysis of the cross-coupling between two conductors over a ground plane.

Figure 4.3.1 illustrates the configuration. It is assumed that the conductors are both 1 mm in diameter, and that each is spaced 1 mm above the plane. The separation between them is set at 4 mm and the length of the assembly is assumed to be 1 m.

Figure 4.3.2 sets out the calculations involved in determining the component values for the circuit model. It is a copy of the first page of a four-page Mathcad worksheet. The first line records the values of the physical constants involved in the derivation; the permeability, permittivity, and the speed of light. It is assumed that the values of relative permittivity and permeability are both unity. The other physical constant is ρ, the resistivity of copper.

The second line of the worksheet defines values for the spatial parameters of the configuration. These correlate with the parameters of Figure 2.11.1.

Steady-state resistance values Rss for the two circular-section conductors are determined using (2.5.11). A guess-value of 5 mΩ is assigned to the ground plane. Above a certain frequency, skin effect will cause an increase in the resistance of the conductors. The cross-over frequency Fx is determined using (2.5.14) and the relationship between Rc and f is defined by (2.5.15).

Inductance values for the circuit model are derived from (2.11.3). Since the two wires are equidistant from the plane, Lc_3 is equal to Lc_1. The three values are stored in the vector Lc. Capacitor values are obtained quite simply, by invoking (2.3.8).

Another significant parameter is Fq, the frequency at which quarter-wave resonance occurs. This is determined using (2.3.9). Knowledge of this frequency allows subsequent analysis to include maximum values of the peaks in the response curves.

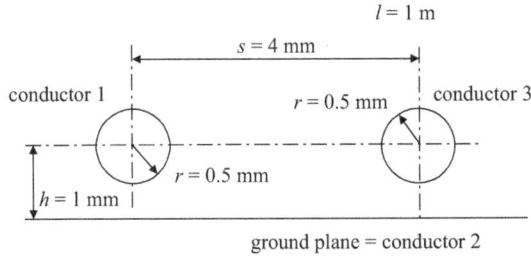

Figure 4.3.1   Two conductors over a ground plane.

Since half-values of resistance and inductance are used in the circuit model of Figure 4.2.1, these are the values printed out in the worksheet. Adding the values of the three capacitors provides sufficient information to draw the representative circuit model for the conductor assembly.

Before the response can be analyzed, it is necessary to include values for the interface components of Figure 4.2.4. Invoking worst-case conditions, it is assumed that the components at the near end are short-circuits and that the wire terminals at the far end are open-circuit. Following normal practice used in SPICE analysis, the open-circuits are represented by 10 MΩ resistors. To allow for the option of simulating losses in the insulation, conductance values Gc are also defined. In this model, all three are assumed to be zero.

Data derived from this first page of the worksheet can be illustrated by the circuit model of Figure 4.3.3. It is assumed that the assembly of Figure 4.3.1 is configured as a pair of transmission lines; with the ground plane acting as the return conductor for both signals. It is also assumed that there is only one voltage source, between conductor 1 and the ground plane. This model represents the situation where the culprit circuit uses conductor 1, while the victim circuit uses conductor 3.

The purpose of the analysis is to simulate the current that would flow in the victim loop over a range of frequencies. Since the far end is open circuit, only the current at the near end will be of any significance. Hence the objective is to simulate the current I_2 when voltage V_1 is applied. If V_1 is held at a constant value of one volt, the result will be the frequency response of the transfer admittance.

Figure 4.3.4 sets out the two program functions derived in section 4.2. It constitutes page 2 of the Mathcad worksheet.

The function Zbranch(f) supplies one input variable; the frequency f. It calculates the values of the two distributed impedances associated with each conductor, and assembles them in the output array Z. As indicated in section 4.2, this contains two rows of three columns. A feature worth noting is the fact that the value of each resistor is calculated at each frequency. This allows skin effect to be incorporated into the analysis.

In addition to supplying frequency as an input variable, the function Zloop(f, Zf) also includes the vector containing component values at the far interface. This allows any of these components to be varied, subsequent to the definition of the function.

The first program step in the Zloop() function is to call up the Zbranch(f) function and place the results in the local variable Z. This allows component values of individual branches to be accessed as required when determining values for the loop impedances. These impedance values are then placed in the four-by-four array at the end of the subroutine.

Worksheet 4.3, page 1

$\mu_o := 4 \cdot \pi \cdot 10^{-7}$ H/m $\qquad$ $\varepsilon_o := 8.854 \cdot 10^{-12}$ F/m $\qquad$ $c := 2.998 \cdot 10^8$ m/s

$\rho := 1.7 \cdot 10^{-8}$ Ω m $\qquad$ $l := 1$ m $\qquad$ $h := 1 \cdot 10^{-3}$ m

$s := 4 \cdot 10^{-3}$ m $\qquad$ $r := 0.5 \cdot 10^{-3}$ m $\qquad$ see Figure 4.3.1

$Rss_1 := \dfrac{\rho \cdot l}{\pi \cdot r^2}$ $\qquad$ $Rss_3 := Rss_1$ $\qquad$ $Rss_2 := 0.005$ Ω $\qquad$ see (2.5.11)

$Fx := \dfrac{4 \cdot \rho}{\mu_o \cdot \pi \cdot r^2} = 6.89 \times 10^4$ Hz $\qquad$ see (2.5.14)

$Lc_1 := \dfrac{\mu_o \cdot l}{2 \cdot \pi} \cdot \ln\left(\dfrac{2 \cdot h \cdot s}{r \cdot \sqrt{s^2 + 4 \cdot h^2}} \right)$ H

$Lc_2 := \dfrac{\mu_o \cdot l}{2 \cdot \pi} \cdot \ln\left(\dfrac{\sqrt{s^2 + 4 \cdot h^2}}{s} \right)$ H $\qquad$ see (2.11.3)

$Lc_3 := Lc_1$

$Cc := \dfrac{1}{Lc} \cdot \left(\dfrac{l}{c} \right)^2$ F $\qquad$ see (2.3.8)

$Fq := \dfrac{1}{4 \cdot \sqrt{Lc_1 \cdot Cc_1}} = 7.495 \times 10^7$ Hz $\qquad$ see (2.3.9)

component values for circuit model of Figure 4.2.3:

$\dfrac{Rss}{2} = \begin{pmatrix} 0.011 \\ 2.5 \times 10^{-3} \\ 0.011 \end{pmatrix}$ $\qquad$ $\dfrac{Lc}{2} = \begin{pmatrix} 1.275 \times 10^{-7} \\ 1.116 \times 10^{-8} \\ 1.275 \times 10^{-7} \end{pmatrix}$ $\qquad$ $Cc = \begin{pmatrix} 4.364 \times 10^{-11} \\ 4.986 \times 10^{-10} \\ 4.364 \times 10^{-11} \end{pmatrix}$

$Zn = \begin{pmatrix} 0 \\ 0 \\ 0 \end{pmatrix}$ Ω $\qquad$ $Zf = \begin{pmatrix} 10^7 \\ 0 \\ 10^7 \end{pmatrix}$ Ω $\qquad$ $Gc = \begin{pmatrix} 0 \\ 0 \\ 0 \end{pmatrix}$ S

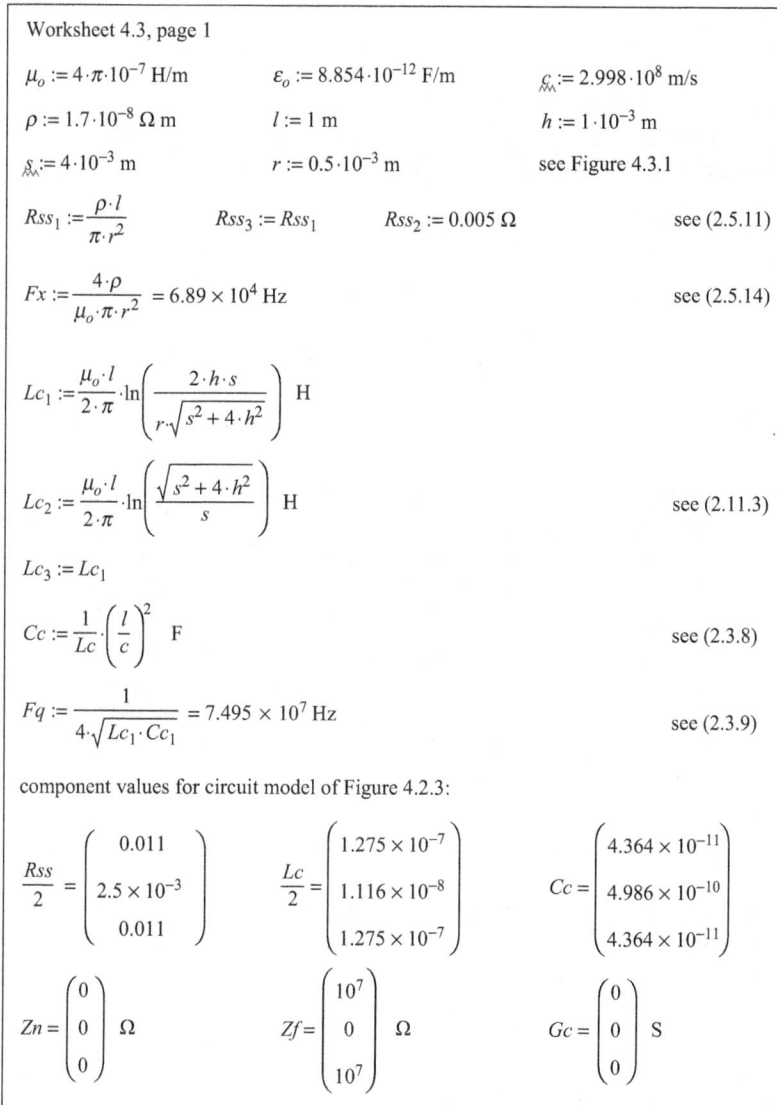

Figure 4.3.2 Deriving circuit parameters.

Creating a graph of the response involves the definition of two vectors; the frequency and the output variable, and this is the task performed by page 3 of the worksheet. Figure 4.3.5 illustrates the process.

A set of spot frequencies is defined in the first line. The integer n is used to set the number of spot frequencies in the range between zero and the frequency of quarter-wave resonance, Fq. To cover the range up to a full wavelength, the control variable s is set to make four times this number of steps. This allows each spot frequency, F_s, to be a multiple of the quarter-wave frequency. Using this method of defining the spot frequencies ensures that

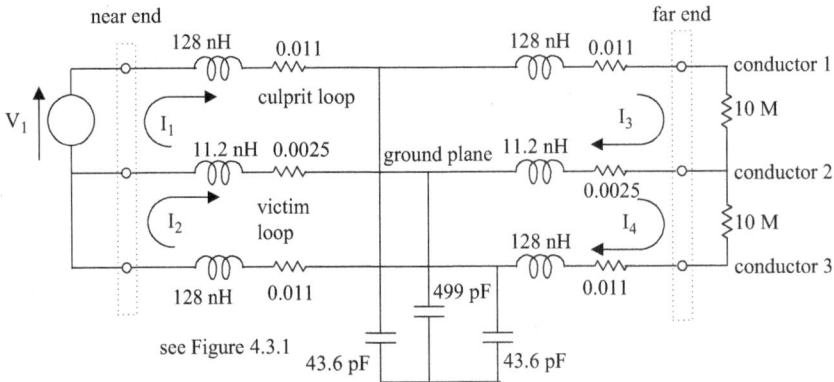

Figure 4.3.3   Representative circuit model of cross-coupling example.

all the resonant frequencies are selected and that the amplitude of every peak and every trough in the response is calculated.

In the model of Figure 4.3.3, the source voltage generator is located in loop 1. Setting the amplitude of loop 1 at 1 V and the other three sources at zero voltage allows the voltage vector V to be defined.

Since all the inputs have now been defined, it becomes possible to create the main program, and this is illustrated on the third line of the worksheet of Figure 4.3.5.

The first program step is to select a spot frequency from the vector F and assign it to the local variable f. The second line calls up the function $Zloop(f, Zf)$. Since Zf has already been defined, (see Figure 4.3.2) both variables are 'visible' to the function. The output is stored in the local variable Z.

The next step is to calculate the values of all four loop currents and place them in the vector I.

The final step in the main subroutine is to select I_2, the current at the near end of the victim loop, and determine its amplitude. Dividing this current by the input voltage V_1 gives the value of the transfer admittance. Since the value of V_1 is unity, the values of current and transfer admittance are identical. Finally, the output is stored in the relevant location in the vector Yoc; the transfer admittance when the line terminations are open-circuit. The graph of this function is shown on Figure 4.3.6.

Although normal convention is to display such a function using logarithmic scale for both parameters, in this case a linear scale is used for frequency. This highlights the symmetrical nature of the response. Symmetry is evident in the fact that the response between 150 MHz and 300 MHz is almost the same as between zero and 150 MHz; almost the same, but not identical. Skin effect causes the resistance to increase, reducing the amplitude of the second peak and increasing the amplitude of the second trough.

By resetting all the components of the impedance vector Zf to zero and re-running the main program, the vector Ysc can be created to define the response of the assembly when the terminations are short-circuited. The response of the assembly when it is critically damped can also be determined. The program steps are illustrated by Figure 4.3.7.

Figure 4.3.8 provides a further illustration of the symmetry of the response, by comparing the open-circuit response with the response when the terminations at the far end are short-circuited. Transfer admittance under short-circuit conditions, Ysc, is illustrated by the dotted curve.

Worksheet 4.3 page 2

$Zbranch(f) :=$ $\omega \leftarrow 2 \cdot \pi \cdot f$

for $i \in 1..3$

$$Rc_i \leftarrow Rss_i \cdot \sqrt{1 + \frac{f}{Fx}}$$ see (2.5.15)

$$\theta \leftarrow \sqrt{(Rc_i + j \cdot \omega \cdot Lc_i) \cdot (Gc_i + j \cdot \omega \cdot Cc_i)}$$ see (4.2.1)

$$Zo \leftarrow \sqrt{\frac{Rc_i + j \cdot \omega \cdot Lc_i}{Gc_i + j \cdot \omega \cdot Cc_i}}$$ see (4.2.2)

$$Z_{1,i} \leftarrow Zo \cdot \tanh\left(\frac{\theta}{2}\right)$$ see (4.2.3)

$$Z_{2,i} \leftarrow Zo \cdot \text{csch}(\theta)$$ see (4.2.4)

Z

$Zloop(f, Zf) :=$ $Z \leftarrow Zbranch(f)$ see (4.2.6)

$Z11 \leftarrow Zn_1 + Z_{1,1} + Z_{2,1} + Z_{2,2} + Z_{1,2} + Zn_2$

$Z12 \leftarrow -(Z_{1,2} + Z_{2,2} + Zn_2)$

$Z13 \leftarrow -(Z_{2,1} + Z_{2,2})$

$Z14 \leftarrow Z_{2,2}$

$Z22 \leftarrow Zn_2 + Z_{1,2} + Z_{2,2} + Z_{2,3} + Z_{1,3} + Zn_3$

$Z23 \leftarrow Z_{2,2}$

$Z24 \leftarrow -(Z_{2,2} + Z_{2,3})$

$Z33 \leftarrow Zf_2 + Z_{1,2} + Z_{2,2} + Z_{2,1} + Z_{1,1} + Zf_1$

$Z34 \leftarrow -(Z_{1,2} + Z_{2,2} + Zf_2)$

$Z44 \leftarrow Zf_3 + Z_{1,3} + Z_{2,3} + Z_{2,2} + Z_{1,2} + Zf_2$

$$\begin{pmatrix} Z11 & Z12 & Z13 & Z14 \\ Z12 & Z22 & Z23 & Z24 \\ Z13 & Z23 & Z33 & Z34 \\ Z14 & Z24 & Z34 & Z44 \end{pmatrix}$$

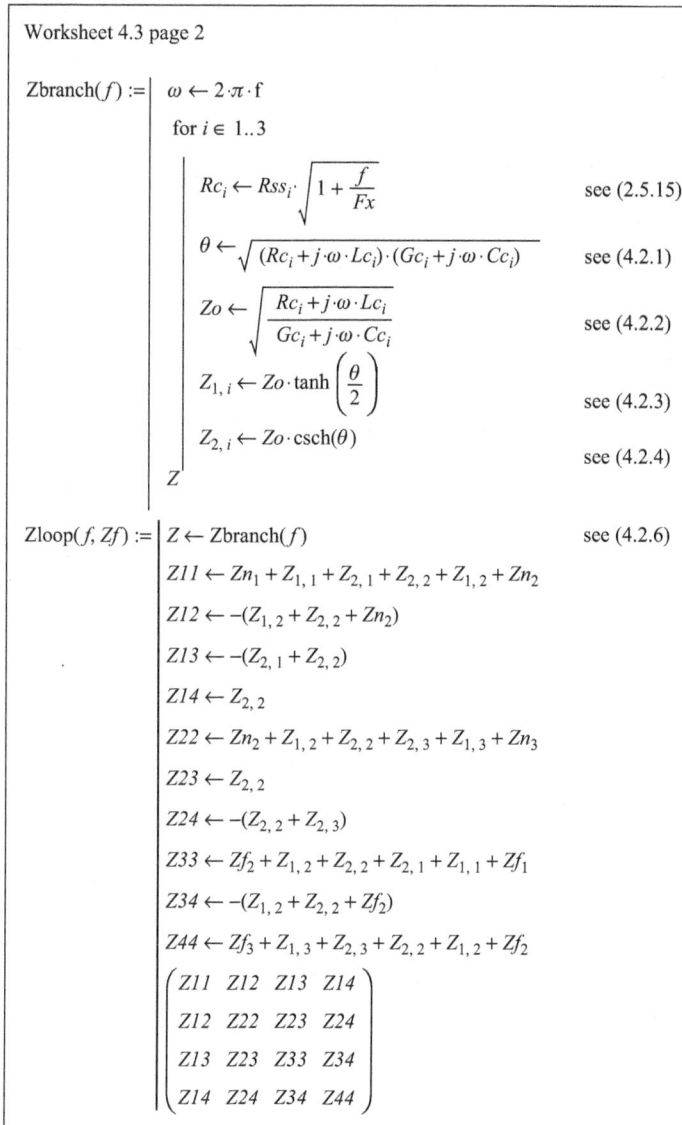

Figure 4.3.4 Calculating branch impedances and loop impedances.

When each conductor is terminated by its characteristic impedance, the response is perfectly flat over the entire frequency range. This is illustrated by the *Ycrit* curve; the dashed line.

The bottom half of the graph, below the *Ycrit* curve, is almost the mirror image of the top half.

Many lessons can be derived from this set of curves, not least among them the fact that the only configuration which is free from peaks in the transfer admittance is also the one which provides the highest efficiency in terms of signal transmission.

Worksheet 4.3, page 3

$n := 100$ $\underset{\sim}{s} := 1..4 \cdot n$ $\underset{\sim}{F_s} := s \cdot \dfrac{Fq}{n}$ defining the spot frequencies

$$\underset{\sim}{V} := \begin{pmatrix} 1 \\ 0 \\ 0 \\ 0 \\ 0 \end{pmatrix} V$$ defining the voltage source

$Yoc_s := \big| f \leftarrow F_s$ main program

$\quad Z \leftarrow Zloop(f, Zf)$ defining the current at the near end of the

$\quad I \leftarrow lsolve(Z, V)$ victim line when the far end is open-circuit;
        see Figure 4.3.3

$\quad \big| I_2 \big|$

Figure 4.3.5   Calculating frequency response of open-circuit transfer admittance.

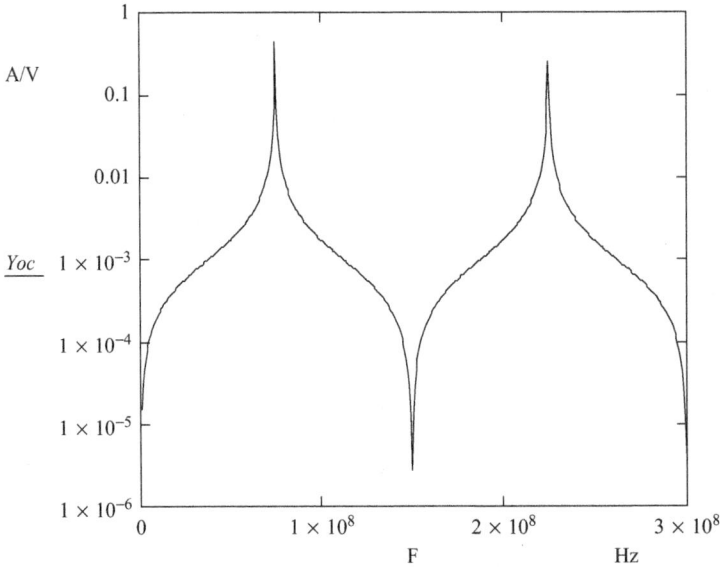

Figure 4.3.6   Transfer admittance with open circuit terminations.

In practice, the peaks will be much lower and much more rounded than those illustrated, since this model does not take account of losses due to radiated emission. Chapter 5 develops the use of models to cater for this effect.

4.4 Bench test models

Bench testing is a vital part of the product development process. It provides assurance that the design requirements are being met. Correlating the observed performance with the predicted performance allows the assumptions used in the theoretical analysis to be verified; or

Worksheet 4.3, page 4

$$Zf := \begin{pmatrix} 0 \\ 0 \\ 0 \end{pmatrix}$$

transfer admittance with short-circuit terminations

$$Ysc_s := \begin{vmatrix} f \leftarrow Fs \\ Z \leftarrow \text{Zloop}(f, Zf) \\ I \leftarrow \text{Isolve}(Z, V) \\ |I_2| \end{vmatrix}$$

$$Zf := \sqrt{\frac{Lc}{Cc}} \qquad Zf = \begin{pmatrix} 76.432 \\ 6.69 \\ 76.432 \end{pmatrix}$$

transfer admittance with critical damping

$$Ycrit_s := \begin{vmatrix} f \leftarrow F_s \\ Z \leftarrow \text{Zloop}(f, Zf) \\ I \leftarrow \text{Isolve}(Z, V) \\ |I_2| \end{vmatrix}$$

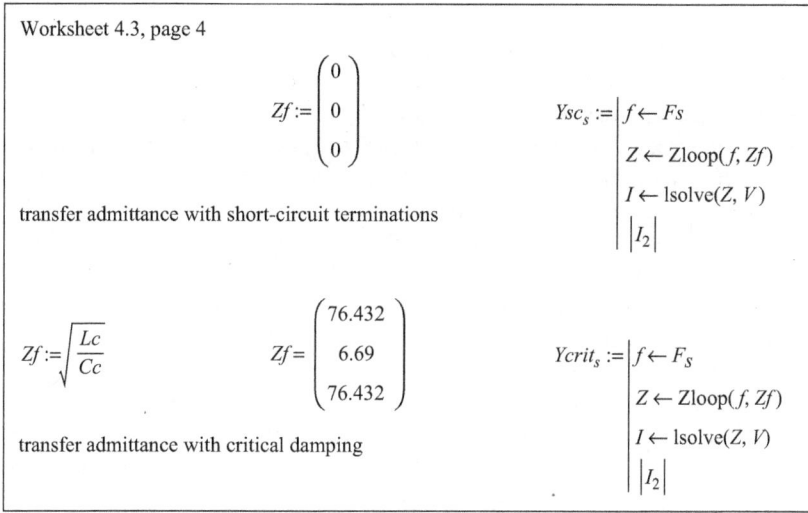

Figure 4.3.7 Calculating frequency response for short-circuited and critically damped lines.

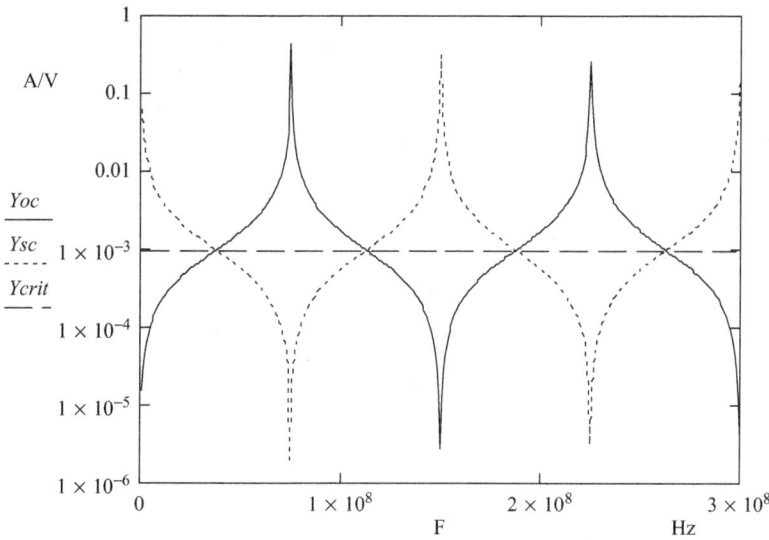

Figure 4.3.8 Transfer admittance for open-circuit, short-circuit, and critically damped lines.

modified in the light of more accurate information. It allows corrections to be made to the design to improve its performance. This is true of requirements such as system function, response time, reliability, size, and cost.

It can also apply to the requirements of electromagnetic compatibility. Since the EMC Test Houses have already acquired a wide experience in meeting the regulatory requirements, it is a logical exercise to tap into that experience to identify the relevant test equipment and test methods.

Two problems immediately emerge. The equipment is extremely expensive, and the test procedures require the services of highly skilled engineers. It would be far too costly for a small organization to attempt to emulate this approach.

However, the product designer also has one advantage; a big advantage. It is possible to design and use interface equipment which allows direct visibility of the signals being processed. This is not an option available to the Test Houses, since any special modification of the equipment-under-test would render the test results invalid.

Another advantage is the fact that General Purpose Test Equipment is already in hand to carry out functional tests of the assembly-under-review. This could also be used for EMC testing. Also, the frequency range of the tests can be tailored to the range over which the equipment-under-test normally function. There is no need to cover the range covered by the Test Houses.

The simplest set of bench tests would be to analyze the performance of the inter-connecting cables; specifically, to measure the frequency response of the transfer admittance. The results could be compared directly with those obtained by circuit modeling. But first, the representative circuit model needs to be created.

Figure 4.4.1 illustrates a setup which can be used to measure the conducted emission of any wiring configuration. In this case, it is assumed that the assembly-under-review is a 10 m length of a conductor pair routed 5 mm above a ground plane.

An input signal can be applied to the near-end terminals via a splitter box. This allows the input voltage to be monitored on channel 1 of the oscilloscope. Channel 2 is used to monitor the common-mode current. The 100 Ω load at the far end of the cable is floating.

The ratio of output to input gives the value of the transfer admittance at the frequency of the signal generator output. Repeating the measurements over a range of spot frequencies enables the frequency response characteristic of the conducted emission to be determined.

If the physical geometry of the signal link is as shown by Figure 4.4.2, then the circuit model can be derived by invoking the calculations recorded on the worksheet of Figure 4.3.2. As with SPICE modeling, it is necessary to simulate the open-circuit as a high value resistance; in this case 10 MΩ.

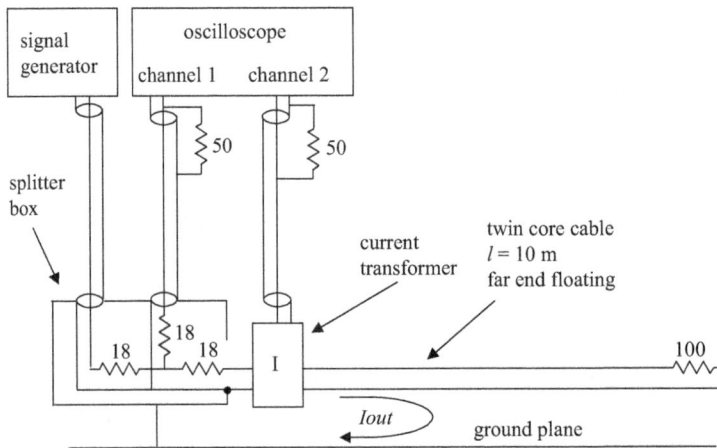

Figure 4.4.1   Setup for conducted emission test – floating configuration.

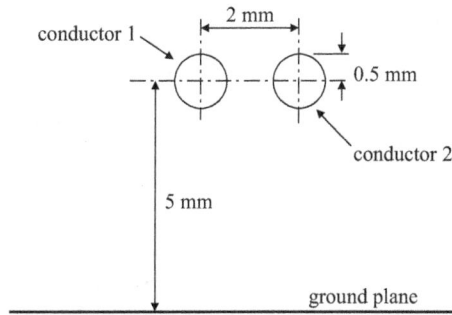

Figure 4.4.2 Cross-sectional view of cable assembly.

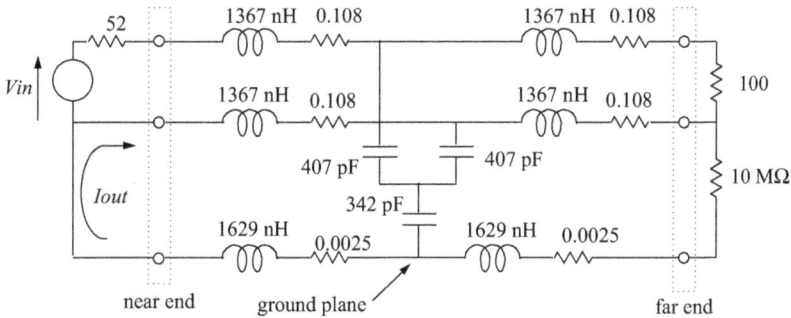

Figure 4.4.3 Representative circuit model of conducted emission test setup.

The resultant model is shown in Figure 4.4.3. It is worth noting that the source voltage is in the differential-mode loop while the output current is in the common-mode loop.

Conducted susceptibility measurements would involve a setup similar to that illustrated by Figure 4.4.4. Here, the input voltage is applied to the common-mode loop via an injection transformer. Since any transformer has an output impedance, the applied voltage can reduce as the load current increases. This effect is catered for by using a separate turn on the transformer to monitor the actual voltage applied. One channel of the oscilloscope provides the means of measurement. This enables the amplitude of the input voltage to be determined.

The second channel of the oscilloscope allows the resultant current in the differential-mode loop to be measured. Again, the ratio of output to input provides a value for the transfer admittance. Measurement over a range of spot frequencies provides the frequency response of the conducted susceptibility characteristic.

A prediction of the response of this configuration can be achieved with the model of Figure 4.4.5. Since the assembly-under-review is exactly the same as that used for the conducted emission test, the passive components of the two models are identical. The only difference lies in the fact that the voltage source is now in the common-mode loop, while the output current is now in the differential-mode loop.

An analysis of the frequency response of either model can be carried out by utilizing the program set out by Figures 4.3.4 and 4.3.5. It is only necessary to redefine the component values, the input voltage, the output current, and the frequency range.

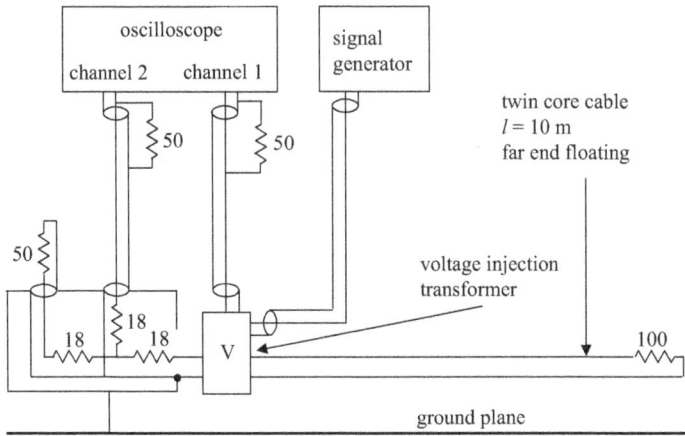

Figure 4.4.4   Setup for conducted susceptibility test.

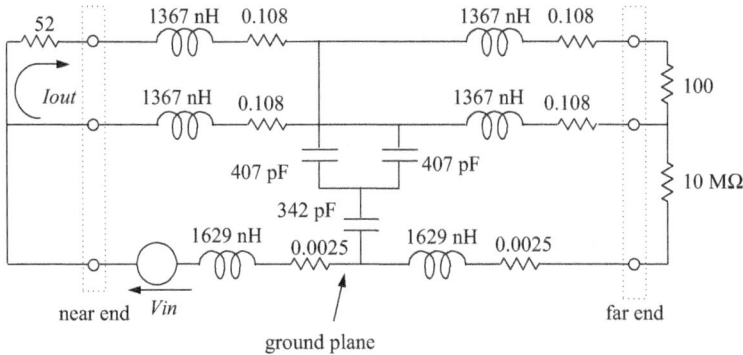

Figure 4.4.5   Representative circuit model of conducted susceptibility test setup.

Analyzing the response of Figure 4.4.3 gives the characteristic shown on Figure 4.4.6. Analyzing the response of Figure 4.4.5 gives exactly the same characteristic. This illustrates the fact that the transfer admittance for conducted susceptibility is exactly the same as for conducted emission. With the circuit models, the responses are identical. This confirms the conclusion of section 2.9.

Actual tests would show some differences, due to the effect of the test equipment and due to radiated emission.

The response curve also illustrates that although the floating configuration gives excellent performance at low frequencies, it actually amplifies the level of interference at the quarter-wave frequency.

If the configuration-under-review is changed, to add a link between the ground plane and the return conductor at the far end, then the circuit model for the susceptibility test changes to that shown on Figure 4.4.7.

The response of this setup is given by Figure 4.4.8. This illustrates the fact that, for the grounded configuration, minimum interference occurs at the quarter-wave frequency, while the peak level occurs at the half-wave frequency.

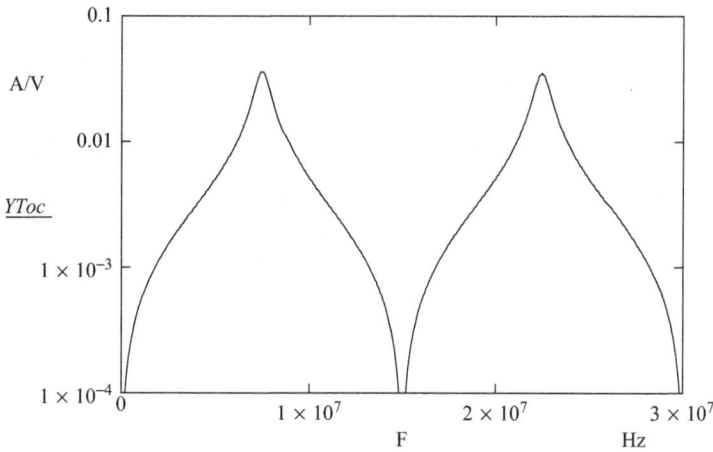

Figure 4.4.6   Transfer admittance of floating configuration.

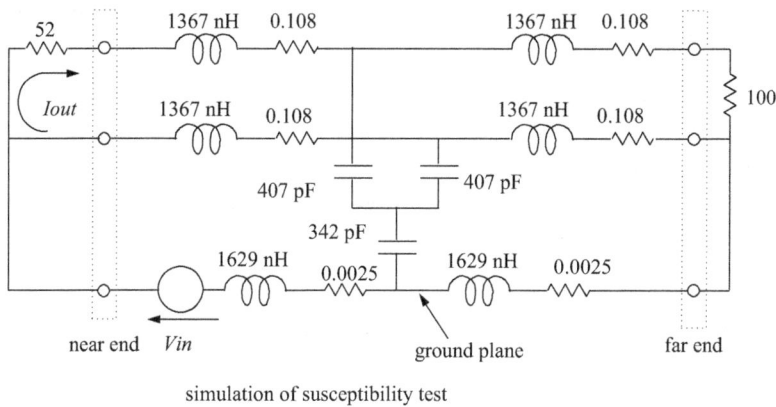

simulation of susceptibility test

Figure 4.4.7   Circuit model for grounded configuration.

This response can be compared with that of Figure 4.4.6, where peak interference with the floating configuration is experienced at quarter-wave frequency, while the minimum level occurs at the half-wave frequency.

There is another significant difference in the responses. The difference between maxima and minima of the response curve of the grounded configuration is much less. This is because the 100-Ω load resistance is damping the amplitude of the reflections in the common-mode loop as well as in the differential mode loop. With the floating configuration, there is no damping whatsoever in the common-mode loop.

It can be concluded that the performances of the grounded and floating configurations tend to complement each other. At the frequency where one gives poor performance, the

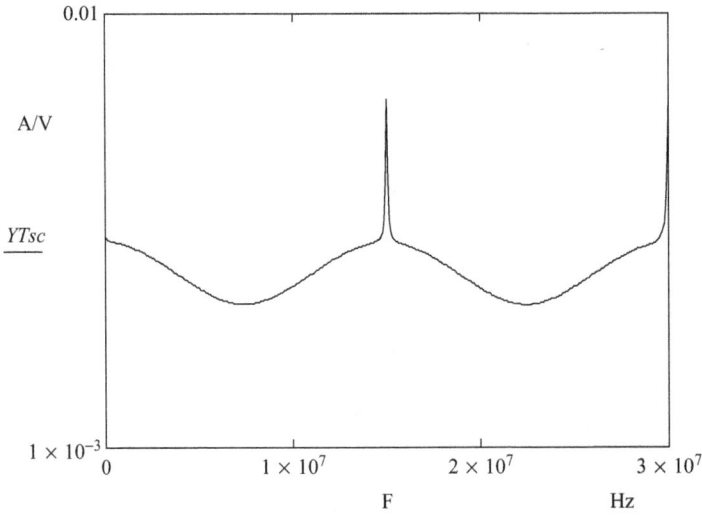

Figure 4.4.8   Transfer admittance of the grounded configuration.

performance of the other is excellent. It is also true that it never gives good performance at all frequencies.

This section has identified the relationships between the bench test equipment, the configuration-under-review and the general circuit model. It has developed a computer program which can be used to analyze the response of that model over the range of frequencies at which interference problem are most critical.

Practical examples of actual tests are provided by Chapter 7 and methods of dealing with interference problems are described in Chapter 8.

CHAPTER 5

Antenna models

EMI extends well beyond the confines of the assembly-under-review. Analysis of radiated emission and radiated susceptibility calls for a review of the properties of antennae. Fortunately, this review allows many simplifying assumptions to be invoked.

Section 5.1 provides a brief overview of the textbook analysis of the half-wave dipole and identifies a very significant parameter; when connected to the output of a radio transmitter which is generating a signal at the half-wave frequency, the dipole exhibits the same characteristics as a resistive load. This has been named the 'radiation resistance' and its value has been calculated to be 73 Ω.

Combining the radiation resistance with the primitive capacitance and primitive inductance of each monopole allows a circuit model to be created; a model which simulates the coupling between the dipole and the environment.

Developing the model to simulate antenna-mode coupling with a twin-conductor cable involves the same process that was used to derive the three-conductor model in section 2.7. This represents each conductor as a T-network and the environment as another T-network. Since the environment must now be assigned values for inductance, capacitance, and resistance, it is fair to say that it behaves like a virtual conductor. Section 5.2 derives formulae for all the components of the circuit model.

Antenna-mode current is induced in any cable which is exposed to external radiation. Simulating the effect of this radiation calls for the introduction of a voltage source in series with the virtual conductor. Section 5.3 relates this 'threat voltage' to the electric field strength of the radiation.

If the power density of the external radiation is constant, but the frequency varies, then the graph of the amplitude of the threat voltage will exhibit a series of peaks and troughs. As far as the analysis of interference coupling is concerned, 'worst-case' conditions can be defined as the envelope curve which touches the peaks.

Section 5.4 provides an illustration of the differential-mode current which would be induced in a twin-conductor cable when the frequency of the external radiation is varied over a wide range.

If the structure is treated as a perfectly reflecting ground plane, then this will represent worst-case conditions. Since a circuit model can be created for this configuration, then such a model can be used to simulate the structure. Section 5.5 provides an example which assigns values to the components of this model. The effect of the external field can be simulated by connecting a voltage source and a 50-Ω resistor in series with the structure. Unlike a dipole receiving antenna which directs input power to a 75 Ω resistor in the receiver, the incoming energy is stored in the assembly. Since more energy is delivered, the effective value of the radiation resistance decreases. The test described in section 7.5 measures this value to be 50 Ω. This value is a good round figure; and can be modified if experience shows it to be too pessimistic.

For radiation susceptibility assessment, it is possible to relate the power output of the test transmitter to the threat voltage applied to the equipment-under-test. Section 5.6 provides the relevant formulae. Since mathematical software is not restricted to the analysis of circuit models, it can be used to compute the response of the equipment to the electromagnetic environment.

It is also possible to relate the antenna-mode current created by the equipment-under-test to the power received by a test receiver during a radiated emission test. Relevant formulae are provided by section 5.7.

Since the approach used in this chapter is always to assume worst-case conditions, the simulated results are likely to err on the side of caution. The most significant benefit derived from the use of worst-case conditions is that the relevant formulae are dramatically simpler than those necessary to derive field distribution patterns.

5.1 The half-wave dipole

There is quite a lot of theory involved in the design of antennae, and there are many textbooks which deal with the subject. Repeating the derivations of the various formulae would be unnecessarily tedious. However, it is useful to summarize the methods used in arriving at the more important equations. Consideration is limited to the half-wave dipole [5.1].

5.1.1 Radiated power

Figure 5.1 illustrates a configuration where a half-wave dipole is driven from a radio-frequency transmitter. This setup will result in an oscillating current in the vertically oriented conductors. Current amplitude will be high at the junction between dipole and transmission line, and will be zero at the tips of the antenna. The relationship between current and time is defined as:

$$I = Io \cdot \cos(\omega \cdot t) \quad (\text{A}) \tag{5.1.1}$$

Current in a small element dz will create a magnetic potential A at the point P. This magnetic potential is a vector, and its direction is parallel to the axis of the conductor. A relationship can be established between A and the current in the element dz.

Spherical co-ordinates, centred at the mid-point of the antenna, are then used to analyze the vectors involved in propagating the field. Figure 5.1.3 illustrates this. Since the vector A is aligned in the same direction as the current, there is no longitudinal component. That is, $A_\phi = 0$. Equations for the radial and latitudinal components A_r and A_θ are derived. It is then possible to define formulae for the magnetic field vector H_ϕ. This is a longitudinal vector; in Figure 5.1.1 it would be directed into the page. Then, using one of Maxwell's equations, the components of the E-field are calculated.

Since the only components of the total field which are capable of propagating power are those which are at right angles to each other, the analysis focusses on the latitudinal electric

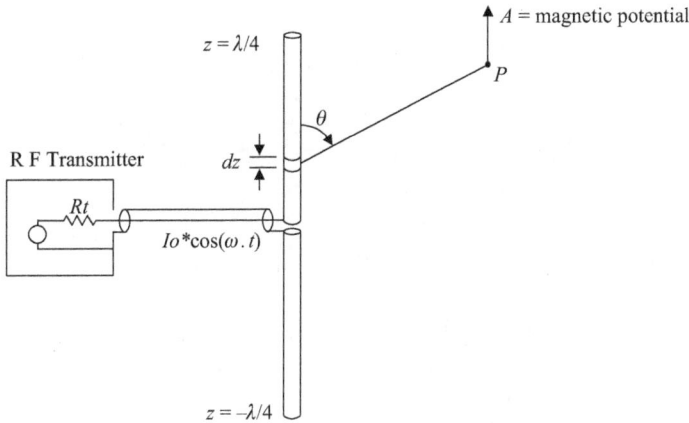

Figure 5.1.1 Analysis of half-wave dipole.

field E_θ and the longitudinal magnetic field H_ϕ. Multiplying the amplitudes of these components together gives an expression for the power density vector S_r. Integrating the value of this vector over the surface of the sphere gives a value for the total radiated power Pt:

$$Pt = \frac{1}{2} \cdot Rrad \cdot Io^2 \quad (\text{W}) \tag{5.1.2}$$

The parameter *Rrad* is named the radiation resistance. It is not a resistor in the conventional sense of the word, but a mathematical constant which happens to have the dimensions of resistance. Its value can be calculated, and for the half-wave dipole:

$$Rrad = 73 \ \Omega \tag{5.1.3}$$

This value for *Rrad* is derived by assuming that the medium is lossless; meaning that the total radiated power does not decrease with distance from the antenna. Nor can it increase with distance. This being so, (5.1.2), with a value of 73 Ω assigned to *Rrad*, defines the maximum power which can be delivered to the environment by a dipole antenna.

It is useful to quote an extract from the Ordnance Board Pillar Proceeding at this point:

At distances closer to about two wavelengths from a single linear aerial, the radiation field is accompanied by further electric and magnetic components the strengths of which fall as the square or cube of the distance from the aerial. These components are not related by Zo and do not radiate power away from the aerial. Although power can be extracted from them by very close receiving aerials, this power is always less than predicted by the extrapolation of far field theory. [5.2]

This means that (5.1.2) will provide a worst-case estimate of the power that would be delivered to a monitor antenna, whether that antenna is located in the near-field or the far-field. For the purposes of EMC analysis, it is desirable to have available a worst-case estimate.

This leads to the circuit model of Figure 5.1.2, where the antenna acts as a resistive load to the co-axial transmission line. It should be emphasized that this relationship applies at only

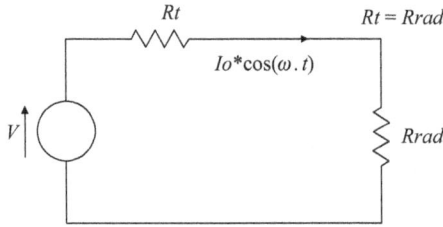

Figure 5.1.2 Equivalent circuit of half-wave dipole at resonance.

one frequency; one which is slightly higher than that of half-wave resonance. At all other non-resonant frequencies, the antenna impedance is a resistance in series with a reactance.

Equation (5.1.2) defines the average power which can be transmitted by the antenna when the frequency is such that this power is at a maximum value. Under all other conditions, the transmitted power is less than this maximum.

If *Irms* is the root-mean-square value of the current, then:

$$Irms = \sqrt{2} \cdot Io \quad \text{(A)} \tag{5.1.4}$$

and

$$Pt = Rrad \cdot Irms^2 \quad \text{(W)} \tag{5.1.5}$$

Since optimum transmission of power is obtained when the output impedance of the transmitter is the same as that of the load, the optimum value for *Rt* and the characteristic impedance of the cable is 73 Ω: making 75 Ω co-axial cable a good choice for signal links between radio-frequency equipment and antennae.

5.1.2 Power density

At the surface of the sphere illustrated in Figure 5.1.3, the relationship between power density, *S*, and the transmitted power, *Pt*, is:

$$S = \frac{Gt \cdot Pt}{4 \cdot \pi \cdot r^2} \quad \text{(W/m}^2\text{)} \tag{5.1.6}$$

The term *Gt* takes into account the fact that power does not radiate uniformly in all directions. It is defined as:

$$Gt = \frac{\dfrac{\text{maximum power radiated}}{\text{unit solid angle}}}{\dfrac{\text{total power delivered to antenna}}{4 \cdot \pi}} \tag{5.1.7}$$

It is described as the antenna gain, and is a measure of the concentration of the power density of the radiated wavefront in a given direction. For a half-wave dipole:

$$Gt = 1.64 \tag{5.1.8}$$

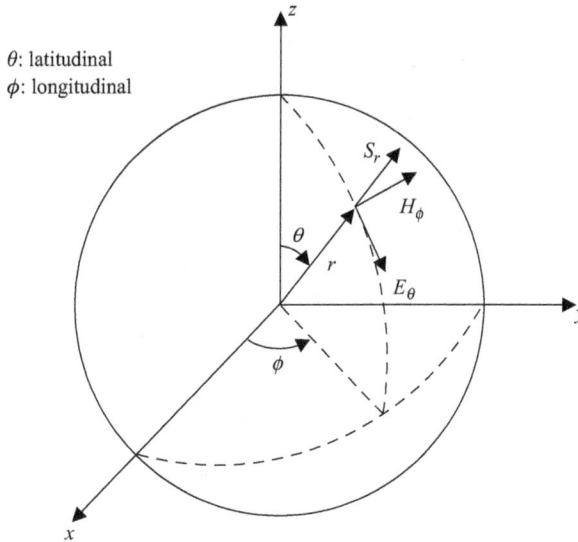

θ: latitudinal
ϕ: longitudinal

Figure 5.1.3   Power density on a spherical surface.

5.1.3 Field strength

Field vectors H and E are tangential to the surface of the sphere, and at right angle to each other. The power density vector S is radial, and the flow of power is outward.

The characteristic impedance of free space is:

$$Z_o = \frac{E}{H} = \sqrt{\frac{\mu_o}{\varepsilon_o}} = 377 \ \Omega \qquad (5.1.9)$$

where E is the electric field strength and H is the magnetic field strength. The relationship between E, H, and the power density is:

$$S = E \cdot H \quad (\text{W/m}^2) \qquad (5.1.10)$$

Hence:

$$S = \frac{E^2}{Z_o} \quad (\text{W/m}^2) \qquad (5.1.11)$$

and

$$S = Z_o \cdot H^2 \quad (\text{W/m}^2) \qquad (5.1.12)$$

This means that if the value of either E or H is used to specify the strength of the electromagnetic field in the far field, it is always possible to determine the value of the power density.

5.1.4 Power received

The dipole can also act as a receiving antenna. For a configuration such as Figure 5.1.4 the power received by the RF receiver is related to the power density Sr of the incoming wavefront by:

$$Pr = Aeff \cdot Sr \quad (\text{W}) \tag{5.1.13}$$

where $Aeff$ is the effective receiving area of the dipole.

A general relationship can be established between the effective receiving area, the gain Gr of a dipole antenna, and the wavelength λ.

$$Aeff = \frac{Gr \cdot \lambda^2}{4 \cdot \pi} \quad (\text{m}^2) \tag{5.1.14}$$

As with the transmitting dipole, Gr is equal to 1.64. Under optimum conditions, the power delivered to the receiver terminals is:

$$Pr = \frac{Gr \cdot \lambda^2 \cdot Sr}{4 \cdot \pi} \quad (\text{W}) \tag{5.1.15}$$

For maximum sensitivity of the receiver, it is necessary for the input resistor to match the output resistance of the antenna. This being so, the voltage along the antenna can be deduced from Figure 5.1.5.

Since power at the receiver input is the product of current in the resistor $Rrec$ and the voltage across that resistor:

$$Pr = \frac{1}{Rrec} \cdot \left(\frac{Vrms}{2} \right)^2 \quad (\text{W})$$

Replacing $Rrec$ with $Rrad$ and rearranging:

$$Vrms = \sqrt{4 \cdot Rrad \cdot Pr} \quad (\text{V}) \tag{5.1.16}$$

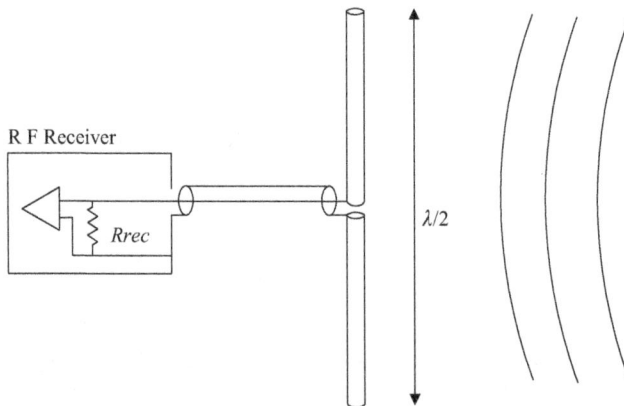

Figure 5.1.4   Receiver assembly.

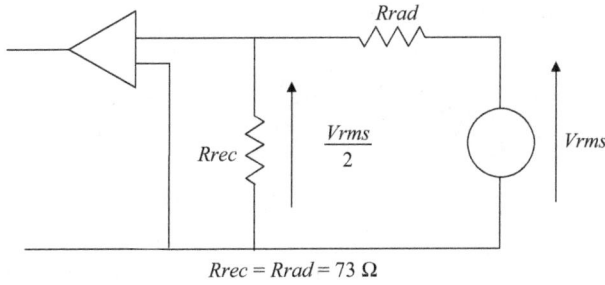

Figure 5.1.5   Simple circuit model of dipole receiver assembly.

For a half-wave dipole receiving antenna, the radiation resistance is 73 Ω, exactly the same as for a transmitting antenna.

5.2 The virtual conductor

In the previous section, the analysis was focused on the behavior of conductors when they are designed to act as antennae. The scope of the analysis was limited to a restricted band-width where the frequency was at or near resonance. Exactly the same mechanisms apply to conductors when antenna action is unintentional. However the scope of the analysis needs to expand to cover situations at all frequencies.

At non-resonant frequencies the reactive components of the conductors play a pre-dominant role. Including these parameters in the model involves the use of primitive inductance and primitive capacitance.

Figure 5.2.1 shows how primitives can be used to create a model of the transmitting dipole. The two monopoles are effectively connected by two capacitors in series with two inductors

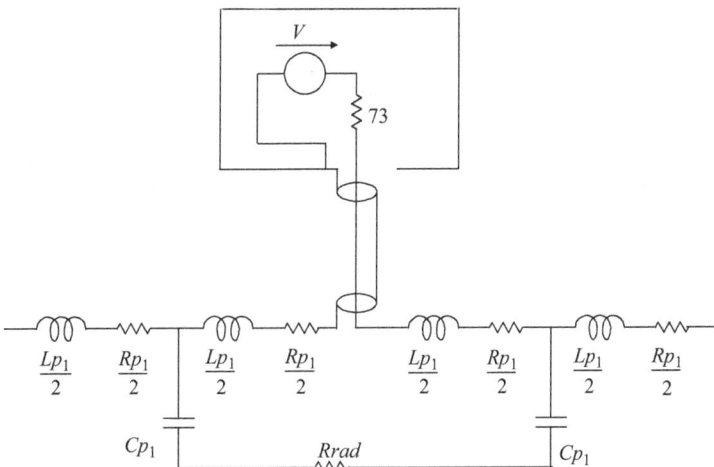

Figure 5.2.1   Circuit model of half-wave dipole.

voltage source located at center of conductor

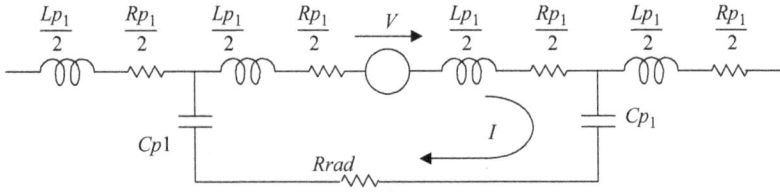

Figure 5.2.2   General circuit model of an isolated conductor.

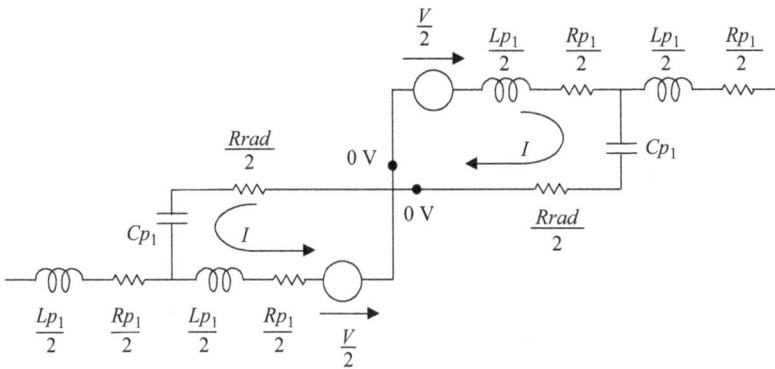

Figure 5.2.3   Balance between two halves of dipole.

and the radiation resistance. At resonance, the circuit acts as a series tuned circuit where the only limit to the current is provided by the resistive components.

Figure 5.2.2 illustrates the situation where the co-axial link to the transmitter has been removed and the two halves of the dipole are shorted together to form a single conductor. Signals can be introduced into this conductor by threading it through a ferrite core which acts as a step-down voltage transformer. In the figure, this is illustrated as a voltage source V.

With such a configuration, there will be a flow of current forward and backward along the conductor. Any current flowing into the right-hand monopole can only come from the left-hand monopole and vice versa. Since the currents and voltages in the two halves of the dipole are balanced it is useful to redraw the model, giving the circuit of Figure 5.2.3.

This is effectively a bridge circuit with voltages in one arm balancing voltages in the other arm. If the voltage at the center of the conductor is defined as zero volts, then the voltage at the mid-point of *Rrad* must also be zero.

Hence it is legitimate to join these two points together in the model. When this is done and attention is focused on the right-hand monopole, the model simplifies to that shown by Figure 5.2.4.

Now, if the single conductor were to be replaced by a pair of parallel conductors connected together at the mid-point and a voltage source inserted in series with both conductors, then the circuit model of the right-hand monopole would be as shown in Figure 5.2.5.

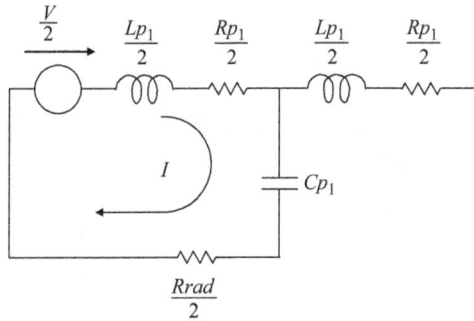

Figure 5.2.4 Focus on right-hand monopole.

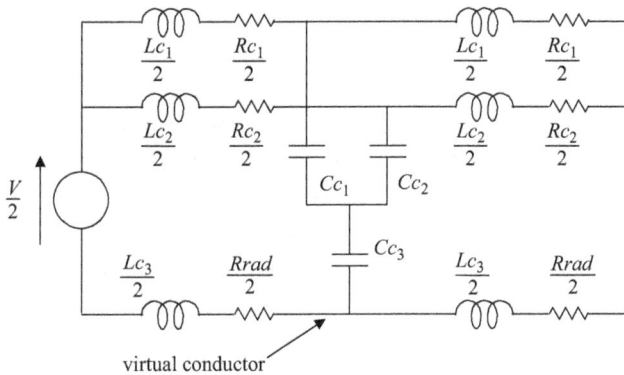

Figure 5.2.5 Circuit model of right-hand section of conductor pair.

This correlates closely with Figure 2.7.6; with the characteristics of the environment replacing the third conductor. The environment is acting as a virtual conductor.

Deriving formulae for the components of this conductor is simply a matter of repeating the procedure introduced in Chapter 2. The starting point is to set out the primitive equations for two conductors:

$$Vp_1 = Zp_{1,1} \cdot Ip_1 + Zp_{1,2} \cdot Ip_2$$
$$Vp_2 = Zp_{2,1} \cdot Ip_1 + Zp_{2,2} \cdot Ip_2 \qquad (5.2.1)$$

These are illustrated in diagram form by Figure 5.2.6, which also relates primitive currents and voltages to loop currents and voltages.

From the diagram:

$$Ip_1 = Il_1$$
$$Ip_2 = Il_2 - Il_1 \qquad (5.2.2)$$

and

$$Vl_1 = Vp_1 - Vp_2$$
$$Vl_2 = Vp_2 \qquad (5.2.3)$$

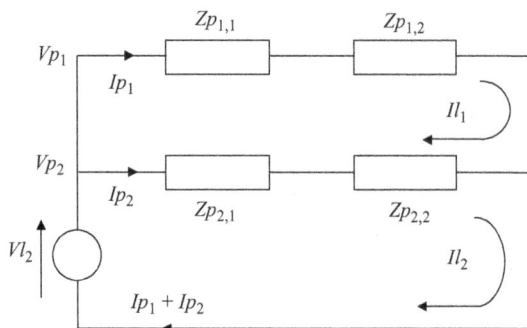

Figure 5.2.6 Diagrammatic representation of primitive equations.

Using (5.2.2) to substitute for Ip_1 and Ip_2 in (5.2.1):

$$Vp_1 = Zp_{1,1} \cdot Il_1 + Zp_{1,2} \cdot (Il_2 - Il_1)$$
$$Vp_2 = Zp_{2,1} \cdot Il_1 + Zp_{2,2} \cdot (Il_2 - Il_1)$$

Rearranging:

$$Vp_1 = (Zp_{1,1} - Zp_{1,2}) \cdot Il_1 + Zp_{1,2} \cdot Il_2$$
$$Vp_2 = (Zp_{2,1} - Zp_{2,2}) \cdot Il_1 + Zp_{2,2} \cdot Il_2$$

Using (5.2.3) to determine loop voltages:

$$Vl_1 = (Zp_{1,1} - Zp_{1,2} - Zp_{2,1} + Zp_{2,2}) \cdot Il_1 + (Zp_{1,2} - Zp_{2,2}) \cdot Il_2$$
$$Vl_2 = (Zp_{2,1} - Zp_{2,2}) \cdot Il_1 + Zp_{2,2} \cdot Il_2$$

Defining the loop impedances as:

$$\begin{aligned} Zl_{1,1} &= Zp_{1,1} - Zp_{1,2} - Zp_{2,1} + Zp_{2,2} \\ Zl_{1,2} &= Zp_{1,2} - Zp_{2,2} \\ Zl_{2,1} &= Zp_{2,1} - Zp_{2,2} \\ Zl_{2,2} &= Zp_{2,2} \end{aligned} \tag{5.2.4}$$

Leads to the loop equations:

$$\begin{aligned} Vl_1 &= Zl_{1,1} \cdot Il_1 + Zl_{1,2} \cdot Il_2 \\ Vl_2 &= Zl_{2,1} \cdot Il_1 + Zl_{2,2} \cdot Il_2 \end{aligned} \tag{5.2.5}$$

If the cross section of the conductor pair is symmetrical:

$$Zp_{1,2} = Zp_{2,1}$$

From (5.2.4):

$$Zl_{1,2} = Zl_{2,1} \tag{5.2.6}$$

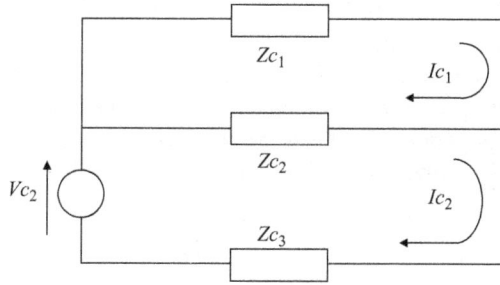

Figure 5.2.7 Circuit model of coupling parameters.

Figure 5.2.7 shows the circuit model which creates a similar pair of equations:

$$Vc_1 = (Zc_1 + Zc_2) \cdot Ic_1 - Zc_{1,2} \cdot Ic_2$$
$$Vc_2 = -Zc_{1,2} \cdot Ic_1 + (Zc_2 + Zc_3) \cdot Ic_2$$

(5.2.7)

Correlating (5.2.5) and (5.2.7) gives:

$$Zl_{1,1} = Zc_1 + Zc_2$$
$$Zl_{1,2} = -Zc_2$$
$$Zl_{2,2} = Zc_2 + Zc_3$$

Defining circuit impedances in terms of loop impedances:

$$Zc_1 = Zl_{1,1} + Zl_{1,2}$$
$$Zc_2 = -Zl_{1,2}$$
$$Zc_3 = Zl_{2,2} + Zl_{1,2}$$

Using equation (5.2.4) to substitute primitive impedances for loop impedances:

$$Zc_1 = (Zp_{1,1} - 2 \cdot Zp_{1,2} + Zp_{2,2}) + (Zp_{1,2} - Zp_{2,2})$$
$$Zc_2 = Zp_{2,2} - Zp_{1,2}$$
$$Zc_3 = Zp_{2,2} + (Zp_{1,2} - Zp_{2,2})$$

This leads to:

$$Zc_1 = Zp_{1,1} - Zp_{1,2}$$
$$Zc_2 = Zp_{2,2} - Zp_{1,2}$$
$$Zc_3 = Zp_{1,2}$$

(5.2.8)

Letting $Zp_{i,j} = j \cdot \omega \cdot Lp_{i,j}$ and invoking (2.3.2) gives:

$$Lc_1 = \frac{\mu_o \cdot \mu_r \cdot l}{2 \cdot \pi} \cdot \ln \frac{r_{1,2}}{r_{1,1}}$$

$$Lc_2 = \frac{\mu_o \cdot \mu_r \cdot l}{2 \cdot \pi} \cdot \ln \frac{r_{1,2}}{r_{2,2}}$$

(5.2.9)

$$Lc_3 = \frac{\mu_o \cdot \mu_r \cdot l}{2 \cdot \pi} \cdot \ln \frac{l}{r_{1,2}}$$

Comparing this set of equations with that of (2.5.5) reveals that the inductance value associated with each conductor is the same, whether the conductor pair is viewed as a transmission line or as an antenna. The new parameter introduced by (5.2.9) is the circuit inductance of the virtual conductor, Lc_3. This is the primitive inductance of a conductor with the same radius as the separation between the conductors.

Letting $Zp_{i,j} = \dfrac{1}{j \cdot \omega \cdot Cp_{i,j}}$ and invoking (2.3.1) gives:

$$Cc_1 = \frac{2 \cdot \pi \cdot \varepsilon_o \cdot \varepsilon_r \cdot l}{\ln \dfrac{r_{1,2}}{r_{1,1}}}$$

$$Cc_2 = \frac{2 \cdot \pi \cdot \varepsilon_o \cdot \varepsilon_r \cdot l}{\ln \dfrac{r_{1,2}}{r_{2,2}}} \qquad (5.2.10)$$

$$Cc_3 = \frac{2 \cdot \pi \cdot \varepsilon_o \cdot \varepsilon_r \cdot l}{\ln \dfrac{l}{r_{1,2}}}$$

Comparing this set of equations with that of (2.5.9) reveals that the capacitance value associated with each conductor is the same, whether the conductor pair is viewed as a transmission line or as an antenna. The new parameter introduced by (5.2.10) is the circuit capacitance of the virtual conductor, Cc_3.

As would be expected, there is a duality between the primitive capacitance and the primitive inductance of the cable assembly.

A necessary condition for the derivation of a value for the radiation resistance $Rrad$ is that the length of the conductor is much greater than its radius. Since the separation between conductors of a transmission line is much less than the length, the same condition is met for a transmission line as for an antenna. At this point, it is reasonable to assume that the radiation resistance of the cable is 73 Ω, the same as defined by (5.1.3).

Having derived formulae for all the parameters associated with the conductor pair, it becomes possible to construct a general circuit model which simulates the coupling between an isolated length of twin-conductor cable and the environment (Figure 5.2.8).

Following on from the creation of this model, it is possible to define the virtual conductor as an imaginary conductor which enables the coupling between cable and environment to be simulated. It behaves as a return conductor for antenna-mode current.

It has the same properties as an actual conductor – capacitance, inductance, and resistance. Numerical values for the reactive parameters can be derived from data on the physical construction of the cable. An initial assumption can be made that the resistance is 73 Ω, the same as that of the half-wave dipole. Tests, such as those described in Chapter 7, can be devised to refine the values of the three parameters.

As far as differential-mode signal transmission is concerned, the values of the reactive components of the circuit model remain exactly the same as those of a pair of conductors. As far as antenna-mode coupling is concerned, the cable can be treated as a single conductor with the same radius as the separation between the conductor pair.

In Figure 5.2.8, it is assumed that two identical voltage sources are used to inject a signal into the cable. Such a configuration could be implemented by clamping a toroidal transformer round the cable at its mid-point.

voltage source transformer coupled to center of cable

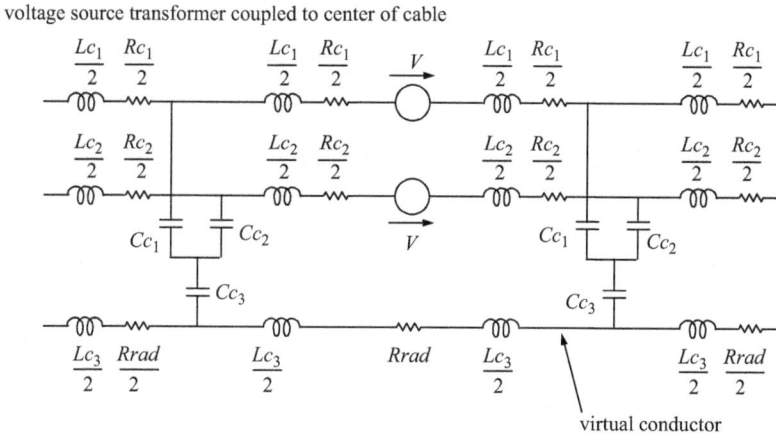

Figure 5.2.8   General circuit model of an isolated cable.

5.3 The threat voltage

The most severe threat to any electronic system will occur in those situations when it is completely exposed to its environment.

In the setup illustrated by Figure 5.3.1, the only connection between the remote transducer and the signal processing unit is a twin-conductor cable. One example could be the link between a proximity sensor and the input circuitry of the signal processing unit. One conductor would carry the signal current whilst the other carries the return current. Normal wiring practice is to connect the return conductor to the zero-volt reference in the processing unit, and for this reference conductor to be connected to the structure.

The conductors of the cable can be represented by the right-hand section of Figure 5.2.8. Since the structure is essentially a single conductor, albeit with a complex cross section, it can be represented by the monopole on the left-hand side of Figure 5.3.2. If the structure were to resonate at the same frequency as the cable, then worst-case conditions would be represented.

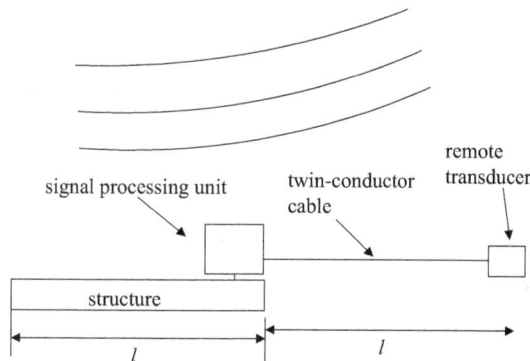

Figure 5.3.1   Cable exposed to external radiation.

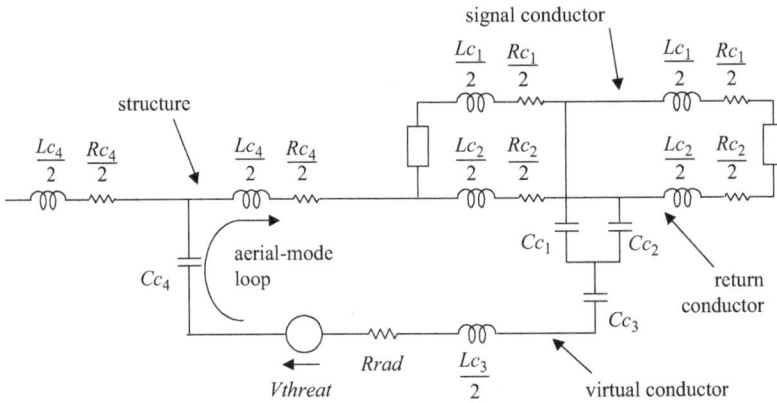

Figure 5.3.2 General circuit model of exposed cable and structure.

This model differs from that shown by Figure 5.2.8 in that the voltage sources in the conductors have been replaced by a single source located in series with the virtual conductor.

The source in Figure 5.3.2 now represents the effect of the external electromagnetic field. The triple-T network representing the left half of the twin-conductor cable has been replaced by a single-T network representing the structure.

Such a representation allows the circuit to be analyzed from the point of view of radiated susceptibility. But before such an analysis can proceed, it is necessary to relate the amplitude of the voltage source to the strength of the field in which the system is immersed.

Figure 5.3.3 depicts a length of conductor which is subjected to an external field and the associated circuit model. An incremental voltage dV is induced in series with each element dz of the conductor.

$$dV = E \cdot dz \qquad (5.3.1)$$

where E is the strength of the electric field at that location.

The velocity of propagation is assumed to be constant. So if the waveform of the electric field is sinusoidal with respect to time, then the distribution in space will also be sinusoidal.

Figure 5.3.4 illustrates the relationship between the strength of the E-field and distance along conductor.

The relationship can be defined as:

$$E = Emax \cdot \cos\left(\frac{2 \cdot \pi}{\lambda} \cdot z\right) \qquad (5.3.2)$$

Over the length of the cable and structure, the voltage is:

$$Vthreat = \int_{-l}^{l} Emax \cdot \cos\left(\frac{2 \cdot \pi}{\lambda} \cdot z\right) \cdot dz$$

This leads to the relationship between threat voltage and electric field strength

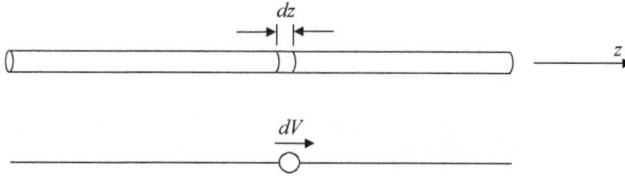

Figure 5.3.3   Effect of electric field.

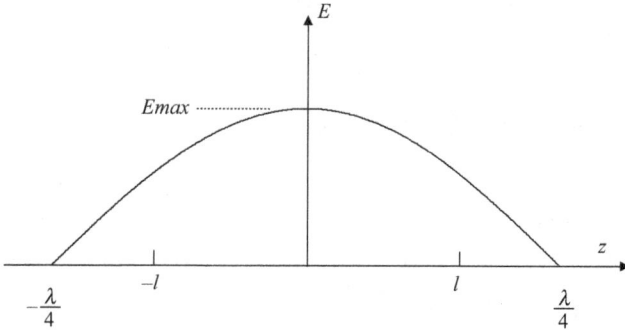

Figure 5.3.4   E-field along conductor when $l < \lambda/4$.

$$Vthreat = \frac{\lambda}{\pi} \cdot Emax \cdot \sin\left(\frac{2 \cdot \pi}{\lambda} \cdot l\right) \qquad (5.3.3)$$

Figure 5.3.5 illustrates the relationship when the length l is less than a quarter wavelength. From (5.1.11):

$$Emax = \sqrt{S \cdot Zo}$$
$$\text{where}\quad Zo = 377 \ \Omega \qquad (5.3.4)$$

The relationship between wavelength and frequency is as defined in (2.3.6):

$$v = \lambda \cdot f$$

If the velocity of propagation is assumed to be the speed of light in vacuum, then:

$$\lambda = \frac{c}{f} \qquad (5.3.5)$$

It is clear from (5.3.5) and (5.3.3) that the threat voltage is a function of both frequency f and length l. Equation (5.3.4) demonstrates that the electric field strength is proportional to the square root of the power density. Combining these equations in the way illustrated by the subroutine $Vthreat_i$ in Figure 5.3.6 allows the relationship between power density and threat voltage to be calculated over a wide range of frequencies. This response is illustrated by the solid curve of Figure 5.3.7. It is assumed that the power density S is 1 W/m^2 and that the length l is 15 m.

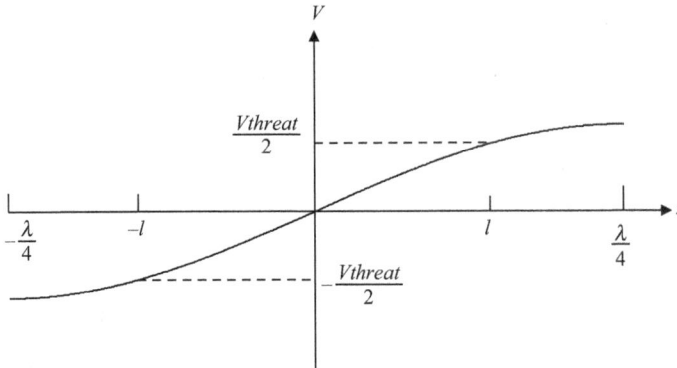

Figure 5.3.5   Variation of threat voltage with length of cable.

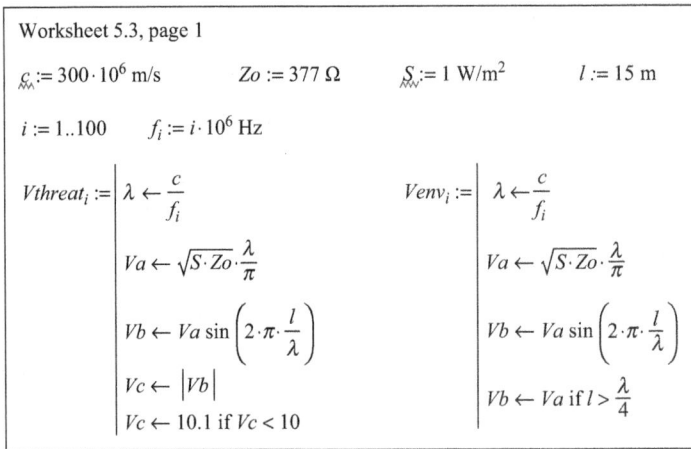

Worksheet 5.3, page 1

$c := 300 \cdot 10^6$ m/s $Zo := 377\ \Omega$ $S := 1$ W/m^2 $l := 15$ m

$i := 1..100$ $f_i := i \cdot 10^6$ Hz

$$Vthreat_i := \begin{vmatrix} \lambda \leftarrow \dfrac{c}{f_i} \\[2mm] Va \leftarrow \sqrt{S \cdot Zo} \cdot \dfrac{\lambda}{\pi} \\[2mm] Vb \leftarrow Va \sin\left(2 \cdot \pi \cdot \dfrac{l}{\lambda}\right) \\[2mm] Vc \leftarrow |Vb| \\[2mm] Vc \leftarrow 10.1 \ \text{if} \ Vc < 10 \end{vmatrix} \qquad Venv_i := \begin{vmatrix} \lambda \leftarrow \dfrac{c}{f_i} \\[2mm] Va \leftarrow \sqrt{S \cdot Zo} \cdot \dfrac{\lambda}{\pi} \\[2mm] Vb \leftarrow Va \sin\left(2 \cdot \pi \cdot \dfrac{l}{\lambda}\right) \\[2mm] Vb \leftarrow Va \ \text{if} \ l > \dfrac{\lambda}{4} \end{vmatrix}$$

Figure 5.3.6   Calculating the relationship between threat voltage and frequency.

This curve is a series of peaks, each peak occurring at a resonant frequency. The first resonance occurs when the length l is equal to a quarter wavelength. At this frequency and at higher resonant frequencies:

$$Vthreat = \frac{\lambda}{\pi} \cdot Emax \qquad\qquad (5.3.6)$$

Joining these points together creates an envelope curve, *Venv*. Since this envelope is of threat voltages which are equal to, or higher than, all voltage points on the *Vthreat* curve, it represents worst-case conditions. This curve is defined by that of the $Venv_i$ function of Figure 5.3.7 and is illustrated by the dashed curve.

When the frequency is lower than that of quarter-wave resonance both curves are co-incident.

Assigning this envelope voltage to the source of the circuit model of Figure 5.3.2 will allow the response of the assembly-under-review to be simulated.

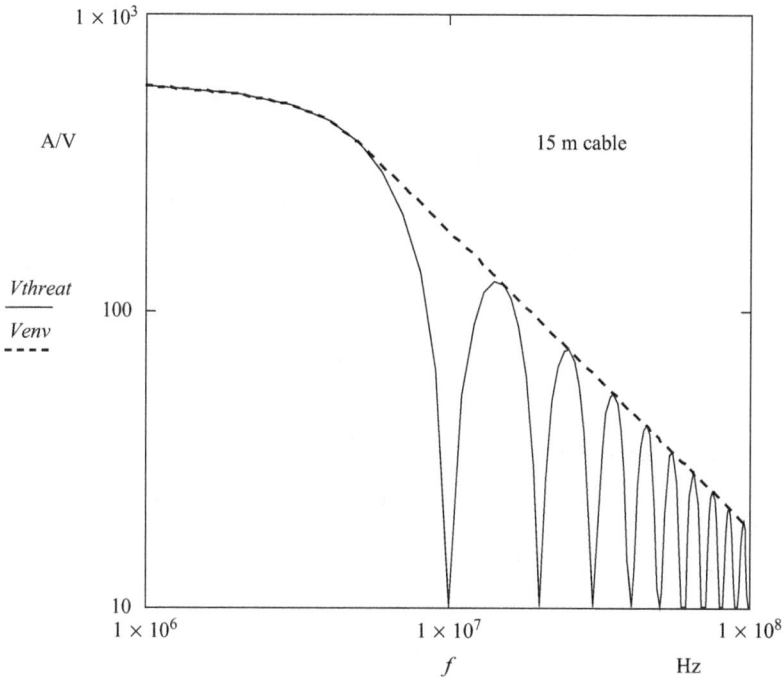

Figure 5.3.7 Relationship between threat voltage and frequency.

5.4 The threat current

In the previous section, the simple assembly of Figure 5.3.1 was used to depict a situation in which an exposed cable can be subjected to an external electromagnetic field. A circuit model was created to simulate the configuration, and the source voltage was related to the power density of the interference. This section develops the analysis to determine the frequency response of the threat current in the signal line.

Under resonance conditions the reactive components representing the structure act as a short-circuit and the only limit to the current is provided by the resistive components. From Figure 5.3.2 these are the radiation resistance and the resistance of the structure. This means that worst-case conditions can be simulated by replacing the reactive components of the structure with a short-circuit. Since the resistance of the structure is minimal compared to the radiation resistance, it is reasonable to represent this as a short-circuit as well.

A welcome consequence of representing the entire structure as a short-circuit is that the model is simplified, as Figure 5.4.1 illustrates.

Determining the response of this model involves the assignment of specific details to the cable and interface circuitry. If it is assumed that the length of the cable is 15 m and that the two conductors are 0.8 mm in diameter with a spacing between centers of 1.2 mm, then all the circuit parameters can be calculated.

Figure 5.4.2 is a copy of the first page of the Mathcad worksheet used to analyze the response. This page determines all the component values. Inductors Lc_1, Lc_2, and Lc_3 are obtained from (5.2.9). The steady-state resistor values Rss_1 and Rss_2 of the two conductors

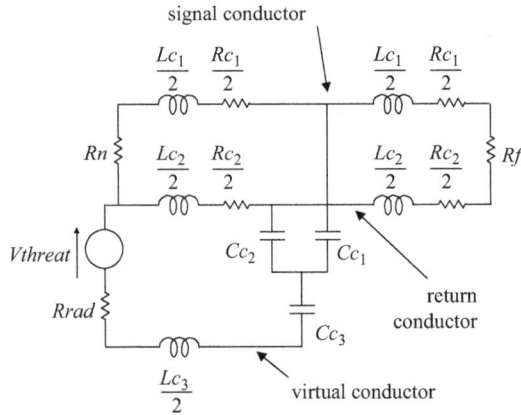

Figure 5.4.1   General circuit model of exposed cable – simplified.

Worksheet 5.4, page 1

$\mu_o := 4 \cdot \pi \cdot 10^{-7}$ H/m $\varepsilon_o := 8.854 \cdot 10^{-12}$ F/m $c := 2.998 \cdot 10^8$ m/s

$\rho := 1.7 \cdot 10^{-8}\ \Omega$ m $l := 15$ m

$r_{1,1} := 0.4 \cdot 10^{-3}$ m $r_{2,2} := r_{1,1}$ $r_{1,2} := 1.2 \cdot 10^{-3}$ m

$$Lc_1 := \frac{\mu_o \cdot l}{2 \cdot \pi} \cdot \ln\left(\frac{r_{1,2}}{r_{1,1}}\right) \qquad Lc_2 := \frac{\mu_o \cdot l}{2 \cdot \pi} \cdot \ln\left(\frac{r_{1,2}}{r_{2,2}}\right) \qquad Lc_3 := \frac{\mu_o \cdot l}{2 \cdot \pi} \cdot \ln\left(\frac{l}{r_{1,2}}\right)$$

$$Rss_1 := \frac{\rho \cdot l}{\pi \cdot (r_{1,1})^2} \qquad Rss_2 := Rss_1 \qquad Rss_3 := 0$$

$$Fx := \frac{4 \cdot \rho}{\mu_o \cdot \pi \cdot (r_{1,1})^2} = 1.077 \times 10^5 \text{ Hz}$$

$$Cc := \left(\frac{l}{c}\right)^2 \cdot \frac{1}{Lc} \qquad Gc := \begin{pmatrix} 0 \\ 0 \\ 0 \end{pmatrix}$$

$$\frac{Lc}{2} = \begin{pmatrix} 1.648 \times 10^{-6} \\ 1.648 \times 10^{-6} \\ 1.415 \times 10^{-5} \end{pmatrix} \qquad Cc = \begin{pmatrix} 7.595 \times 10^{-10} \\ 7.595 \times 10^{-10} \\ 8.846 \times 10^{-11} \end{pmatrix} \qquad \frac{Rss}{2} = \begin{pmatrix} 0.254 \\ 0.254 \\ 0 \end{pmatrix} \qquad Rrad := 73$$

$$Zo := \sqrt{\frac{Lc}{Cc}} \qquad Zo := \begin{pmatrix} 65.873 \\ 65.873 \\ 565.632 \end{pmatrix} \qquad \begin{array}{l} Rn := Zo_1 + Zo_2 \qquad Rf := Rn \\ Rn = 131.746\ \Omega \end{array}$$

$$Fq := \frac{1}{4 \cdot \sqrt{Lc_1 \cdot Cc_1}} = 4.997 \times 10^6 \text{ Hz}$$

Figure 5.4.2   Calculating values for the circuit components.

are obtained from (2.5.11). Since the virtual conductor does not have the properties of a conventional conductor, Rss_3 is set to zero. The crossover frequency Fx is obtained from (2.5.14) and the capacitor values are derived from (2.3.8). Conductance values are set at zero, since it is assumed that there are no losses in the insulation.

Values for inductance, capacitance, and resistance of the representative circuit model are then displayed on the sheet. To minimize reflections at the near and far ends of the line, the values of Rn and Rf are set at a value equal to the characteristic impedance. If the values of the terminating impedances are any other value, the peaks in the response curve would be higher. Terminating the line with perfectly matched impedances does not eliminate interference.

The final parameter to be calculated on this page is the value of the quarter-wave frequency Fq. This invokes the use of (2.3.9).

Assigning these values to the general circuit model of Figure 5.4.1 leads to the representative circuit model of Figure 5.4.3.

Re-drawing Figure 5.4.1 in terms of distributed parameters results in Figure 5.4.4.

Inspection of this model allows the equations of the Zloop(f) function to be defined, and this function is listed in Figure 5.4.5. It invokes the Zbranch(f) function, which is identical to that in Figure 4.3.4.

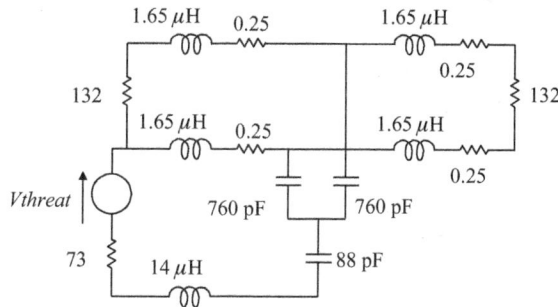

Figure 5.4.3   Representative circuit model of exposed cable – simplified.

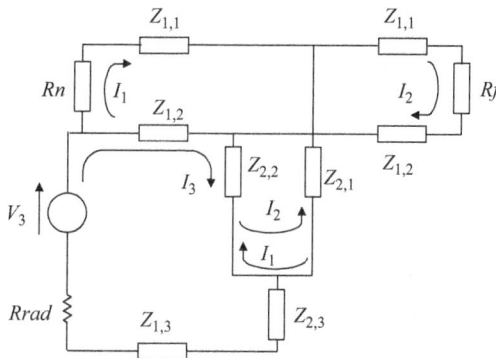

Figure 5.4.4   Distributed parameter model for figure 5.4.3.

The Vthreat(f) function is essentially a copy of the envelope function *Venv* of Figure 5.3.6. It is assumed that the power density of the interference signal is 1 W/m^2.

These three functions are then used to calculate the frequency response of the threat current *Ithreat* in the signal loop. Figure 5.4.6 defines the main program and Figure 5.4.6 illustrates the response curve.

Worksheet 5.4, page 2

$$Zbranch(f) := \begin{vmatrix} \omega \leftarrow 2 \cdot \pi \cdot f \\ \text{for } i \in 1..3 \\ \qquad \begin{vmatrix} Rc_i \leftarrow Rss_i \sqrt{1 + \dfrac{f}{Fx}} \\ \theta \leftarrow \sqrt{(Rc_i + j \cdot \omega \cdot Lc_i) \cdot (Gc_i + j \cdot \omega \cdot Cc_i)} \\ Zo \leftarrow \sqrt{\dfrac{Rc_i + j \cdot \omega \cdot Lc_i}{Gc_i + j \cdot \omega \cdot Cc_i}} \\ Z_{1,i} \leftarrow Zo \cdot \tanh\left(\dfrac{\theta}{2}\right) \\ Z_{2,i} \leftarrow Zo \cdot \operatorname{csch}(\theta) \end{vmatrix} \\ Z \end{vmatrix}$$

$$Zloop(f) = \begin{vmatrix} Z \leftarrow Zbranch(f) \\ Z11 \leftarrow Z_{1,1} + Z_{2,1} + Z_{2,2} + Z_{1,2} + Rn \\ Z12 \leftarrow -(Z_{2,2} + Z_{2,1}) \\ Z13 \leftarrow -(Z_{1,2} + Z_{2,2}) \\ Z22 \leftarrow Z_{1,2} + Z_{2,2} + Z_{2,1} + Z_{1,1} + Rf \\ Z23 \leftarrow Z_{2,2} \\ Z33 \leftarrow Z_{1,2} + Z_{2,2} + Z_{2,3} + Z_{1,3} + Rrad \\ \begin{pmatrix} Z11 & Z12 & Z13 \\ Z12 & Z22 & Z23 \\ Z13 & Z23 & Z33 \end{pmatrix} \end{vmatrix}$$

$$S := 1 \qquad Zo := 377 \qquad Vthreat(f) := \begin{vmatrix} \lambda \leftarrow \dfrac{c}{f} \\ Va \leftarrow \sqrt{S \cdot Zo} \cdot \dfrac{\lambda}{\pi} \\ Vb \leftarrow Va \sin\left(2 \cdot \pi \cdot \dfrac{l}{\lambda}\right) \\ Vb \leftarrow Va \text{ if } l > \dfrac{\lambda}{4} \end{vmatrix}$$

Figure 5.4.5 Calculating values for impedance matrix and threat voltage at each frequency.

Worksheet 5.4, page 3

$n := 100$ $\qquad s := 1..20 \cdot n$ $\qquad F_s := s \cdot \dfrac{Fq}{n}$

$$Ithreat_s := \begin{vmatrix} f \leftarrow F_s \\ Z \leftarrow \text{Zloop}(f) \\ V \leftarrow \begin{pmatrix} 0 \\ 0 \\ \text{Vthreat}(f) \end{pmatrix} \\ I \leftarrow \text{lsolve}(Z, V) \\ Iout \leftarrow |I_1| \\ Iout \leftarrow 10^{-2} \text{ if } Iout \le 10^{-2} \end{vmatrix}$$

$\max(Ithreat) = 1.031$

Figure 5.4.6 Calculating the frequency response of the threat current.

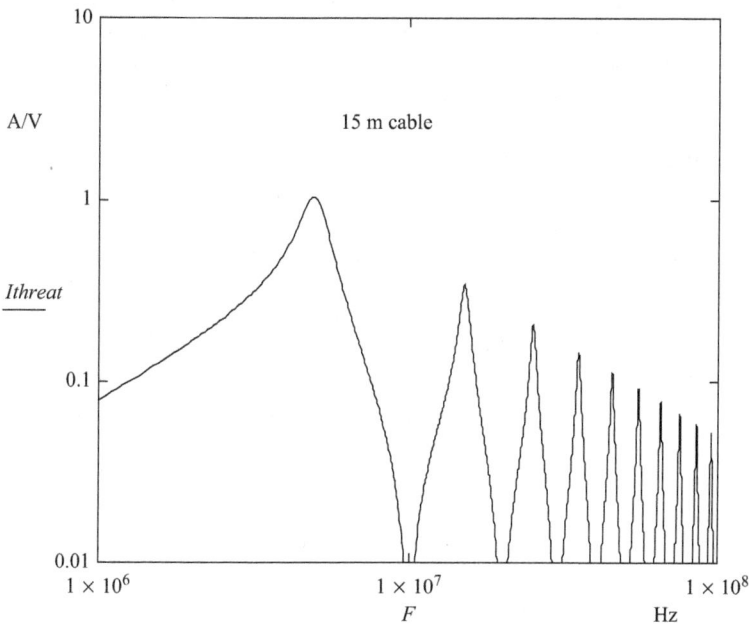

Figure 5.4.7 Frequency response of threat current.

The curve itself is very similar to that of Figure 5.3.7, the main difference being that the response falls sharply as the frequency reduces below that of quarter-wave resonance.

This means that the configuration of Figure 5.3.1 is most susceptible to interference at about 5 MHz. At that frequency, an incoming wave with a power density of 1 W/m^2 will generate a threat current of about 1 A in the signal circuitry. This current could trigger an electro-explosive device, or could cause overheating in transducers or semiconductor devices – with consequential damage to the system.

The subsequent peaks at higher frequencies could also pose a threat; they could induce false signals into the system, or prevent actual signals from reaching the intended input. System upset could occur. However, the designer is now in possession of information not previously available. Given a detailed understanding of both the effect of the interference on the signal-under-review and the bandwidth, amplitude, and function of that signal, it should be possible to design the system to be immune to radiation in any defined EMI environment.

5.5 Coupling via the structure

In the vast majority of configurations, the signal link between two units of equipment is carried by a cable routed along the structure, the signal processing circuitry is mounted on printed circuit boards, the boards are supported by a conducting framework, and that framework is connected to a conducting structure via low-resistance joints. This is the case for most vehicles, whether they are cars, lorries, aircraft, spacecraft, ships, or submersibles.

The same is true of fixed installations. Any equipment unit with a conductive outer surface which is connected to the mains supply must also be connected to the earth conductor – for safety reasons. This applies to most signal generators, spectrum analysers, oscilloscopes, refrigerators, and computer towers. This earth wire provides a low resistance path between all equipment units in the system. In industrial situations, the cables are routed along metal conduits which are connected to the earth wires, to the structure of the buildings, and to lightning conductors. The earthed conductors effectively form a mesh which encloses the equipment. Section 8.7 provides a brief description of the shielding effect this provides.

Military vehicles and many land, sea, and air vehicles are not permitted to bond the zero-volt reference conductors of the printed circuit boards directly to the chassis, because the chassis is used as the battery return. In these cases, the signal link between two units of equipment is usually carried by a single conductor. Such configurations rely on either a return path via the structure or a return conductor routed to some star point in the system. Sections 8.1, 8.2, and 8.3 show how the concepts of 'equipotential ground', the 'single earth point', and the need to 'avoid earth loops' lead to extremely poor EMC. There is no point in attempting to analyse the myriad interference coupling paths that exist in these systems.

The only hope of controlling the EMC of any system in an effective and efficient way is to invoke the guidelines of section 1.8.1. If the system is designed along these lines, then a typical mode of interference coupling would be as shown by Figure 5.5.1.

This is an interface diagram which identifies the essential components of a simple signal link. A 15 m length of twin conductor cable is used to carry the signal. The output of the near-end unit is a voltage source in series with a 132 Ω resistance; the resistance value corresponding to the characteristic impedance of the cable. The load presented to the signal by the far-end unit is also 132 Ω. Such a configuration should reduce the amplitude of any reflections. (The only way of eliminating all reflections is to terminate each conductor with its characteristic impedance. The flat line of *Ycrit* in Figure 4.3.8 illustrates the frequency response when there are no reflections.)

It is assumed that the return conductor is connected to the framework of each unit via a short length of conductor. That is, the configuration is 'grounded'.

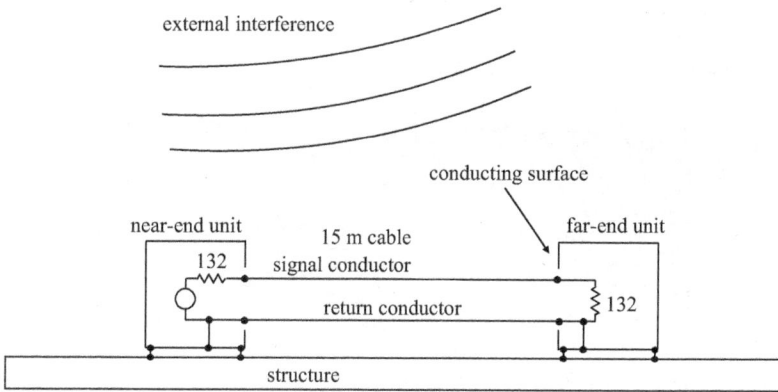

Figure 5.5.1   Interface diagram for the signal link under review.

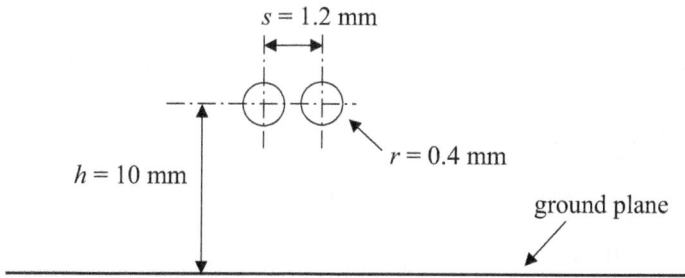

Figure 5.5.2   Cross-sectional view of two conductors over a ground plane.

In most situations, the routing of the cable is fairly convoluted. So, the geometry of the assembly becomes complicated and the task of defining the various cross sections becomes onerous.

One simple way of overcoming this problem is to invoke the concept of the 'ground plane' in its intended sense; as a purely reflecting surface. The ground plane does not need to be infinite. It just needs to be able to create an image of the conductors routed above it. A glass mirror does not need to be infinite in size to provide a good quality image of the object in front. An estimate can be made of the average separation between cable and structure and that value used to define the cross section. Figure 5.5.2 illustrates the geometry of a section of such an assembly.

Given knowledge of the cross section of the cable assembly and the details of the interface circuitry, enough information now exists to create a circuit model. Since a representative circuit model of a similar assembly has already been derived in the section on cross-coupling, a Mathcad program is available to calculate the component values. Figure 5.5.3 is essentially a copy of the worksheet page from that section, modified slightly to utilize input data from the link-under-review. This constitutes the first page of worksheet 5.5.

Worksheet 5.5, page 1

$\mu_o := 4 \cdot \pi \cdot 10^{-7}$ H/m $\varepsilon_o := 8.854 \cdot 10^{-12}$ F/m $c := 2.998 \cdot 10^8$ m/s

$\rho := 1.7 \cdot 10^{-8}$ Ω m $l := 15$ m $h := 10 \cdot 10^{-3}$ m

$s := 1.2 \cdot 10^{-3}$ m $r := 0.4 \cdot 10^{-3}$ m   see Figure 4.3.1

$Rss_1 := \dfrac{\rho \cdot l}{\pi \cdot r^2}$ Ω $Rss_2 := Rss_1$ $Rss_3 := 0.005$ Ω see (2.5.11)

$Fx := \dfrac{4 \cdot \rho}{\mu_o \cdot \pi \cdot r^2} = 1.077 \times 10^5$ Hz   see (2.5.14)

$Lc_1 := \dfrac{\mu_o \cdot l}{2 \cdot \pi} \cdot \ln\left(\dfrac{2 \cdot h \cdot s}{r \cdot \sqrt{s^2 + 4 \cdot h^2}}\right)$ $Lc_2 := Lc_1$

$Lc_3 := \dfrac{\mu_o \cdot l}{2 \cdot \pi} \cdot \ln\left(\dfrac{\sqrt{s^2 + 4 \cdot h^2}}{s}\right)$ see (2.11.3)

$Cc := \dfrac{1}{Lc} \cdot \left(\dfrac{l}{c}\right)^2$ see (2.3.8)

$Fq := \dfrac{1}{4 \cdot \sqrt{Lc_1 \cdot Cc_1}} = 4.997 \times 10^6$ Hz   see (2.3.9)

Component values for three-conductor assembly of Figure 5.5.2:

$\dfrac{Rss}{2} = \begin{pmatrix} 0.254 \\ 0.254 \\ 2.5 \times 10^{-3} \end{pmatrix}$ $\dfrac{Lc}{2} = \begin{pmatrix} 1.645 \times 10^{-6} \\ 1.645 \times 10^{-6} \\ 4.223 \times 10^{-6} \end{pmatrix}$ $Cc = \begin{pmatrix} 7.60\,8 \times 10^{-10} \\ 7.608 \times 10^{-10} \\ 2.964 \times 10^{-10} \end{pmatrix}$

$Zn := \begin{pmatrix} 132 \\ 0 \\ 0 \end{pmatrix}$ Ω $Zf := \begin{pmatrix} 132 \\ 0 \\ 0 \end{pmatrix}$ Ω $Gc := \begin{pmatrix} 0 \\ 0 \\ 0 \end{pmatrix}$ S $Rrad := 50$ Ω

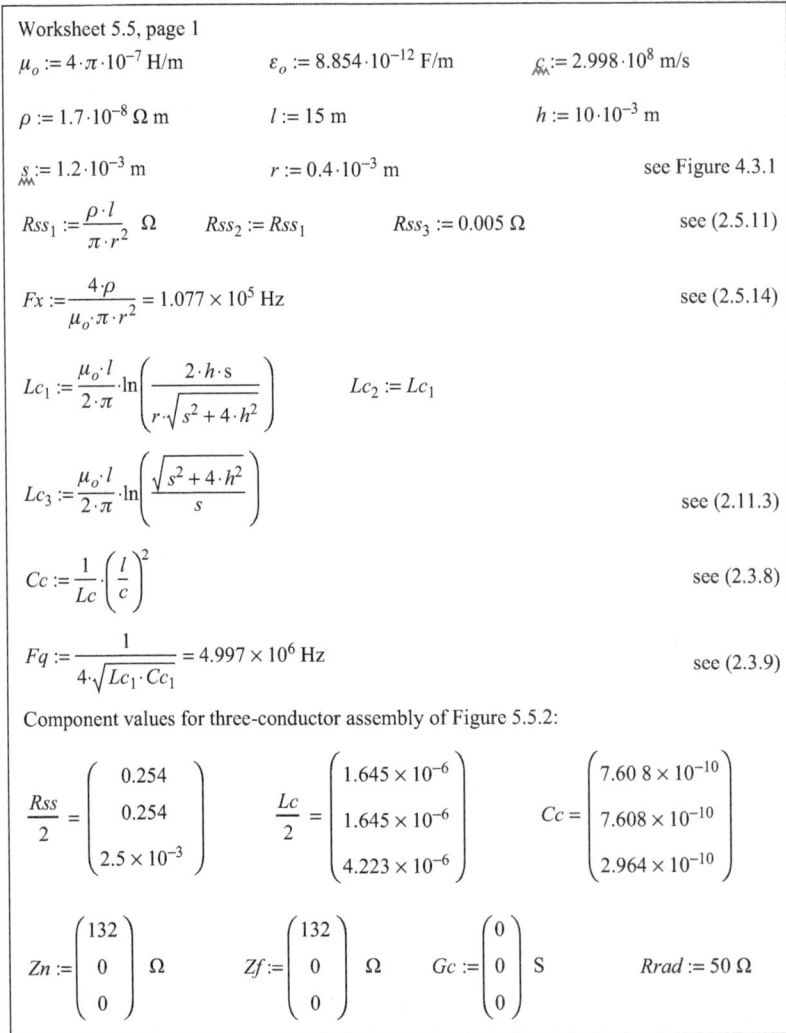

Figure 5.5.3 Calculating parameter values for the three-conductor circuit model.

In this setup, the objective is to analyze the response of the circuitry to an external field rather than the cross-coupling between two signals. So the program identifies signal, return, and structure as conductors 1, 2, and 3, respectively.

Data from this first page of the worksheet can be used to assign values to most of the components of the triple-T model; most, but not all. It is still necessary to relate the model to the external field.

The action of the incoming electromagnetic field is to create a current in the structure, and this current develops a voltage in the common-mode loop. If it is assumed that the shielding effectiveness of the structure is zero, then it can also be assumed that all the power

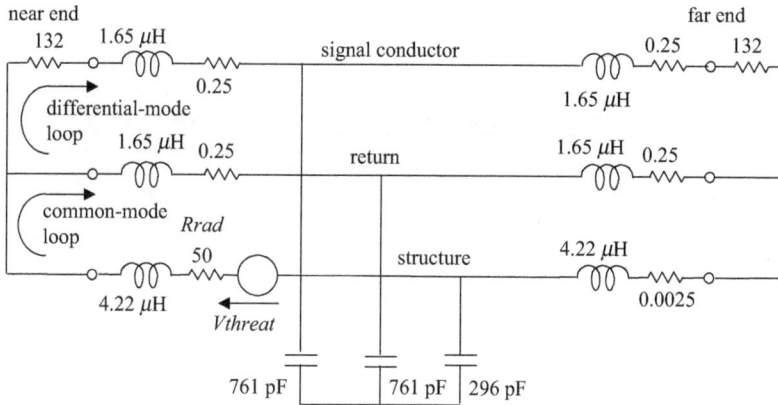

Figure 5.5.4   Representative circuit model of radiation coupling via the structure.

in the interference field is transferred to the structure. Such a transfer of power can be represented by a voltage source *Vthreat* in series with the radiation resistance *Rrad* and with the structure. Under worst-case conditions the amplitude of the voltage source *Vthreat* can be defined by the envelope curve of Figure 5.3.6.

In section 5.2 on the virtual conductor it is assumed that the value of the radiation resistance is 73 Ω; the same as that of a half-wave dipole. However, an actual test on a single length of conductor revealed that, with an assumed value of 73 Ω, the response of the model was much lower than the actual response. Figure 7.4.8 illustrates this discrepancy. It was reasoned that this discrepancy was due to storage of energy in the conductor rather than the transfer of power to a 73 Ω load.

A subsequent test on a twin-conductor cable revealed that the measured value of the radiation resistance was indeed much less than 73 Ω. Figure 7.5.10 illustrates the response of the cable-under-test on the same graph as the response of the model, and shows that the peak values coincide when it is assumed that the radiation resistance is 50 Ω. This is the value assumed for *Rrad* in Figure 5.5.4.

Having established a rationale for a method of linking the circuit model to the external field, it becomes possible to complete the model. Figure 5.5.4 shows the representative circuit model for a twin conductor cable routed along the structure.

It is a fairly simple step to convert this lumped parameter model to one which uses distributed parameters. This is illustrated by Figure 5.5.5. This distributed parameter model also defines the four circuit loops used in the subsequent mesh analysis.

There is now enough data to compile the second page of the worksheet, and this page is illustrated by Figure 5.5.6. The *Zbranch(f)* function calculates numerical values for the six distributed parameters of Figure 5.5.5. This data is provided as input to the *Zloop(f)* function which compiles a four-by-four array of loop impedance values for the circuit at the spot frequency *f*. This page of the worksheet is a slightly modified version of page 2 of the worksheet 4.3.

Figure 5.5.7 illustrates the final set of computations on page 3 of the worksheet. The power density *S* of the incoming interference is set at a constant level of 1 W/m^2.

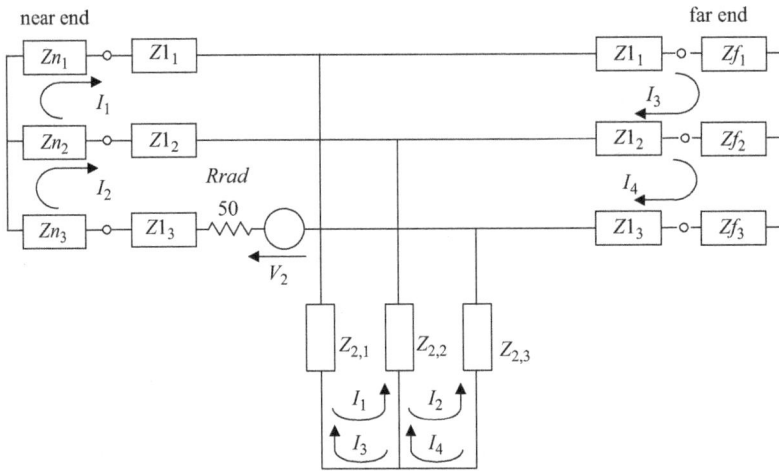

Figure 5.5.5 Distributed parameter model for figure 5.5.4.

The *Vthreat(f)* function then calculates the value of the threat voltage at the given frequency *f*. This function is a copy of the envelope function for the threat voltage, derived in section 5.3 and recorded in Figure 5.3.6.

The integer n is used to set the number of spot frequencies in the frequency range between zero and the frequency of quarter-wave resonance Fq. It has already been calculated (in page 1 of the worksheet) that the value of Fq is approximately 5 MHz. The integer s is used to set the total range of frequencies to twenty times the quarter-wave frequency. The variable F is a vector which defines every spot frequency in the range.

The main program is defined by the *Iout* function. This picks out a value for the frequency f from row s of the F vector, assigns a four-by-four matrix of loop impedance values to the variable Z, sets the voltage V_2 to the value of the threat voltage at that frequency, calculates values for the four loop currents, provides the magnitude of the current I_2 as an output variable, and stores the results in the vector *Iout*.

Since I_2 is a measure of the current at the near end of the common-mode loop, the vector *Iout* is effectively a table of the common-mode current at every spot frequency. The two vectors *Iout* and F can then be used to display a graph which defines the frequency response of the common-mode current. This is replicated by Figure 5.5.8.

This response follows the expected pattern; a null at the quarter-wave frequency followed by a peak at the half-wave frequency. This is followed by a series of peaks and nulls of ever-decreasing amplitude.

By selecting the value of I_1 as the output variable of the main program, a graph can be created which defines the frequency response of the differential-mode current. This is illustrated by Figure 5.5.9.

Several features of this response are worth noting:

Between 100 kHz and 1 MHz, the amplitude of the current in the differential-mode loop rises. This is because the current in the common-mode loop is relatively constant. Constant current at increasing frequency flowing in the inductance of the return conductor causes the

Worksheet 5.5, page 2

$$\text{Zbranch}(f) := \begin{vmatrix} \omega \leftarrow 2 \cdot \pi \cdot f \\ \text{for } i \in 1..3 \\ \quad \begin{vmatrix} Rc_i \leftarrow Rss_i \cdot \sqrt{1 + \dfrac{f}{Fx}} \\ \theta \leftarrow \sqrt{(Rc_i + j \cdot \omega \cdot Lc_i) \cdot (Gc_i + j \cdot \omega \cdot Cc_i)} \\ Zo \leftarrow \sqrt{\dfrac{Rc_i + j \cdot \omega \cdot Lc_i}{Gc_i + j \cdot \omega \cdot Cc_i}} \\ Z_{1,i} \leftarrow Zo \cdot \tanh\left(\dfrac{\theta}{2}\right) \\ Z_{2,i} \leftarrow Zo \cdot \text{csch}(\theta) \end{vmatrix} \\ Z \end{vmatrix}$$

$$\text{Zloop}(f) := \begin{vmatrix} Z \leftarrow \text{Zbranch}(f) \\ Z11 \leftarrow Zn_1 + Z_{1,1} + Z_{2,1} + Z_{2,2} + Z_{1,2} + Zn_2 \\ Z12 \leftarrow -(Z_{1,2} + Z_{2,2} + Zn_2) \\ Z13 \leftarrow -(Z_{2,1} + Z_{2,2}) \\ Z14 \leftarrow Z_{2,2} \\ Z22 \leftarrow Zn_2 + Z_{1,2} + Z_{2,2} + Z_{2,3} + Z_{1,3} + Zn_3 + Rrad \\ Z23 \leftarrow Z_{2,2} \\ Z24 \leftarrow -(Z_{2,2} + Z_{2,3}) \\ Z33 \leftarrow Zf_2 + Z_{1,2} + Z_{2,2} + Z_{2,1} + Z_{1,1} + Zf_1 \\ Z34 \leftarrow -(Z_{1,2} + Z_{2,2} + Zf_2) \\ Z44 \leftarrow Zf_3 + Z_{1,3} + Z_{2,3} + Z_{2,2} + Z_{1,2} + Zf_2 \\ \begin{pmatrix} Z11 & Z12 & Z13 & Z14 \\ Z12 & Z22 & Z23 & Z24 \\ Z13 & Z23 & Z33 & Z34 \\ Z14 & Z24 & Z34 & Z44 \end{pmatrix} \end{vmatrix}$$

Figure 5.5.6 Calculating the values of the branch and loop parameters.

voltage across that inductance to rise. Since this voltage is applied to the differential-mode loop, the current in that loop increases.

As the frequency rises above 2.5 MHz, the response flattens out. This is because the 132 Ω resistors act to minimize the amplitude of the resonant peaks. It is worth noting that 2.5 MHz corresponds to one eighth of a wavelength and that the first crossover point in Figure 4.3.8 also occurs at one eighth of a wavelength.

Worksheet 5.5, page 3

$S := 1$ $\text{Vthreat}(f) := \Bigg|$ $\lambda \leftarrow \dfrac{c}{f}$ Calculating the threat voltage
for a power density S
at a frequency f.

$Va \leftarrow \sqrt{S \cdot 377} \cdot \dfrac{\lambda}{\pi}$ see Figure 5.19

$Vb \leftarrow Va \sin\left(2 \cdot \pi \cdot \dfrac{l}{\lambda}\right)$

$Vb \leftarrow Va \text{ if } l > \dfrac{\lambda}{4}$

$n := 100$ $s := 1..20 \cdot n$ $F_s := s \cdot \dfrac{Fq}{n}$ Defining the frequency range

$Iout_s := \Bigg|$ $f \leftarrow F_s$ Main program

$Z \leftarrow \text{Zloop}(f)$

$V \leftarrow \begin{pmatrix} 0 \\ \text{Vthreat}(f) \\ 0 \\ 0 \end{pmatrix}$

$I \leftarrow \text{lsolve}(Z, V)$

$|I_2|$

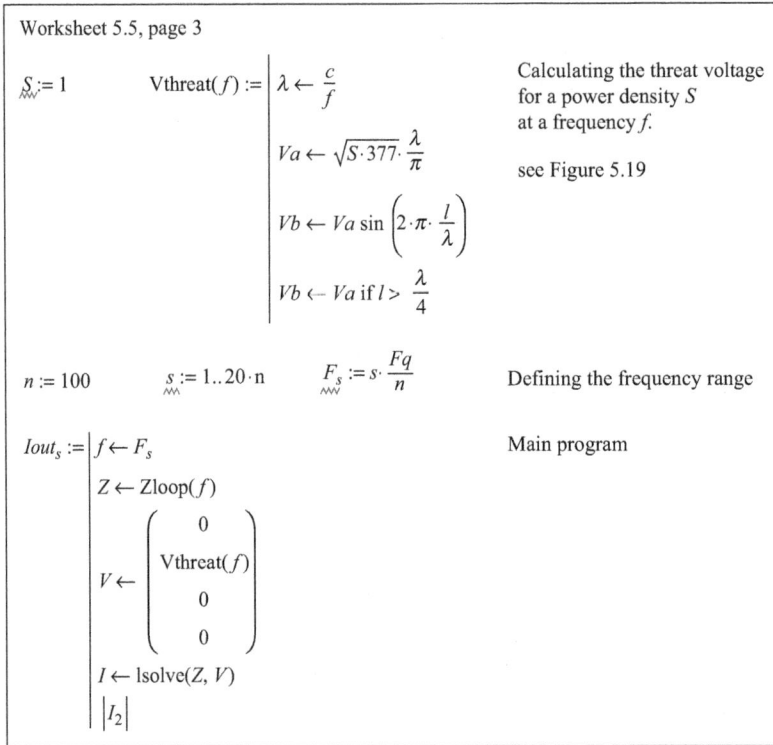

Figure 5.5.7   Calculating the frequency response of the common-mode current.

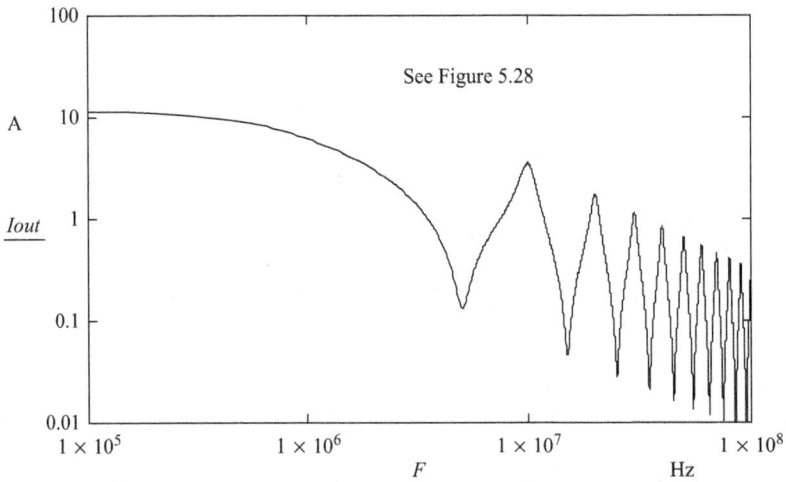

Figure 5.5.8   Frequency response of current in the common-mode loop.

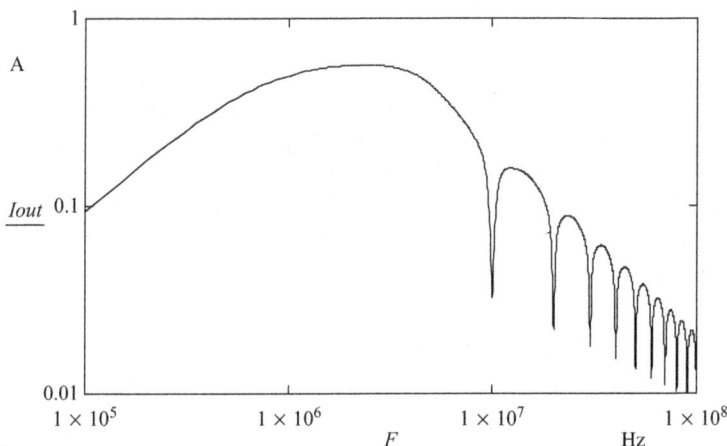

Figure 5.5.9 Frequency response of current in the differential-mode loop.

Although the common-mode current drops to a minimum at the quarter-wave frequency of 5 MHz, the amplitude of the differential-mode current remains relatively constant. This also is a feature of the use of impedance-matching resistors at the terminations. The *Ycrit* curve in Figure 4.3.8 illustrates the response of an ideal configuration where the use of the characteristic impedance at each interface prevents both nulls and peaks in the response.

Probably the most interesting feature is the fact that, although the common-mode current reaches a peak value at the half-wave frequency (10 MHz), the differential-mode current drops to a null. At this frequency, the common-mode loop currents at the near end and at the far end both reach a resonant peak. But they are of opposite polarity. Whilst the voltage developed along one half of the length of the return conductor is positive, the voltage developed along the other half is negative. Since the common-mode voltages across the return conductor cancel out, the net voltage along the complete length of the return conductor is minimal. Since this is the voltage which appears in the differential-mode loop, the current induced by that voltage also dips to a minimum value.

The waveform pattern repeats itself every 10 MHz, but the maximum amplitude drops at 20 dBA per decade. This identifies the fact that, as the frequency increases, the maximum power which can be delivered to the victim circuit reduces. The shorter the link-under-review, the less unwanted power it can pick up from an interference field of defined power density. This is a significant consideration when designing printed circuit boards; or firing circuits for electro-explosive devices. This conclusion is supported by (5.1.15), which shows that the maximum power is proportional to the square of the wavelength. Maximum power is picked up when the length of the conductor is equal to a half-wavelength of the interfering signal. If the conductor is longer, then the half-wavelength is longer and the maximum power delivered is higher.

It can be concluded that it would be better to use a screened twisted pair to provide some measure of shielding for the signal link depicted in Figure 5.5.1. Further analysis and testing would be needed to check the integrity of any modification to the design.

This example illustrates an analysis of a uniform cable routed over a perfectly flat structure with very simple terminations at each interface. This could be regarded as the easiest simplest most idealised solution. It also represents worst-case conditions, since it does not take into account the fact that other equipment in the system is also absorbing interference energy.

If the cross section of the cable is not uniform, it would be necessary to analyse the properties of each section individually and invoke the concept of equivalent circuits, as described in section 1.3.3. The general circuit model of Figure 5.5.5 can cater for impedances of any value at the terminations. If any doubt exists as to the accuracy of any particular simulation, then electrical measurements can be carried out on a test rig.

5.6 Radiation susceptibility

In the previous section, a model was created which can simulate the response of the equipment-under-review to the presence of an external electromagnetic field at the surface of the assembly. To extend the scope of the simulation, it is necessary to invoke the use of electromagnetic theory, and this can involve some very complicated mathematics.

Even so, it is possible to identify a few simple relationships which can be used to obtain a first estimate of the response.

For the analysis of the susceptibility of the equipment to external radiation, the need is to relate the threat voltage in the antenna-mode loop to the output of a remote transmitter. Since the formal EMC Requirements usually specify the setup and test levels for Radiation Susceptibility, the immediate objective is to simulate the output of the Test Transmitter.

Figure 5.6.1 illustrates the type of setup used for this form of testing. An antenna is located at a defined distance r from the equipment under test, oriented to deliver maximum field strength at the surface of the equipment, and a specified power Pr is applied to the test antenna over a specified range of frequencies. The antenna gain, Gt, is also specified.

The power density S at the surface of the equipment can be calculated using (5.1.6), and the intensity of the electric field E can be derived by using (5.1.11). The threat voltage *Vthreat* is obtained from (5.3.6) if the length l of the cable is greater than a quarter wavelength. If the length is less than a quarter wavelength, (5.3.3) applies. This gives the response of the envelope curve of Figure 5.3.6.

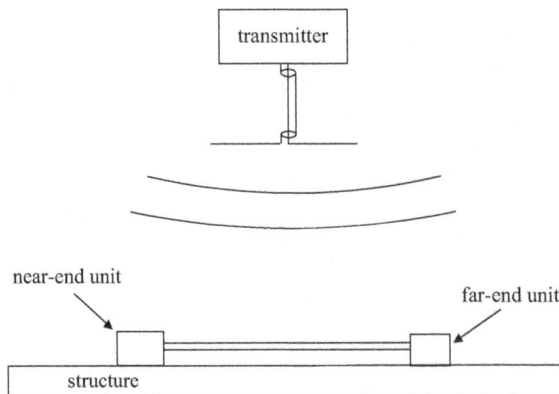

Figure 5.6.1 Setup for radiation susceptibility test.

Grouping these equations together allows the threat voltage to be related to the transmitter power:

$$S = \frac{Gt \cdot Pt}{4 \cdot \pi \cdot r^2}$$
$$E = \sqrt{Z_o \cdot S}$$
$$Vthreat = \frac{\lambda}{\pi} \cdot E \quad \text{if} \quad l > \frac{\lambda}{4}$$
$$Vthreat = \frac{\lambda}{\pi} \cdot E \cdot \sin\left(\frac{2 \cdot \pi \cdot l}{\lambda}\right) \quad \text{if} \quad l \le \frac{\lambda}{4}$$

(5.6.1)

The maximum power which can be delivered to the victim loop is limited only by the radiation resistance and the threat voltage. From Figure 5.5.4:

$$Pthreat = \frac{Vthreat^2}{Rrad}$$

(5.6.2)

Using (5.3.6) to substitute for *Vthreat*:

$$Pthreat = \left(\frac{\lambda}{\pi}\right)^2 \cdot \frac{E^2}{Rrad}$$

Substituting for *E*:

$$Pthreat = \frac{\lambda^2 \cdot S}{\pi} \cdot \frac{Z_o}{\pi \cdot Rrad}$$

(5.6.3)

Hence, if *Ge* is defined as:

$$Ge = \frac{Z_o}{\pi \cdot Rrad}$$

(5.6.4)

then, using (5.1.9) and (5.1.3) to substitute for *Zo* and *Rrad* in (5.6.4), the value of the gain *Ge* can be calculated to be 1.64. This correlates precisely with (5.1.8). Equation (5.6.3) can be re-written as:

$$Pthreat = \frac{Ge \cdot \lambda^2 \cdot S}{\pi}$$

(5.6.5)

Comparing (5.6.5) with (5.1.15) leads to the conclusion that the power delivered to the victim loop can, in theory, be four times that which appears at the receiver terminals of a matched antenna/load assembly. Such a conclusion would imply that the assumed value of the radiation resistance should be 36.5 Ω, rather than the 73 Ω defined in (5.1.3).

Although the tests described in sections 7.4 and 7.5 measure the value of the radiation resistance to be much less than 73 Ω, neither of them achieve the theoretically minimum value of 37.5 Ω. It can be surmised that the discrepancy is due to re-radiation of the stored energy.

It can also be reasoned that the test of section 7.5 represents worst-case conditions, since all the antenna-mode power is delivered to the cable. This would not be the case in a practical situation. Some of the power of the input radiation would be absorbed by the adjacent wiring in the system. Since the test on the twin-conductor cable results in a measured value of 50 Ω for *Rrad*, this would seem to be a reasonable starting point for any analysis of radiation susceptibility.

5.7 Radiated emission

The circuit model derived in section 5.5 can also be used to determine the radiated emission of the equipment under test. Figure 5.7.1 illustrates the setup.

If the equipment being tested is the same as that of Figure 5.5.1, then the model used to predict emissions can be based on Figure 5.5.4. The only changes necessary are to replace *Vthreat* and *Rrad* with a short-circuit and to insert a voltage source *Vdiff.*

In Figure 5.7.2 the voltage source *Vdiff* is located in the differential-mode loop, to simulate the output signal generated by the near-end equipment. This will create a common-mode loop current; and the amplitude of this current can be calculated. Under worst-case conditions it can be assumed that the shielding effectiveness of the structure is zero. In which case, the common-mode current will be the radiation source. In the figure, it is identified as *Irad*.

The power *Prad* transmitted into the environment by the current *Irad* would be:

$$Prad = Irad^2 \cdot Rrad \tag{5.7.1}$$

From (5.1.6), the power density at a monitor antenna located a distance r away from the assembly would be:

$$Sm = \frac{Ge \cdot Prad}{4 \cdot \pi \cdot r^2} \tag{5.7.2}$$

Since the assembly under test is the same as that which had been subjected to the radiation susceptibility test, then the gain *Ge* would be the same as defined by (5.6.4). Substituting for *Ge* and *Prad* gives:

$$Sm = \frac{Z_o}{\pi \cdot Rrad} \cdot \frac{Irad^2 \cdot Rrad}{4 \cdot \pi \cdot r^2}$$
$$= Z_o \cdot \frac{Irad^2}{4 \cdot \pi^2 \cdot r^2}$$

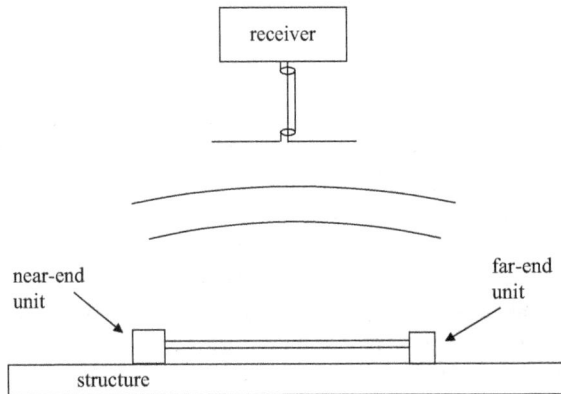

Figure 5.7.1 Setup for radiated emission test.

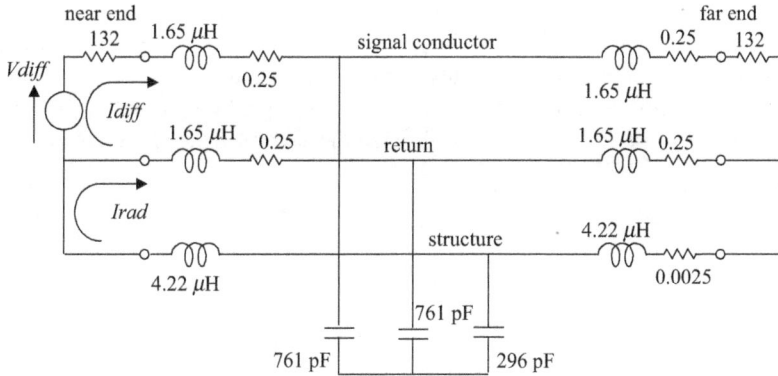

Figure 5.7.2 Representative circuit model for analysis of radiated emission.

It is worth noting that this relationship does not depend on the value of *Rrad*. From (5.1.12) the magnetic field strength at the surface of the monitor antenna is:

$$Hm = \sqrt{\frac{Sm}{Z_o}}$$

Substituting for *Sm*:

$$Hm = \frac{Irad}{2 \cdot \pi \cdot r} \tag{5.7.3}$$

Even though there are no simplifying assumptions in its derivation, this relationship is about as simple as one could wish for. (It could have been derived from (2.2.1) by assuming an infinitely long conductor. But it wasn't.)

It is also worth noting that this derivation is based on the assumption that the conductor is acting as a half-wave dipole at maximum efficiency in a lossless medium. When the length of the conductor is anything other than $\lambda/2$, the power delivered to the environment will be significantly less than *Prad*. In turn, this means that (5.7.3) defines the maximum magnetic field strength *Hm* that can be created by the current *Irad*.

Using (5.7.3) to define the magnetic field strength at the monitor antenna, (5.1.12) to relate power density to magnetic field, and (5.1.15) to relate power density to the power delivered to the receiver input terminals, leads to the set of equations:

$$Hm = \frac{Irad}{2 \cdot \pi \cdot r}$$
$$Sm = Z_o \cdot Hm^2 \tag{5.7.4}$$
$$Pm = \frac{Gm \cdot \lambda^2 \cdot Sm}{4 \cdot \pi}$$

where *Gm* is the gain of the monitor antenna used by the Test House.

This establishes a relationship between the current radiated into the environment by the equipment under test and the power delivered to the input of the monitor receiver.

Adding this set of equations to the subroutine which calculates the value of the common-mode current will provide a clear relationship between signal output voltage of the near-end unit of Figure 5.7.1 and the level of interference measured by the test receiver.

CHAPTER 6

Transient analysis

Transients are an ever-present source of glitches in electronic systems. Sources such as relays, switches, motors, and power supplies can cause the sudden release of stored magnetic energy when the current maintaining that magnetization is interrupted. Electrostatic discharges can cause a momentary creation of high-amplitude current spikes. Whatever the source, brief bursts of high intensity electromagnetic radiation can occur almost anywhere in an electrical system.

Such transients can easily corrupt the data streams handled by microprocessors. In fact, any analogue or digital circuit link can be upset. Depending on the susceptibility of the processing circuitry, such events could be inconsequential, annoying, dangerous, or catastrophic.

This section describes the use of time-step analysis to simulate transient behavior. It derives a model which can simulate the emission produced by a step pulse propagated along a twin-conductor line. Given an insight into the mechanisms involved, it should be possible for the designer to devise circuits which minimize this emission. Such circuits should also be capable of minimizing susceptibility. Chapter 8 identifies a few of the techniques deriving from this approach.

As with frequency response analysis, the effects of an incoming field can be represented by a voltage source, while the outgoing field can be defined in terms of the antenna-mode current. The amplitude of the voltage source is a function of the E-field of the external radiation, while the H-field of the emitted radiation is a function of the antenna-mode current. To a first approximation, it can be assumed that the amplitude of both fields is inversely proportional to the separation between transmitter and receiver. All the passive parameters (resistance, inductance, capacitance, and conductance) remain unchanged. The electrical properties of a particular cable assembly are independent of the waveform of the signal or power carried by that assembly.

Section 6.1 provides an introduction to the basics of time-step analysis, and contains two examples of how to develop programs to simulate transient waveforms. An essential feature of time-step analysis is that the input voltage is defined and current is calculated instant-by-instant. The computations involve absolute real values. Unlike frequency response analysis, time-step analysis never invokes the concepts of phasors, phase angles, imaginary numbers, reactance, or susceptance. A further implication is that impedances can only be resistive. So, to avoid any misunderstanding of the terms used during the course of this type of analysis, the variable Ro is used to define the characteristic impedance.

The equations in section 6.2 are based on the concepts of partial currents and partial voltages used in textbook analysis of transients in transmission lines. A program is described which simulates the response of a twin conductor line when a step pulse is applied to the near end. It is possible to develop this model to include the response of interface circuitry at the terminations.

Section 6.3 examines the characteristics of the delay-line model, and shows that the properties of inductance and capacitance are still present.

When a sinusoidal source at the quarter-wave frequency is switched on to a twin-conductor cable which is open-circuit at the far end, the current will gradually build up as the energy in successive half-cycles is stored in the cable. This is illustrated by Figure 6.3.10.

If a signal at half-wave frequency is connected, the current delivered to the cable will gradually decrease. This is because the stored energy in the cable increases to a level where energy delivered from source to cable is balanced by energy delivered from cable to source. Figure 6.3.11 illustrates this.

These two illustrations show that an analysis of the transient response of a system to a sinusoidal signal helps us to better understand frequency response analysis.

Section 6.4 tackles the problem of analyzing antenna-mode transients. An experiment is described where a step pulse was injected into a twin-conductor cable which was open-circuit at the far end. After a trial-and-error process, a circuit model was identified; one which provided a fair reproduction of the response. The existence of such a model raised several questions about the mechanisms involved, questions which led to a further development of the model.

In section 6.5, the picture emerges of wavefront propagating along the signal conductor in the form similar to the bow wave of a ship. Given a separation of 2 mm between the conductors, the wavefront does not arrive at the return conductor until the forward edge has progressed a further 2 mm along the signal conductor. Electric charge is deposited on the return conductor. Induced voltage causes a current to flow back toward the source. This current creates an electromagnetic field that couples with the signal conductor and that also radiates outwards. Since the current in the return conductor does not completely balance that in the signal conductor, there is an unbalanced charge which propagates along the cable, just behind the wavefront.

Since the return current is always less than the signal current, there must be a net flow of aerial-mode current out toward the far end. This outgoing current must be balanced by an inflow of current from the ground connections. Effectively, the system is behaving like a dipole, with the cable acting as one monopole and the ground conductors acting as the other.

This means that there are at least three current components involved

- that which carries differential-mode signal,
- that which deposits unbalanced charge on the cable, and
- that which flows out radially in the form of electromagnetic radiation.

Section 6.6 derives a general circuit model which allows the amplitude of all three of these components to be calculated. A program is described which carries out the computations. Comparing the response of the model with the waveform displayed on an oscilloscope allows the two waveforms to be aligned. The values so derived for component values allow a representative model to be derived for the cable-under-test; Figure 6.6.5. The tests themselves are described in section 7.6.

6.1 Time-step analysis

6.1.1 Basic concept

The classical approach to transient analysis is to use transforms – either Laplace transforms or Heaviside transforms. However, this approach becomes extremely complex when dealing with the type of circuits commonly encountered. Hence, the approach adopted here is essentially the same as that used by Simulation Programs with Integrated Circuit Emphasis (SPICE) software – time-step analysis [6.1].

This form of analysis is similar to the iterative method used in mathematics to solve non-linear equations. The essential difference is that each calculation defines the condition of the system at a later time. If the time steps are sufficiently small, then the variation in the amplitude of each current can be assumed to be linear.

Two factors are involved in each set of calculations:

- The state of the system immediately prior to the time of the calculation
- Incremental changes in the applied voltages

Either nodal analysis or mesh analysis is possible. Since mesh analysis has been used for calculations in the frequency domain, it is logical to continue with that type of analysis in the time domain.

6.1.2 Basic equations

The basic equations are the simplest possible:

For an inductor:

$$V = L \cdot \frac{dI}{dt} \qquad (6.1.1)$$

For a resistor:

$$V = R \cdot I \qquad (6.1.2)$$

For a capacitor:

$$V = \frac{Q}{C} \qquad (6.1.3)$$

where:

$$Q = \int_0^t I \cdot dt \qquad (6.1.4)$$

given that

$t =$ elapsed time
$dt =$ finite increment of time
$dI =$ finite increment of current

6.1.3 Series LCR circuit

For the circuit illustrated by Figure 6.1.1, the total voltage at any instant is the sum of the instantaneous voltages across each component. This gives the loop equation:

$$V = L \cdot \frac{dI}{dt} + R \cdot I + \frac{Q}{C} \tag{6.1.5}$$

Analysis, based on the application of a Heaviside transformation, leads to an equation that defines the relationship between current and time when a step voltage is applied to the input terminals of the circuit. If the circuit is initially in a quiescent state, then:

$$I = \frac{V}{L} \cdot \frac{1}{a - b} \cdot (e^{-b \cdot t} - e^{-a \cdot t})$$

where $a = \alpha + \beta$, $b = \alpha - \beta$, $\alpha = \frac{R}{2 \cdot L}$, and $\beta = \sqrt{\frac{R^2}{4 \cdot L^2} - \frac{1}{L \cdot C}}$.

A less elegant, but simpler approach would be to use time-step analysis to formulate a short computer program. The steps necessary to compile such a program are outlined below.

During the time increment dt, the current changes by the increment dI, where:

$$dI = \frac{dt}{L} \cdot \left(V - R \cdot I - \frac{Q}{C} \right) \tag{6.1.6}$$

At the end of that increment of time the value of the current will have changed. In a Mathcad computer program such as that illustrated in Appendix A, the new value for the current would be defined as:

$$I \leftarrow I + dI \tag{6.1.7}$$

During this time, the charge on the capacitor will have changed, and the program statement which updates the value of Q would be:

$$Q \leftarrow Q + I \cdot dt \tag{6.1.8}$$

These last three equations form the basis of a simple subroutine 'next(D)' in the Mathcad worksheet of Figure 6.1.2. The variable D is a two-element vector containing the value of the current and the charge in the circuit of Figure 6.1.1 at any particular instant.

The subroutine picks out these two values, uses (6.1.6)–(6.1.8) to update them, and then returns the new value of I and Q. The value of the time increment dt has been set at 1 µs on the top line of the worksheet, together with the other circuit parameters.

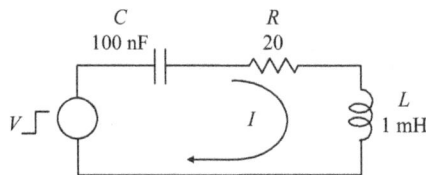

Figure 6.1.1   Series LCR circuit.

Worksheet 6.1.1

$L := 1 \cdot 10^{-3}$ H $\qquad\qquad$ $C := 100 \cdot 10^{-9}$ F $\qquad\qquad$ $R := 20\ \Omega$

$V := 1$ V $\qquad\qquad\qquad$ $dt := 10^{-6}$ s $\qquad\qquad\qquad$ $N := 100$

$$D := \begin{pmatrix} 0 \\ 0 \end{pmatrix} \qquad \text{next}(D) := \begin{vmatrix} I \leftarrow D_1 \\[4pt] Q \leftarrow D_2 \\[4pt] dI \leftarrow \dfrac{dt}{L} \cdot \left(V - R \cdot I - \dfrac{Q}{C} \right) \\[8pt] I \leftarrow I + dI \\[4pt] Q \leftarrow Q + I \cdot dt \\[6pt] \begin{pmatrix} I \\ Q \end{pmatrix} \end{vmatrix}$$

$$i := 2..N \qquad Iout_i := \begin{vmatrix} D \leftarrow \text{next}(D) \\[4pt] D_1 \end{vmatrix} \qquad t_i := (i-1) \cdot dt$$

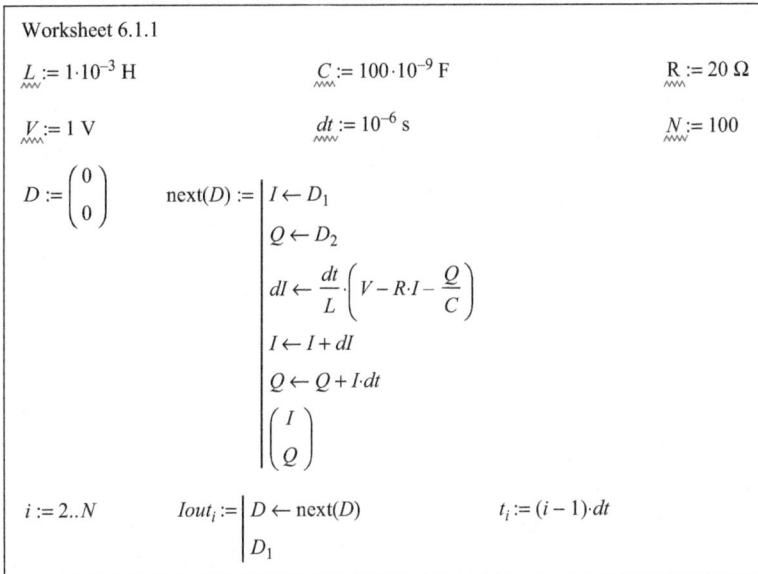

Figure 6.1.2 Calculating the transient response of a series LCR circuit.

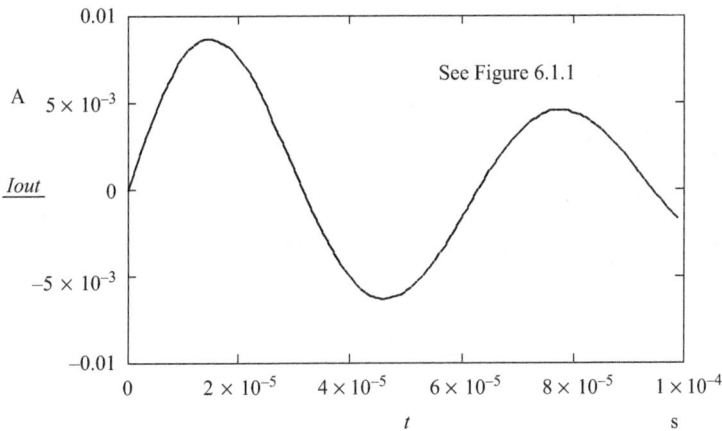

Figure 6.1.3 Transient response of series LCR circuit.

The control variable i for the main program is set to vary from 2 to N, where N has been preset at 100. The main program itself is defined immediately below the subroutine. It simply updates the data variable D at each time step, picks out the value of D_1 (the current I), and transfers this to the element $Iout_i$, where the vector $Iout$ is the record of the output current.

The variable t records the time of each calculation, and allows the graph of $Iout$ versus time to be displayed; Figure 6.1.3. Not surprisingly, this is a damped sine wave.

6.1.4 Parallel LCR circuit

When the L, C, and R components are in parallel, three separate loops are formed as illustrated by Figure 6.1.4.

The loop equations are:

$$Vgen = (R1 + R2) \cdot I_1 + \frac{Q_1}{C1} - \frac{Q_2}{C1} - R2 \cdot I_2 \qquad (6.1.9)$$

$$0 = - R2 \cdot I_1 - \frac{Q_1}{C1} + \frac{Q_2}{C1} + (R2 + R3) \cdot I_2 - R3 \cdot I_3 \qquad (6.1.10)$$

$$0 = -R3 \cdot I_2 + R3 \cdot I_3 + L1 \cdot \frac{dI_3}{dt} \qquad (6.1.11)$$

In formulating the program which calculates the values of the variables at each time step, it is useful to take into consideration the characteristics of the step pulse. Theoretically, it changes from zero volts to $Vgen$ in zero time. As far as the initial step is concerned, $C1$ behaves as a short-circuit and $L1$ behaves as an open-circuit. Most of the initial surge current I_1 flows through $C1$ and $R2$ in series. The resultant voltage across $R2$ creates a current I_2. Current I_2 flowing in $R3$ creates a voltage across $L1$.

This means that, initially, I_1 is the most significant current, I_2 is less significant, while the value of I_3 is inconsequential. This effectively determines the order in which parameter values should be calculated.

Since the charge on the capacitor is always due to $I_1 - I_2$, (6.1.9) and (6.1.10) can be simplified by defining:

$$Q = Q_1 - Q_2 \qquad (6.1.12)$$

It is assumed that the initial amplitudes of I_1, I_2, and I_3 are zero.

Current I_1 can be calculated by rearranging (6.1.9). This gives:

$$I_1 = \frac{1}{R1 + R2} \cdot \left(Vgen + R2 \cdot I_2 - \frac{Q}{C1} \right) \qquad (6.1.13)$$

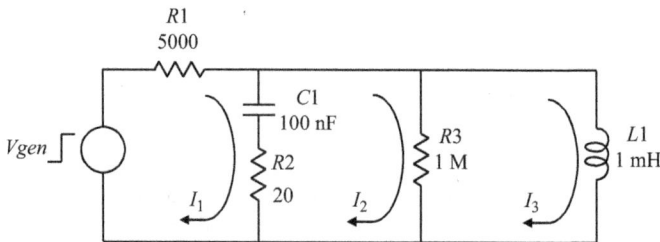

Figure 6.1.4 Parallel LCR circuit.

Since the value of I_1 is now known, it can be used to calculate a new value of I_2:

$$I_2 = \frac{1}{R2 + R3} \cdot \left(R2 \cdot I_1 + R3 \cdot I_3 + \frac{Q}{C1} \right) \tag{6.1.14}$$

The new value of I_2 can be used to calculate the rate of change of I_3:

$$dI_3 = \frac{dt}{L1} \cdot R3 \cdot (I_2 - I_3) \tag{6.1.15}$$

Worksheet 6.1.2

$R1 := 5000\ \Omega$ $\quad R2 := 20\ \Omega$ $\quad R3 := 10^6\ \Omega$

$L1 := 1 \cdot 10^{-3}\ \mathrm{H}$ $\quad C1 := 100 \cdot 10^{-9}\ \mathrm{F}$ $\quad Vgen := 1\ \mathrm{V}$

$dt := 10^{-8}\ \mathrm{s}$

$$D := \begin{pmatrix} 0 \\ 0 \\ 0 \\ 0 \end{pmatrix}$$

$$\mathrm{next}(D) := \left| \begin{array}{l} I2 \leftarrow D_2 \\ I3 \leftarrow D_3 \\ Q \leftarrow D_4 \\ I1 \leftarrow \dfrac{1}{R1 + R2} \cdot \left(Vgen + R2.I2 - \dfrac{Q}{C1} \right) \\ I2 \leftarrow \dfrac{1}{R2 + R3} \cdot \left(R2.I1 + R3.I3 + \dfrac{Q}{C1} \right) \\ dI3 \leftarrow \dfrac{dt}{L1} \cdot R3 \cdot (I2 - I3) \\ I3 \leftarrow I3 + dI3 \\ Q \leftarrow Q + (I1 - I2) \cdot dt \\ \begin{pmatrix} I1 \\ I2 \\ I3 \\ Q \end{pmatrix} \end{array} \right.$$

$T := 100 \cdot 10^{-6}\ \mathrm{s}$

$N := \mathrm{ceil}\left(\dfrac{T}{dt} \right)$

$i := 2..N$ $\quad Iout_i := \left| \begin{array}{l} D \leftarrow \mathrm{next}(D) \\ D_3 \end{array} \right.$ $\quad t_i := (i - 1).dt$

Figure 6.1.5 Calculating the amplitude of the transient current in the parallel LCR circuit.

The new value of I_3 becomes:

$$I_3 \leftarrow I_3 + dI_3 \qquad (6.1.16)$$

Since the values of I_2 and I_3 have also been updated:

$$Q \leftarrow Q + (I_1 - I_2) \cdot dt \qquad (6.1.17)$$

It is then a routine matter to assemble (6.1.13)–(6.1.17) into a subroutine which updates the values of current and charge at every time step. This is illustrated in Figure 6.1.5.

The simplest way of checking the accuracy of the calculations is to vary the time step dt and recalculate. If a reduction in the amplitude of the time step results in no perceptible change to the output waveform, then the simulation can be assumed to be reasonably accurate.

In the worksheet of Figure 6.1.5, the simulation time T is set at 100 μs, and the number of iterations is defined as:

$$N = \text{ceil}\left(\frac{T}{dt}\right) \qquad (6.1.18)$$

(In Mathcad, the function ceil(z) returns the smallest integer greater than or equal to z.) No matter what value is selected for dt, the simulation will always run for 100μs.

The waveform displayed in Figure 6.1.6 is that of the current in the inductance $L1$. This is a ringing transient, similar to that of Figure 6.1.3. The main difference between this and the previous response is the fact that it approaches a steady-state amplitude of 200 μA. That is, the current in a 5 kΩ resistor when a voltage of 1 V is applied.

In formulating programs similar to the two illustrated in this section, the first objective is to identify the parameter with the greatest change during the first time increment. In the first example, this is the voltage across the inductor; in the second example it is the current in the capacitor.

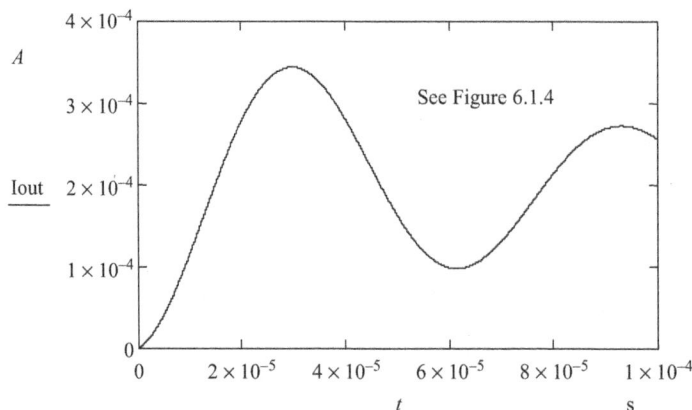

Figure 6.1.6 Waveform of the current in the inductor of the parallel LCR circuit.

6.2 Delay-line model

Transmission-line theory invokes the concept of electromagnetic waves traveling forwards and backwards along a path bounded by a pair of parallel conductors. Figure 6.2.1 illustrates the concept.

Although the velocity of an electromagnetic wave in free space is approximately 300 m/μs, the presence of dielectric material in the cable reduces this velocity. If it is assumed that the velocity of propagation is v, the time taken for an electromagnetic wave to propagate from one end of the line to the other is the time constant T, where:

$$T = \frac{l}{v} \tag{6.2.1}$$

Initial assumptions are that there are no radiation losses from a transmission line and that the line does not pick up external radiation. It is worth noting that these assumptions are not identified in conventional textbooks.

If it is assumed that an electromagnetic signal has been delivered to the line by the sender at the near end, then it will arrive at the far end after an elapsed time of T seconds. Some of it will be absorbed by the load in the receiver, while some of it will be reflected. Figure 6.2.2 illustrates the components of the signal at any instant in terms of currents and voltages.

Conventional theory depicts this situation in terms of current in one conductor of a two-conductor line, and assumes that the properties of the line are concentrated in this one conductor. To enable the analysis to be developed to deal with three-conductor lines, it is necessary to view all currents in terms of loop currents. In the convention adopted here, positive loop current is defined as clockwise.

It is useful to use a shorthand method of identifying each parameter to keep track of all these currents and voltages during the computation process. The method adopted here is to use the letter f to indicate that the parameter is at the far end, the letter n to indicate that it is at the near end, and to use the first letter of the relevant word to indicate whether the

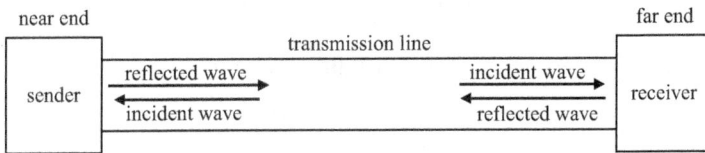

Figure 6.2.1   Transmission-line reflections.

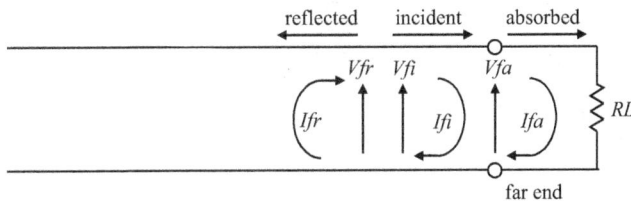

Figure 6.2.2   Currents and voltages at the far end of the line.

parameter is absorbed, reflected, or incident. Hence, Ifa is the current which is absorbed by the load resistor in the receiver at the far end.

A noteworthy feature of the parameters Ifi and Ifr is that they are 'partial currents'. The total current in the line at any particular location is the sum of Ifr and Ifi at that location. Similarly, Vfr and Vfi can be described as 'partial voltages'.

Currents and voltages in Figure 6.2.2 can be related:

$$Vfi + Vfr = Vfa \tag{6.2.2}$$

$$Ifi + Ifr = Ifa \tag{6.2.3}$$

Voltages can be related to currents:

$$Vfi = Ro \cdot Ifi \tag{6.2.4}$$

$$Vfr = -Ro \cdot Ifr \tag{6.2.5}$$

$$Vfa = RL \cdot Ifa \tag{6.2.6}$$

where Ro is the characteristic impedance of the line and RL is the load resistor. Subtracting (6.2.5) from (6.2.4) gives:

$$Vfi - Vfr = Ro \cdot (Ifi + Ifr)$$

Using (6.2.3):

$$Vfi - Vfr = Ro \cdot Ifa \tag{6.2.7}$$

Using (6.2.6) to substitute for Vfa in (6.2.2):

$$Vfi + Vfr = RL \cdot Ifa \tag{6.2.8}$$

Adding (6.2.7) and (6.2.8):

$$2 \cdot Vfi = (Ro + RL) \cdot Ifa \tag{6.2.9}$$

Rearranging (6.2.9) and using (6.2.4) to substitute for Vfi gives:

$$Ifa = \frac{2 \cdot Ro \cdot Ifi}{Ro + RL} \tag{6.2.10}$$

This relates the current delivered to the receiver to the incident current arriving from the transmission line. The reflected current is determined by rearranging (6.2.3).

$$Ifr = Ifa - Ifi \tag{6.2.11}$$

Using (6.2.10) to substitute for Ifa in (6.2.11) leads to the standard equation for the reflection coefficient found in any textbook on electromagnetic theory.

$$K = \frac{Ifr}{Ifi} = \frac{Ro - RL}{Ro + RL}$$

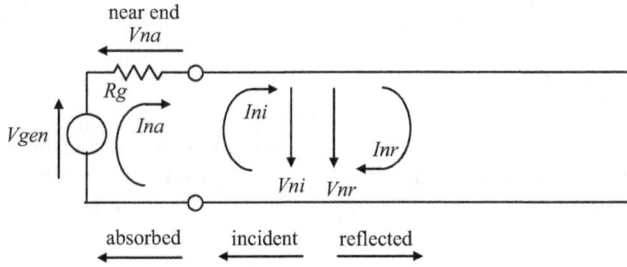

Figure 6.2.3   Currents and voltages at the near end of the line.

This coefficient is not utilized here because the load at the far end could be any mixture of resistors, inductors, and capacitors. It is important to know the value of the current actually delivered to the receiver, *Ifa*. The magnitude of this current is a function of the state of the interface circuitry at the far end. Once *Ifa* is known, the value of *Ifr* can be obtained from (6.2.11).

Current reflected by the far-end termination, *Ifr*, travels back along the line and manifests itself as incident current *Ini* at the terminals at the near end, as shown in Figure 6.2.3. Since positive loop current is defined as clockwise on all diagrams, and since power is delivered by the line to the near-end terminals, the incident voltage *Vni* must necessarily be inverted with respect to the voltage *Vfr*.

At any instant, the voltages at the near end are:

$$Vni + Vnr + Vgen = Vna \tag{6.2.12}$$

$$Ini + Inr = Ina \tag{6.2.13}$$

$$Vni = Ro \cdot Ini \tag{6.2.14}$$

$$Vnr = -Ro \cdot Inr \tag{6.2.15}$$

$$Vna = Rg \cdot Ina \tag{6.2.16}$$

Using the same process that was used to analyze currents and voltages at the far end, the relationships at the near end become:

$$Ina = \frac{2 \cdot Ro \cdot Ini + Vgen}{Ro + Rg} \tag{6.2.17}$$

and

$$Inr = Ina - Ini \tag{6.2.18}$$

Unlike the computations involving lumped parameters, where the status of the system at time t is dependent on its status at time $t - dt$, the status of the equipment interface at each end of the transmission line is dependent on the status of the interface at the other end of the line at time $t - T$.

Since the current at each end of the line is time-dependent, it is necessary to store data on instantaneous values for a large number of time steps. Such a requirement calls for a table of values to be recorded, each record holding data on parameter values for a defined instant in time. An array needs to be created to hold this data.

Fortunately, it is unnecessary for this array to store a large number of variables for each instant. If the interface circuits are purely resistive, then only two parameters are involved, the reflected current at the far end, *Ifr*, and the reflected current at the near end, *Inr*. This limits the number of columns of the array to two. The number of rows can also be limited, since the state of the transmission line before $t - T$ plays no part in the computation. If the duration T is divided into N steps of duration dt then:

$$dt = \frac{T}{N} \tag{6.2.19}$$

Defining the value for dt in this way ensures that a computation is carried out for every instant a transient arrives at a termination.

This means that the table should contain N records and that each record should contain two values. Since it is necessary to scan the table several times during any particular simulation, there needs to be some way of relating the time of each event to the appropriate record.

If the time interval between each sample is always dt, then the relationship between computation number n and time t is given by:

$$t_n = n \cdot dt \tag{6.2.20}$$

Table 6.2.1 illustrates the necessary correlation between computation number, n, and the relevant column, p, of the table, given the assumption that the number of columns, N, is 10.

The diagram of Figure 6.2.4 illustrates a configuration in which a signal is transmitted from one end of a twin-conductor line to the other. The time taken for a transient to propagate from one end to the other is 100 ns and the characteristic impedance is 100 Ω. The signal source at the near end is a voltage generator *Vgen* with an output impedance of 10 Ω. The resistor *RL* at the far end provides a load impedance of 1000 Ω.

Determining the response of this signal link to a step voltage is simply a matter of assembling the relevant equations into an ordered sequence. The worksheet of Figure 6.2.5 illustrates one method.

The top two lines of the worksheet is derived directly from the characteristics of the configuration-under-review. On the third line, the number *N1* is set at 100. This defines the number of calculations to be carried out during a single traverse of a transient pulse. The value of *N1* is selected to allow the simulation of 30 such sweeps.

Table 6.2.1 Relating each data sample to the computation number

n	1	2	3	4	5	6	7	8	9	10
n	11	12	13	14	15	16	17	18	19	20
n	21	22	23	24	25	26	27	...	...	...
p	1	2	3	4	5	6	7	8	9	10
Inr										
Ifr										

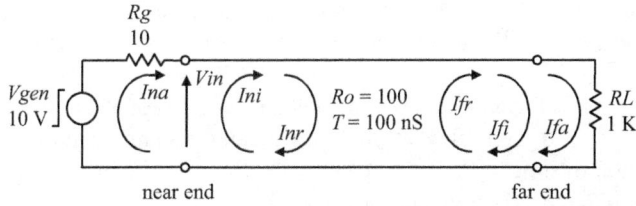

Figure 6.2.4 Delay-line model.

Worksheet 6.2, page 1   See Figure 6.2.1

$Ro := 100 \, \Omega$ $Rg := 10 \, \Omega$ $RL := 1000 \, \Omega$

$Vgen := 10 \, V$ $T := 100 \cdot 10^{-9} \, s$

$N1 := 100$ $dt := \dfrac{-T}{N1}$ $N2 := 30 \cdot N1$

$$send(INPUT, Vgen) := \left| \begin{array}{l} Ini \leftarrow INPUT_2 \\[4pt] Ina \leftarrow \dfrac{2 \cdot Ro \cdot Ini + Vgen}{Rg + Ro} \\[8pt] Inr \leftarrow Ina - Ini \\[4pt] \begin{pmatrix} Ina \\ Inr \end{pmatrix} \end{array} \right. \qquad recv(INPUT) := \left| \begin{array}{l} Ifi \leftarrow INPUT_1 \\[4pt] Ifa \leftarrow \dfrac{2 \cdot Ro \cdot Ifi}{RL + Ro} \\[8pt] Ifr \leftarrow Ifa - Ifi \\[4pt] \begin{pmatrix} Ifa \\ Ifr \end{pmatrix} \end{array} \right.$$

$$point(n) := \left| \begin{array}{l} m \leftarrow mod(n, N1) \\[4pt] m \leftarrow N1 \text{ if } m = 0 \end{array} \right.$$

$$Ina := \left| \begin{array}{l} data_{2, N1} \leftarrow 0 \\[4pt] \text{for } i \in 1..N2 \\ \quad \left| \begin{array}{l} p \leftarrow point(i) \\[4pt] INPUT \leftarrow data^{\langle p \rangle} \\[4pt] \begin{pmatrix} Ina \\ Inr \end{pmatrix} \leftarrow send(INPUT, Vgen) \\[8pt] \begin{pmatrix} Ifa \\ Ifr \end{pmatrix} \leftarrow recv(INPUT) \\[8pt] OUTPUT \leftarrow \begin{pmatrix} Inr \\ Ifr \end{pmatrix} \\[8pt] data^{\langle p \rangle} \leftarrow OUTPUT \\[4pt] I_i \leftarrow Ina \end{array} \right. \\[4pt] I \end{array} \right.$$

$n := 1..N2 \qquad t_n := (n-1).dt$

Figure 6.2.5 Calculating the waveform of the current at the near end of the line.

The function send(*INPUT,Vgen*) carries out the computations which define the currents existing at the near end of the line at any particular instant – the time *t*. It responds to two input variables.

The voltage *Vgen* is the amplitude of the voltage source. For a step pulse, its value is zero at all time before *t* = 0, and 10 V at all times thereafter. For a sine wave, it would be a sinusoidal function of time.

The variable *INPUT* is a two element vector containing the values of *Inr* and *Ifr* at time *t* − *T*.

The first line of this subroutine picks out the value of *Ifr*. Since it is assumed that there are no losses in the line, then this value is the amplitude of the current *Ini* arriving at the near-end terminals at time *t*.

Given that the values of *Vgen* and *Ini* are now available to the subroutine, the values of *Ina* and *Ini* can be calculated using (6.2.17) and (6.2.18). The output of the subroutine is a two-element vector containing the values of these two parameters at time *t*.

The subroutine recv(*INPUT*) performs a similar set of calculations for currents at the far end of the line and returns a two element vector with the values of *Ifa* and *Ifr*.

The point(*n*) function takes as input the number *n* of the sequence of computations and uses it to calculate a value for *p*, a pointer to the appropriate record in Table 6.2.1.

In Mathcad, the function mod(*n,N*) returns the remainder of *n* when divided by *N*.

The main program is used to calculate a set of values for *Ina*, the current delivered to the input terminals of the transmission line.

The first action of the main program is to define the array *data* as having two rows and *N1* columns. Initially, all the values are set to zero.

A control variable *i* is set to run from 1 to *N2*, the total number of computations involved.

For each computation, the value of the integer *p* is calculated. This points to the appropriate column in the data array, and the record stored in that column is defined as the vector *INPUT*. This is a two-element vector holding the values of *Inr* and *Ifr* at time *t* − *T*.

These values are treated as *Ifi* and *Ini* at time *t* by the two subroutines send(*INPUT,Vgen*) and recv(*INPUT*), and values for *Inr* and *Ifr* are calculated. These are placed in a two element vector, *OUTPUT*, and the contents of this vector are used to overwrite the record in column *p* of the *data* array.

The final action of each computation is to select the local variable *Ina* and place the value in element *i* of the vector *I*.

The output of the main program is a vector containing all the values computed for *Ina*. This parameter can be regarded as the current delivered to the input terminals of the transmission line.

The waveform of this current is illustrated by Figure 6.2.6. The amplitude of the initial step current is due to a 10 V supply loaded by 10 Ω and 100 Ω in series. It is only after several reflections have occurred that the current settles down to its steady-state value – the current due to a 10 V supply loaded by 1000 Ω and 10 Ω in series. During the settling-in period, there is a burst of high frequency oscillation in the system.

The configuration depicted by Figure 6.2.4 is indicative of many signal links in the average system. In many configurations, the output impedance of the sender is less than 10 Ω and the input impedance of the load is greater than 1000 Ω. This means that there is a brief burst of oscillation on the line every time a voltage step occurs.

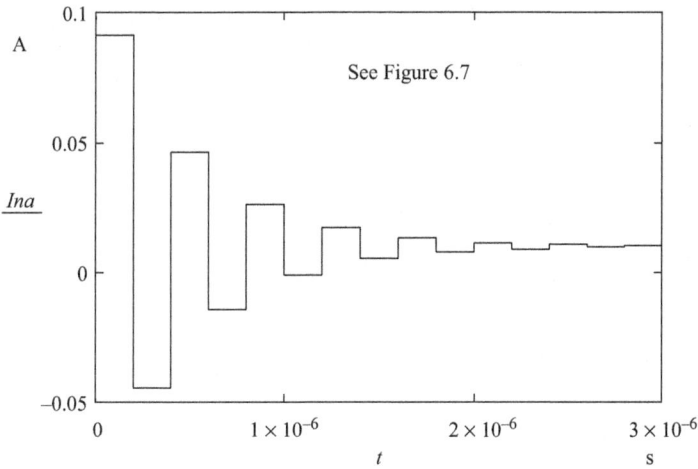

Figure 6.2.6 Waveform of the current delivered to input of transmission line.

Step changes in voltage occur every time a logic signal changes state and every time a device is switched on or off. If the interface circuitry is not matched to the line, then it will also carry a great number of high frequency transient currents. Moreover, the frequency will be close to that of quarter-wave resonance, the ideal frequency for creating maximum emission.

6.3 Line characteristics

It would seem that the model for the delay line has lost two of the essential features of any circuit model; there are no inductors and there are no capacitors. However, these parameters have not really disappeared.

The definitive parameters for a loss-free line are the characteristic impedance Ro and the transition time T. (In transient analysis, the concept of reactance does not exist, since the parameters L and C are treated separately. Impedance parameters can only be resistive.) Since these have been derived from the values for the inductance and capacitance of the line, it should be possible to reverse the process.

Equations (4.1.7) and (4.1.18) define the basic relationships for any transmission line. For a loss-free line, frequency response analysis gives:

$$Ro = \sqrt{\frac{La}{Ca}} \tag{6.3.1}$$

and

$$\theta = j \cdot \omega \cdot \sqrt{La \cdot Ca} \tag{6.3.2}$$

where La and Ca are the loop inductance and the loop capacitance of the line. From (2.3.8):

$$\sqrt{La \cdot Ca} = \frac{l}{v} = T \tag{6.3.3}$$

It follows that:

$$\theta = j \cdot \omega \cdot T \tag{6.3.4}$$

From (6.3.1):

$$Ca = \frac{La}{Ro^2} \tag{6.3.5}$$

Substituting for Ca in (6.3.3):

$$T = \frac{La}{Ro}$$

This removes the j operator and the ω parameter from the equations and allows inductance to be defined in terms of T and Ro:

$$La = T \cdot Ro \tag{6.3.6}$$

Substituting for La in (6.3.5):

$$Ca = \frac{T}{Ro} \tag{6.3.7}$$

Hence, (6.3.6) and (6.3.7) allow the values of La and Ca to be derived from Ro and T.

The inductance of a transmission line can be determined by shorting the far-end terminations and applying a voltage to the near end via a resistance of known value. Figure 6.3.1 illustrates the configuration and Figure 6.3.2 shows the corresponding model when lumped parameters are used.

Analyzing the response of Figure 6.3.1 is simply a matter of copying the Mathcad worksheet of Figure 6.2.5, changing the value of RL to zero, and re-running the program. This constitutes page 1 of a new worksheet.

Analyzing the response of Figure 6.3.2 is a matter of following the process described in section 6.1.1. The loop equation is:

$$Vgen = Rg \cdot I + L \cdot \frac{dI}{dt} \tag{6.3.8}$$

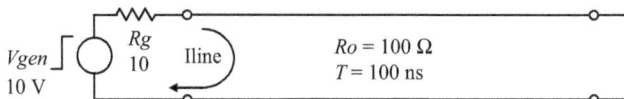

Figure 6.3.1 Transmission line configured as an inductance.

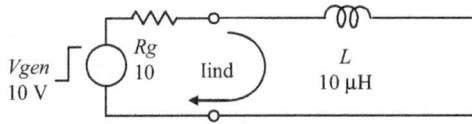

Figure 6.3.2 Lumped parameter model of short-circuited line.

Worksheet 6.3.1 page 2
(page 1 of worksheet is a copy of Figure 6.2.5 with the value of *RL* set at zero)

$Rg := 10$ $\qquad$ $RL := 0$ $\qquad$ $dt := 1 \cdot 10^{-9}$ $\qquad$ $\underset{\sim}{L} := T \cdot Ro$ $\qquad$ $L := 1 \cdot 10^{-5}$

$I := 0$ $\qquad$ $\text{next}(I) := \left| \begin{array}{l} dI \leftarrow \dfrac{dt}{L} \cdot (Vgen - Rg \cdot I) \\[2mm] I \leftarrow I + dI \end{array} \right.$ $\qquad\qquad$ $Iind_n := \left| \begin{array}{l} I \leftarrow \text{next}(I) \\[1mm] I \end{array} \right.$

Figure 6.3.3 Calculating response of model of Figure 6.3.2.

Rearranging this equation gives:

$$dI = \frac{L}{dt} \cdot (Vgen - Rg \cdot I) \tag{6.3.9}$$

Equation (6.3.9) forms the core calculation of the program illustrated by Figure 6.3.3. This is page 2 of the new worksheet.

The response of the transmission-line model is illustrated by the stepped curve *Iline* of Figure 6.3.4. The response of the lumped parameter model is illustrated by the solid curve *Iind*. As expected, the current rises exponentially toward 1 A; the steady-state current in 10 Ω due to a constant voltage of 10 V.

Correlation between the responses of the two models is quite evident. It illustrates the fact that the transmission-line model indeed has inductive properties, and that (6.3.6) defines the value of that inductance.

The capacitance of the line can be revealed by open circuiting the terminations at the far end of the line and by increasing the value of *Rg*.

In Figure 6.3.5, the value of *RL* is changed to 10 MΩ, which is an open-circuit compared to *Ro*. The value of *Rg* is changed to 1000 Ω, 10 times the value of *Ro*.

For this configuration, the lumped parameter model would be as shown by Figure 6.3.6, and the value of *C* can be calculated using (6.3.7).

The loop equation for this model is:

$$Vgen = Rg \cdot I + \frac{Q}{C} \tag{6.3.10}$$

and this leads to:

$$I = \frac{1}{Rg} \cdot \left(Vgen - \frac{Q}{C} \right) \tag{6.3.11}$$

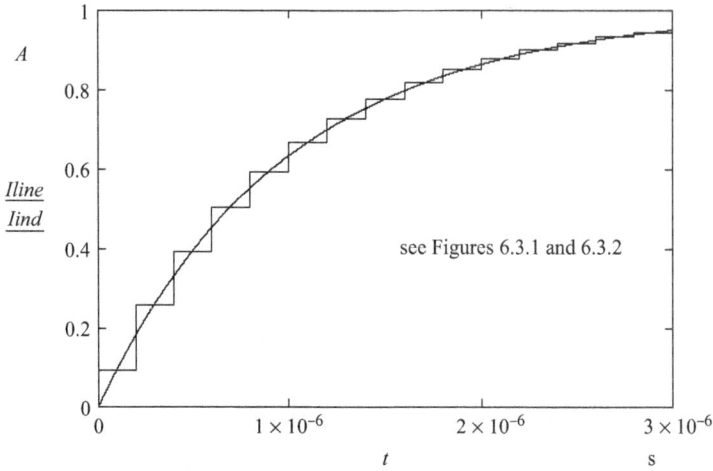

Figure 6.3.4 Response of short-circuited line and 10 μH inductor.

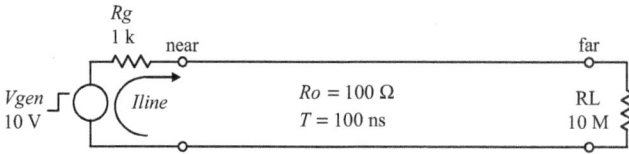

Figure 6.3.5 Transmission line configured as a capacitance.

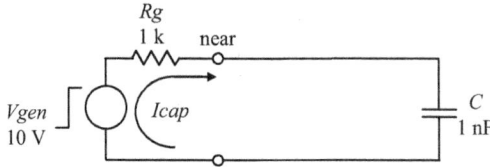

Figure 6.3.6 Lumped parameter model of open-circuit line.

Comparing the responses of the open-circuit line with that of the R–C circuit can be carried out by invoking a similar procedure to that employed with the assessment of inductance. Figure 6.3.7 is a copy of the second page of the relevant worksheet. The new values of Rg and RL are recorded at the top of the page, and the core calculation of the program is derived from (6.3.11).

Figure 6.3.8 shows the results of the computations. As expected, the initial amplitude of the current into the capacitor C is 10 mA, the current flowing in 1 kΩ due to a voltage of 10 V. The stepped curve which tracks this response is due to current delivered to the open-circuit transmission line of Figure 6.3.5. The correlation between the two curves could not be closer. This demonstrates that the transmission line possesses capacitive properties, and that the value of the capacitance is given by (6.3.7).

Normally, a capacitor is constructed by winding a length of two closely spaced conductors into a tight spiral. Coupling between the turns of the spiral results in multiple

Worksheet 6.3.2 page 2
(page 1 of this worksheet is a copy of Figure 6.2.5 with the value of RL set to 10^7)

$Rg := 1 \cdot 10^3$ $\qquad RL := 1 \cdot 10^7$ $\qquad dt := 1 \cdot 10^{-9}$ $\qquad C := \dfrac{T}{Ro}$ $\qquad C := 1 \cdot 10^{-9}$

$Icap := \Big| \begin{array}{l} Q \leftarrow 0 \\ \\ \text{for } i \in 1..N2 \\ \quad \Big| \begin{array}{l} Ic \leftarrow \dfrac{1}{Rg} \cdot \left(Vgen - \dfrac{Q}{C} \right) \\ \\ Q \leftarrow Q + Ic \cdot dt \\ \\ I_i \leftarrow Ic \end{array} \\ \\ I \end{array}$ $\qquad\qquad T := 100 \cdot 10^{-9}$

Figure 6.3.7   Calculating the transient current in the 1 nF capacitor.

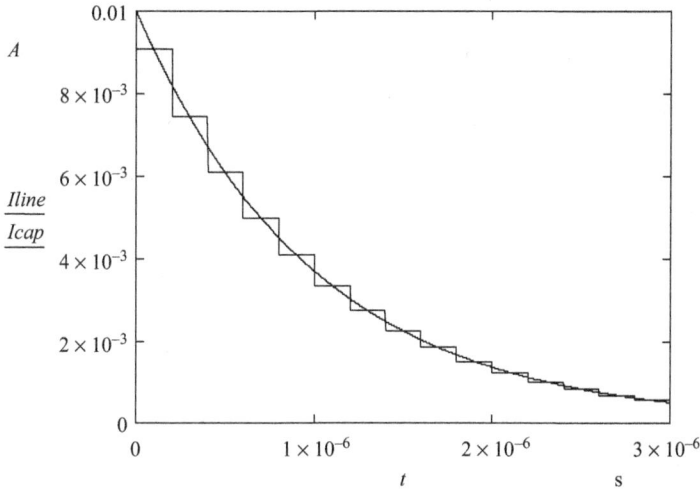

Figure 6.3.8   Response of open-circuited line compared to that of a capacitor.

reflections in the assembly, the net result being a curve that is indistinguishable from that of the lumped parameter model.

Changing *Vgen* from a step-function generator to a sinusoidal voltage source enables the response of the open-circuit line to be assessed at any frequency. Figure 6.3.9 illustrates the configuration. The quarter-wave frequency of the line is:

$$f_q = \frac{1}{4 \cdot T} = 2.5 \text{ MHz} \tag{6.3.12}$$

If this is the frequency of the source *Vgen*, then the current delivered would be as shown in Figure 6.3.10. This illustrates the fact that the amplitude of the current gradually builds up. At every half-cycle, the amplitude is increased by a small increment. A limit is reached when the peak current is 1 A, the current in *Rg* when a peak voltage of 10 V is applied.

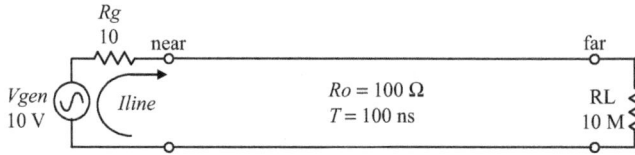

Figure 6.3.9 Transmission line acting as a tuned circuit.

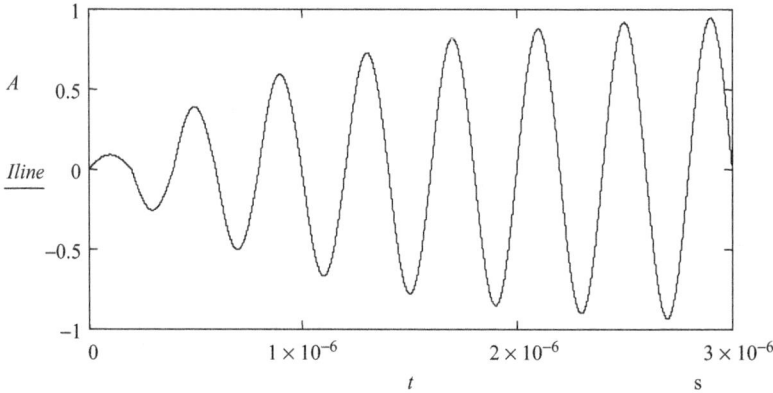

Figure 6.3.10 Response of open-circuit line at quarter-wave frequency.

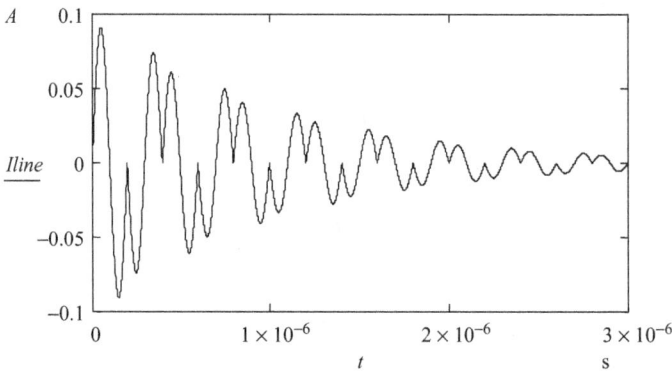

Figure 6.3.11 Response of open-circuit transmission line at half-wave frequency.

Hence, when an alternating current at the quarter-wave frequency is applied to an open-circuit line, the input terminals appear to be short-circuited. Such a conclusion is exactly the same as that obtained with AC analysis. However, there is a significant difference between the two types of analysis; AC analysis does not predict the gradual build-up of current amplitude.

If the frequency of the source generator were to be changed to 5 MHz, the half-wave frequency, then the current waveform at the near end of the line, would be as illustrated by Figure 6.3.11.

For the first cycle of the waveform the current is due to a 10 V sinusoidal signal in a resistance of 110 Ω. That is, the resistance value is equal to $Rg + Ro$. Subsequent cycles of the waveform are progressively reduced. After about 100 cycles, the current delivered to the line is effectively zero. Such a response is due to the fact that the incident current at the near end of the line generates a voltage across Rg that precisely balances the voltage delivered by the source generator $Vgen$.

This means that, as far as the source generator is concerned, a transmission line that is open-circuit at the far end will appear as an open-circuit at the near end. Again, this conclusion is the same as that predicted by ac analysis. However ac analysis does not predict the existence of a brief burst of current into the line, and that a standing wave exists along the line all the time an input signal is applied.

6.4 Antenna-mode current

Having developed a delay-line model to simulate the action of a twin-conductor cable, the next obvious step was to set up an experiment and observe how an actual line responds to a step input. Comparing the actual waveform with that of the simulation should provide some indication of the accuracy of the model. So this was done.

A 15 m length of two-core mains cable was purchased and a signal generator used to inject a square wave of about 8 μs duration into one end of the line. Figure 6.4.1 is an illustration of the setup.

The interface circuitry at the near end was designed to provide low source impedance, while the far end of the cable was open-circuit. This provided a configuration in which several reflections could be observed for each step of the input waveform.

The input voltage was monitored by channel 1 of the oscilloscope via a simple potentiometer network, while the output current was monitored on channel 2 via a current transformer. Waveforms were recorded as accurately as possible.

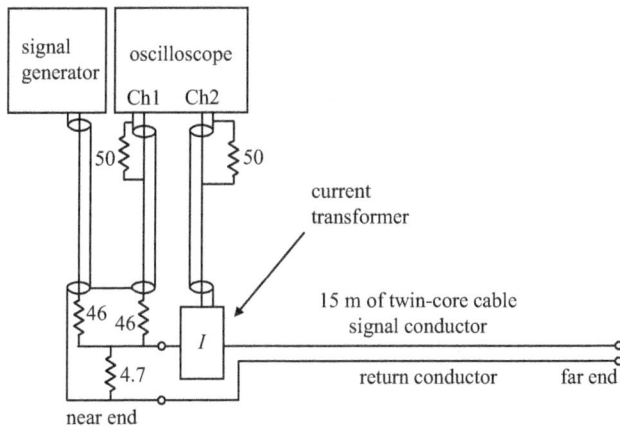

Figure 6.4.1   Set up for radiated transient test.

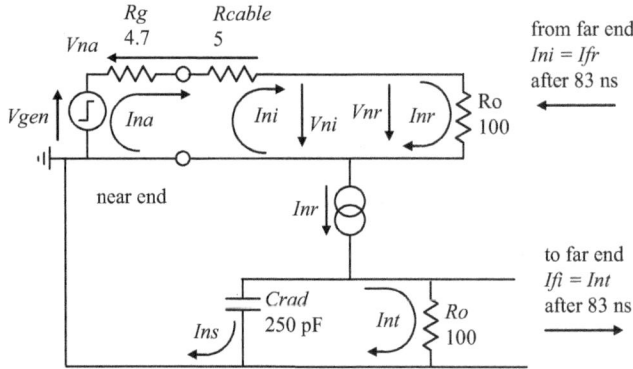

Figure 6.4.2 Delay-line model of 15 m two-core cable.

The first circuit model created to simulate this waveform was found to be wildly inac-curate. A modified model was found to be equally unrepresentative. This was followed by a trial-and-error process that eventually resulted in the model of Figure 6.4.2. This produced a waveform that was a fair representation of the trace displayed on channel 2 of the oscilloscope.

What follows is a description of the circuit model and an attempt to relate its operation to the behavior of the system-under-review. The fact that this approach also seems to be very 'trial and error' is because that's just what it was. However, the end result is the identifi-cation of features of electromagnetic coupling which could not be predicted by frequency response analysis.

In this model, *Rg* represents the source impedance of the step-function generator *Vgen,* while *Rcable* is a resistor that allows copper losses and dielectric losses to be catered for, and *Ro* represents the characteristic impedance of the cable.

Equations for the near end of the cable are:

$$Vni + Vnr + Vgen = Vna \tag{6.4.1}$$

$$Ini + Inr = Ina \tag{6.4.2}$$

$$Vni = Ro \cdot Ini \tag{6.4.3}$$

$$Vnr = -Ro \cdot Inr \tag{6.4.4}$$

$$Vna = (Rg + Rcable) \cdot Ina \tag{6.4.5}$$

These are essentially the same as those defined in section 6.2, with the exception that the parameter *Rcable* is included in (6.4.5). Using the process described in section 6.2, it can be deduced that:

$$Ina = \frac{2 \cdot Ro \cdot Ini + Vgen}{Ro + Rg + Rcable} \tag{6.4.6}$$

Rearranging (6.4.2):

$$Inr = Ina - Ini \tag{6.4.7}$$

This current transient travels down the line at a velocity approaching that of light. During this transit, most of the current flows from the signal conductor, through the characteristic impedance Ro, and back via the return conductor. (Most of the current delivered to the signal conductor is derived from current flow via the conductors of the structure and the source resistance Rg. Section 6.5 deals with this aspect of the phenomenon in more detail.) However, not all of the differential-mode current is delivered to the terminations at the far end. The current that does not arrive is represented by Ins flowing in the radiation capacitor $Crad$. The energy apparently lost to the environment due to current flow in the resistor Ro is effectively stored as a voltage across the capacitor.

The amplitude of Ins can be calculated using Figure 6.4.3, since a current source in parallel with a resistor can be represented by a voltage source in series with that resistor. The loop equation for this circuit is:

$$Inr \cdot Ro = Ins \cdot Ro + \frac{Qns}{Crad} \tag{6.4.8}$$

Hence:

$$Ins = Inr - \frac{Qns}{Ro \cdot Crad} \tag{6.4.9}$$

After each time step, the value of Qns changes. Using Mathcad terminology:

$$Qns \leftarrow Qns + Ins \cdot dt \tag{6.4.10}$$

The current actually transmitted along the cable must be:

$$Int = Inr - Ins \tag{6.4.11}$$

After a delay of T seconds, this transmitted current arrives at the far end. Again using Mathcad terminology:

$$Ifi \leftarrow Int \tag{6.4.12}$$

In the model, it is assumed that the open-circuit is represented by a high value resistor RL, and a value of 10 MΩ is assigned to this parameter.

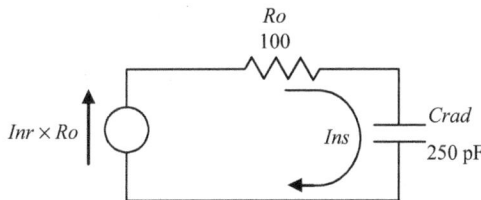

Figure 6.4.3 Calculating the value of the stored current.

The current absorbed in *RL* is:

$$Ifa = \frac{2 \cdot Ro \cdot Ifi}{Ro + RL} \qquad (6.4.13)$$

and the current reflected from the far end is:

$$Ifr = Ifa - Ifi \qquad (6.4.14)$$

After a further time delay of *T* seconds, the incident current arriving at the near end is given by:

$$Ini \leftarrow Ifr \qquad (6.4.15)$$

Assembling the relevant equations into a software program leads to the creation of a Mathcad worksheet. The first page is illustrated in Figure 6.4.4.

The top two lines of the program are derived from component values of the circuit model.

The function send(*near, INPUT, Vgen*) is a modified version of send(*near, Vgen*) of Figure 6.2.5. The modifications are essentially the inclusion of (6.4.9)–(6.4.11).

Since the states of the currents and voltages at the near end of the line are now a function of *Ins* and *Qns*, the values of these parameters need to be available as an input to the subroutine. This is done through the use of the vector *near*. Since this vector is also used to provide updated values of *Ina* and *Int* to the main program, these parameters are also included as input variables (even though they are not actually used by the subroutine).

The function *recv(INPUT)* is essentially the same as that defined in the worksheet of Figure 6.2.5.

Figure 6.4.5 illustrates the main program. It is a development of the main program of Figure 6.2.5. To correlate the simulated waveform with the actual waveform displayed on channel 2 of the scope, it was necessary to define the time of the leading edge *T1* and the sweep time *T2*. Since all the time steps in the program are equal in value, it is a simple matter to identify the step counts *N1* and *N2* at which they occur.

Since the vector *near* is used in the near-end calculations, this vector needs to be declared at the start of the main program. It contains four variables. So the number of rows is four.

The first step of the iterative subroutine of the main program is to set the initial value of *Vgen* to zero. At time *T1* it is switched to the value *Vg*. The value of *Vg* is defined on the first page of the worksheet. The most significant modification to the subroutine of Figure 6.2.5 is due to the need to update the values of the parameters *Ins* and *Qns*. So the number of rows in the *near* vector changes from two to four.

After all the values have been updated the final action of the iterative subroutine is to place the value of *Ina* in the appropriate row of the vector *Out*. When the iterations have been completed, the contents of this vector are transferred to the vector *Idiff*.

Figure 6.4.6 shows the waveform of the current *Ina* of the circuit model of Figure 6.4.2. This simulates the current that is being monitored by the current transformer in the setup of Figure 6.4.1. This waveform is extremely informative.

The first rising edge is exactly as would be predicted by the simple delay-line model of section 6.2. After this first step, the source delivers a constant current to the line. The amplitude is due to the application of the voltage *Vgen* to *Rg*, *Rcable*, and *Ro* in series.

Worksheet 6.4, page 1

$Ro := 100\ \Omega$ $\quad\quad\quad\quad\quad\quad T := 83 \cdot 10^{-9}\ s$ $\quad\quad\quad\quad Crad := 250 \cdot 10^{-12}\ F$

$Rg := 4.7$ $\quad\quad\quad\quad\quad\quad\quad\quad Rcable := 5$ $\quad\quad\quad\quad\quad\quad RL := 10^{7}$

$Vg := 1\ V$ $\quad\quad\quad\quad\quad\quad\quad N := 100$ $\quad\quad\quad\quad\quad\quad\quad dt := \dfrac{T}{N}$

$$send(near, INPUT, Vgen) := \begin{Vmatrix} \begin{pmatrix} Ifi \\ Ini \end{pmatrix} \leftarrow INPUT \\[10pt] \begin{pmatrix} Ina \\ Int \\ Ins \\ Qns \end{pmatrix} \leftarrow near \\[20pt] Ina \leftarrow \dfrac{2 \cdot Ro \cdot Ini + Vgen}{Rg + Ro + Rcable} \\[10pt] Inr \leftarrow Ina - Ini \\[8pt] Ins \leftarrow Inr - \dfrac{Qns}{Ro \cdot Crad} \\[10pt] Int \leftarrow Inr - Ins \\[8pt] Qns \leftarrow Qns + Ins \cdot dt \\[8pt] \begin{pmatrix} Ina \\ Int \\ Ins \\ Qns \end{pmatrix} \end{Vmatrix}$$

$$recv(INPUT) := \begin{Vmatrix} \begin{pmatrix} Ifi \\ Ini \end{pmatrix} \leftarrow INPUT \\[10pt] Ifa \leftarrow \dfrac{2 \cdot Ro \cdot Ifi}{RL + Ro} \\[10pt] Ifr \leftarrow Ifa - Ifi \\[8pt] \begin{pmatrix} Ifa \\ Ifr \end{pmatrix} \end{Vmatrix}$$

Figure 6.4.4   Subroutines for near-end and far-end calculations.

This constant current is maintained throughout the time taken by the transient edge to travel to the far end and for the reflected pulse to travel back to the near end. During this period, current delivered to the capacitance *Crad* is totally invisible to the source generator at the near end. It does not appear in the waveform monitored by channel 2 of the oscilloscope.

When the first leading edge reaches the far end, all the current is reflected straight back into the line. Its amplitude is unchanged, but its sign is reversed. Positive current leaving the near end appears as a negative current on its return.

Worksheet 6.4, page 2

$$point(n) := \begin{vmatrix} m \leftarrow mod(n, N) \\ m \leftarrow N \text{ if } m = 0 \end{vmatrix} \qquad\qquad\qquad \text{see Table 6.2.1}$$

$T1 := 100 \cdot 10^{-9}$ s   start time

$T2 := 2 \cdot 10^{-6}$ s   end time $N1 := \dfrac{T1}{dt}$ $N2 := \dfrac{T2}{dt}$

$$Idiff := \begin{vmatrix} data_{2,N} \leftarrow 0 \\ near_4 \leftarrow 0 \\ \text{for } i \in 1..N2 \\ \quad \begin{vmatrix} Vgen \leftarrow Vg \text{ if } i > N1 \\ p \leftarrow point(i) \\ INPUT \leftarrow data^{\langle p \rangle} \\ near \leftarrow send(near, INPUT, Vgen) \\ \begin{pmatrix} Ina \\ Int \\ Ins \\ Qns \end{pmatrix} \leftarrow near \\ \begin{pmatrix} Ifa \\ Ifr \end{pmatrix} \leftarrow recv(INPUT) \\ OUTPUT \leftarrow \begin{pmatrix} Int \\ Ifr \end{pmatrix} \\ data^{\langle p \rangle} \leftarrow OUTPUT \\ Out_i \leftarrow Ina \end{vmatrix} \\ Out \end{vmatrix}$$

$n := 1..N2$ $t_n := (n-1) \cdot dt$

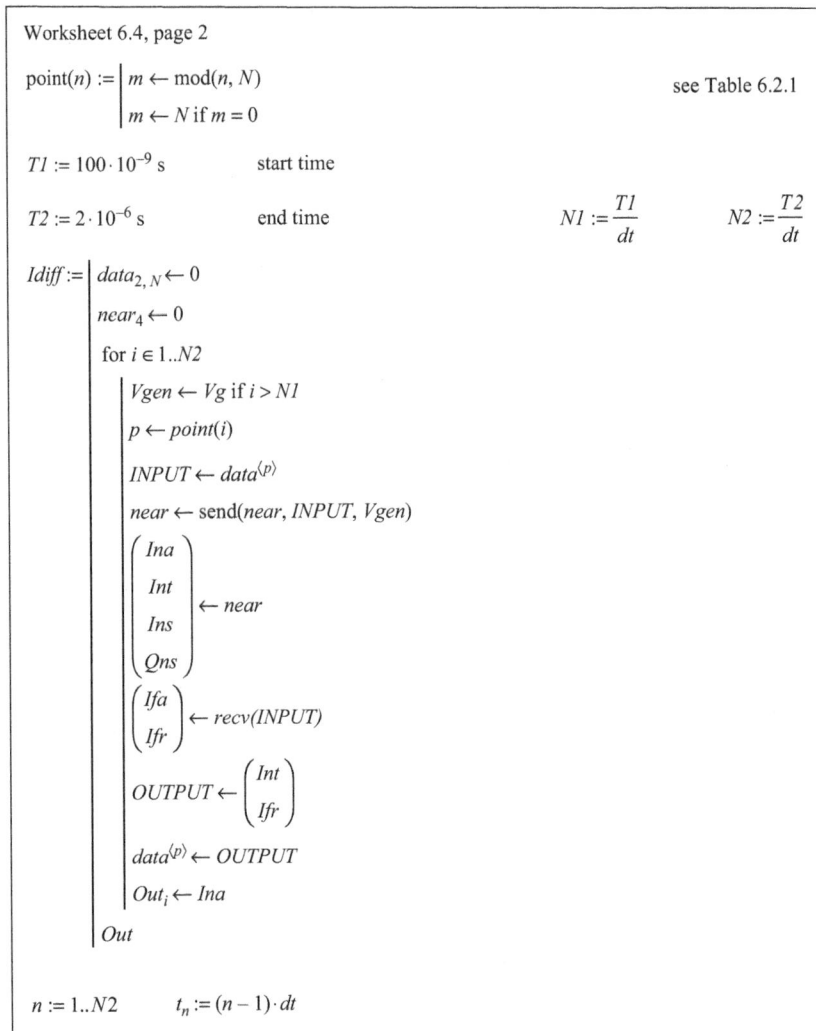

Figure 6.4.5 Calculating the waveform of the current at the near end of the line.

Since the load at the near end is much less than Ro, the amplitude of the reflected current is almost doubled.

If the simple delay-line model had been used, then this current would have caused a step change in the line current from positive to negative. This does not happen in practice. The trailing edge is an exponential curve, providing a clear indication that some of the current did not reach the far end. The lost energy is effectively stored as a voltage across $Crad$.

When the voltage step reaches the far end, current Inr of Figure 6.4.3 ceases to flow, and the amplitude of the voltage source $Inr \cdot Ro$ drops to zero. Antenna mode current Ins now flows back from the cable into the terminals at the near end.

Since the edge of the waveform is now distributed in time, and since subsequent traverses of the transients are subjected to the same physical phenomena, the result is a gradual

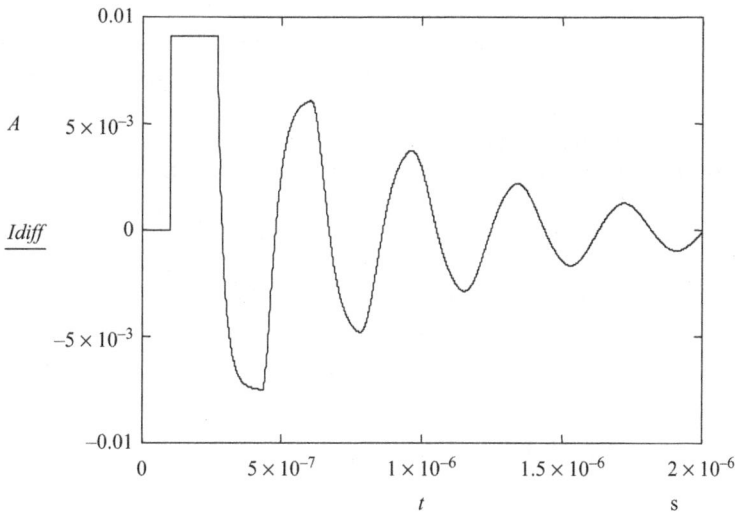

Figure 6.4.6   Waveform of current at near end of transmission line.

transition from a square waveform to a sinusoidal waveform. After only a few cycles, the mismatched line is resonating at its natural frequency.

The amplitude of the waveform gradually decays and after about 3 μs is virtually undetectable.

It should be emphasized at this point that the component values of the circuit model were adjusted iteratively until the simulated waveform bore a close resemblance to the waveform monitored on the scope. In fact, the development of the model and the explanation of how it simulated the physical phenomena went hand-in-hand. Although the explanation so far provided is reasonably plausible, there are still some puzzling questions raised by the test results. These questions are posed and answered in the next section.

6.5 Radiated emission

Although the analysis of section 6.4 created a circuit model that replicated the observed phenomena, several puzzling features of the test results remain. Chief among these is the fact that the antenna-mode current is not visible to the current transformer. So the question 'why does the transformer not detect this current?' needs an answer.

6.5.1 Current linking the transformer

There can be no doubt that antenna-mode current is flowing in the signal conductor, since that is the only conductor being energized. The return conductor is held at ground potential at the near end. The fact that the transformer of Figure 6.4.1 does not detect this component of the current means that there must be antenna-mode current in the return conductor as well.

Earlier tests on the same cable, using a sinusoidal source, had demonstrated quite clearly that the coupling between the conductors is close enough to guarantee that most of the current emanating from the signal conductor is picked up by the return conductor and flows in the opposite direction. (See section 7.5.)

Also, the test of a single conductor, described in section 7.4, illustrates the fact that all the current transmitted along one monopole is provided by current delivered from the other monopole. With the setup shown on Figure 7.6.1, the signal conductor can be regarded as one monopole, while the other monopole exists in the form of the assembly of earthed conductors of the test equipment.

It is not plausible to assume that the send and return components of the differential-mode current propagate simultaneously along the cable. It is true that if the signal source in Figure 7.6.1 had been completely isolated from structural conductors, then the voltages applied to the input terminals of the cable would have been equal in magnitude and opposite in sign. But it is not isolated.

Since the return terminal is connected to both the return conductor and to ground, it follows that the voltages applied to signal and return conductors are unbalanced. Since the primitive capacitance of the ground conductors is much greater than that of the return conductor, that unbalance must be significant. The amplitude of the current flowing from the return conductor into the ground conductors has to be much less than that delivered into the signal conductor.

However, the fact that the antenna-mode current must emanate from the ground conductor does not explain why it is flowing in the same direction as the signal current. A process of elimination means that the energy causing this outward flow can only come from the voltage source, via the signal conductor.

The picture emerges of a current transient propagating along the surface of the signal conductor and creating the wavefront of an electromagnetic field. This spreads out in the same way as the bow wave of a ship. Since the conductors of the transmission line are 2 mm apart, the wavefront does not reach the return conductor until the current pulse has propagated at least 2 mm along the signal conductor.

When the wavefront does reach the return conductor, it induces a voltage on the surface. This voltage creates a current that flows back toward the near end. Current in the return conductor creates an electromagnetic field of its own, a field that spreads out to enclose the signal conductor. This field tends to neutralize the field emanating from the signal conductor; but not all of it. The coupling cannot be 100 per cent efficient. So the amplitude of the induced current is less.

By the time the field from the return conductor reaches the signal conductor, the advancing wavefront has progressed a further 2 mm. Since this happens for every increment of length, it must happen along the whole length.

Behind the wavefront, there is an increasing length of cable in which current is flowing in one direction in the signal cable, and in the opposite direction in the return conductor. Since the current in the return conductor is slightly less, there must be an unbalanced current flow in the cable. Since both conductors are now acting as transmitting antennae, the net result is that there is a residual current flow from the near end. Mutual coupling ensures that this antenna-mode current is shared equally between the two conductors.

Hence, there are two current components flowing in each conductor; the differential-mode current and the antenna current. Figure 6.5.1 illustrates the current flow, viewed from

signal conductor ——————$Idiff + \dfrac{Iaerial}{2}$——————→——————

return conductor ——————$-Idiff + \dfrac{Iaerial}{2}$——————→——————

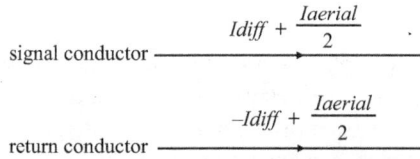

Figure 6.5.1 Unidirectional current in cable.

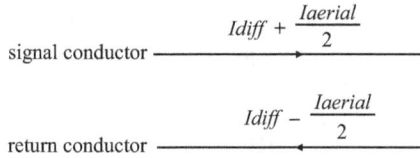

signal conductor ——————$Idiff + \dfrac{Iaerial}{2}$——————→——————

return conductor ——————$Idiff - \dfrac{Iaerial}{2}$——————←——————

Figure 6.5.2 Loop current in cable.

the point of view of unidirectional current. This shows that there is a net flow outward of *Iaerial*. At the wavefront, antenna-mode current is flowing in the forward direction along both conductors.

Figure 6.5.2 views the same two conductors, this time from the point of view of loop current. Magnetic material in the transformer responds only to the current which encloses the core. Since the antenna-mode current in one conductor balances the antenna-mode current in the other, this part of the current flow is 'invisible' to the transformer.

In classical transmission-line theory, the concept of partial currents was introduced to explain the behavior of reflections at the terminations. The above reasoning indicates that there is at least one more partial current to add to incident current and reflected current – antenna-mode current.

To summarize: Electric charge is deposited on the return conductor. Induced voltage causes a current to flow back along the return conductor toward the source. This current creates an electromagnetic field that couples with the signal conductor and that also radiates outwards. Since the current in the return conductor does not completely balance that in the signal conductor, there is an unbalanced charge that propagates along the cable, just behind the wavefront. In addition, there is antenna-mode current which propagates outwards along both conductors and is converted into an electromagnetic wave. This means that there are at least three current components involved:

- that which carries differential-mode signal,
- that which deposits unbalanced charge on the cable, and
- that which flows out radially in the form of electromagnetic radiation.

It is useful at this point to make a distinction between common-mode current and antenna-mode current:

- Common-mode current is that which flows in the loop formed by the return conductor and the adjacent structure.
- Antenna-mode current is that which flows between the cable and the environment when there is no nearby structure.

6.5.2 Line voltage

The reasoning also leads to an explanation for the use of lumped parameters in the delay-line model of Figure 6.4.2. Conventional delay-line theory caters for the fact that, as the wave-front progresses, the voltage between the conductors undergoes a step change. It does not cater for a situation where the step change on one conductor is less than the step change on the other.

If such a situation is taken into account, it means that there is also an unbalanced voltage component that progresses along the cable.

Conductive material behind the step is charged; conductive material in front of the step is uncharged. Effectively, the charged portion of the system capacitance is increasing, instant by instant. For an elemental length of cable:

$$dQ = V \cdot dC \tag{6.5.1}$$

where dC is the capacitance of that element. Since the cross section of the cable is constant, the capacitance is proportional to the length which has been charged. Since the propagation velocity is constant, the rate of change of length with time is constant. So the apparent rate of change of capacitance with time is constant. From (6.5.1):

$$\frac{dQ}{dt} = V \cdot \frac{dC}{dt} \tag{6.5.2}$$

Equation (6.5.2) is applicable to a situation where a constant voltage is being applied to a capacitance that varies with time.

Conventional circuit theory makes the assumption that the capacitance is constant and that the parameter which varies with time is the voltage. That is:

$$\frac{dQ}{dt} = C \cdot \frac{dV}{dt} \tag{6.5.3}$$

It follows, from (6.5.2) and (6.5.3), that:

$$C \cdot \frac{dV}{dt} = V \cdot \frac{dC}{dt} \tag{6.5.4}$$

This relationship justifies the existence of lumped parameters in the delay-line model of Figure 6.4.2.

6.5.3 Source current and voltage

From the point of view of the source generator, another part of the picture emerges. Since *Vgen* is the only source of power in the system, it must supply both the differential-mode current and the antenna-mode current. Figure 6.5.3 shows that the current in the resistor *Rg* must be the sum of *Idiff* and *Iaerial*.

As far as the generator is concerned, the source of *Iaerial* must be the ground conductor. Such a deduction is entirely plausible, since the conductor contains a large surface area of

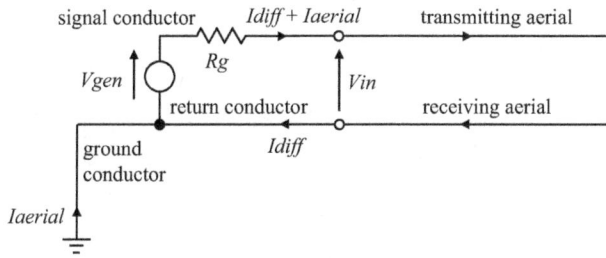

Figure 6.5.3 Current sharing at source generator.

conducting material. This being so, it must have a high capacitance, a low inductance, and hence, a low characteristic impedance.

This means that, at the output terminals of the supply, all the antenna-mode current is flowing in the signal conductor. The signal conductor is behaving as a transmitting antenna and the return conductor as a receiving antenna. Differential-mode current is delivered to the return conductor, while the rest is converted into electromagnetic radiation.

From Figure 6.5.3, the voltage applied to the input terminals of the cable is *Vin*, where:

$$Vin = Vgen - Rg \cdot (Idiff + Iaerial)$$

Comparing Figure 6.5.3 with Figure 6.4.2 allows the parameters of the two models to be correlated. That is, the current *Ina* used in the calculations corresponds to *Idiff*, and the antenna-mode current *Ins* corresponds to *Iaerial*. This gives:

$$Vin = Vgen - Rg \cdot (Ina + Ins) \tag{6.5.5}$$

6.5.4 Radiated current

Current used to deliver unbalanced charge to the transmission line is only one component of antenna-mode current.

Developing the model to simulate current which is actually being radiated involves one further step in the reasoning. In conventional textbook analysis of transmission-line transients, it is necessary for the inductive and capacitive parameters to undergo a metamorphosis and reappear as a resistance *Ro* and a time delay *T*. The relationships are set out in section 6.3. For a two-conductor cable, normal practice is to represent the characteristic impedance as a single resistor.

However, the modeling process for alternating signals has revealed that this single resistor provides an inadequate representation of the observed phenomenon. Three conductors are necessary; the signal conductor Ro_1, the return conductor Ro_2, and the virtual conductor Ro_3. Figure 6.5.4 provides a more revealing picture of the impedance presented to the output terminals of the signal source.

This figure makes it clear that the source current flows along the signal conductor. Most returns via the return conductor, but some flows into the environment.

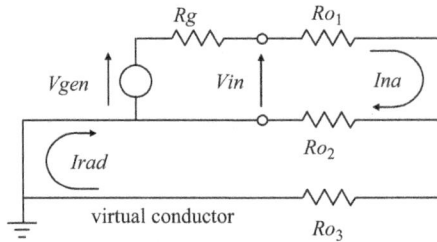

Figure 6.5.4 Simulating the radiated current, *Irad*.

Calculating the value for these three components is relatively simple, using the equation:

$$Ro_i = \sqrt{\frac{Lc_i}{Cc_i}}$$ (6.5.6)

where i is an integer identifying the conductor, and Lc_i and Cc_i are defined by (5.2.9) and (5.2.10).

The resistance Ro as 'seen' by the output terminals of the source at the near end is:

$$Ro = Ro_1 + \frac{Ro_2 \cdot Ro_3}{Ro_2 + Ro_3}$$ (6.5.7)

The relationship between *Irad* and *Ina* is:

$$Irad = \frac{Ro_2}{Ro_2 + Ro_3} \cdot Ina$$ (6.5.8)

Since the far end is isolated from ground, there is no third conductor available to deliver extra charge to the cable. So the primitive voltages existing at the far-end terminals are perfectly balanced with respect to the environment. Even so, both conductors act as transmitting antennae, as far as returning current from the far end is concerned.

6.5.5 Cable losses

Any transmission line provides a highly efficient means of carrying an electromagnetic signal from one location to another. Even so, there are losses:

- Differential-mode energy is stored in the cable. This is released into the environment when the line is disconnected.
- Some energy is used to heat up the conductors, due to copper losses. This can be simulated by inserting a resistor in series with the source resistance. Since transient currents are involved, skin effect will cause the value of this resistor to be much higher than the steady-state resistance of the conductors.

- There could be losses in the dielectric material of the insulation. These could be simulated by placing a high-value resistor between the two conductors at the generator end of the line.
- Some of the transient differential-mode current radiates away. When the polarity of the radiation from one conductor is positive, the polarity of the radiation from the other is negative. If the cable is twisted, this differential-mode radiation tends to cancel itself out at a short distance from the line.
- There is also a transient antenna-mode current *Irad* which emanates from the cable in the form of electromagnetic radiation. Figure 6.5.4 shows how this can be simulated. Current flows into the cable from the ground conductor to replace this loss. (This is the same phenomenon that was analyzed in section 5.2, using sinusoidal waveforms.)

Section 6.4 has shown that, as well as propagating down the line, the current transient leaves a residual charge on the line. The amplitude of the stored charge can be determined by the model of Figure 6.4.2. If a step voltage is applied to the input terminals, current *Ins* flows into *Crad* for $2 \cdot T$ seconds, that is, the time it takes for the step to be reflected back to the input terminals. Since this charge is eventually converted back into a current and the energy dissipated in the resistors *Rg* and *Rcable*, this stored energy also represents a loss.

6.5.6 Line parameter measurements

Section 6.3 has shown that when the delay-line model is used to simulate the response of a lossless line, the inductive and capacitive parameters disappear from the equations. They are replaced by the characteristic impedance and the propagation time. That is, L and C are replaced by *Ro* and *T*.

If a step pulse is applied to a line with a mismatched termination at the far end, then a steady current I will flow into the line for $2 \cdot T$ seconds. Figure 7.6.5 shows how this time period and current can be measured for an open-circuit line.

From (6.3.7):

$$C = \frac{T}{Ro}$$

Substituting $\frac{V}{I}$ for *Ro* in this equation gives:

$$C = \frac{I \cdot T}{V} \tag{6.5.9}$$

Similarly, from (6.3.6):

$$L = T \cdot Ro$$

Substituting $\frac{V}{I}$ for *Ro*:

$$L = \frac{V \cdot T}{I} \tag{6.5.10}$$

Since V, I, and T are known, both L and C can be calculated.

6.6 Transient emission model

The reasoning of the previous section points the way to a further development of the model used to simulate transient radiation; development which allows all the current and voltage components to be simulated.

In Figure 6.6.1, the resistor representing the characteristic impedance of the transmission line is replaced by a potentiometer network. Resistors $Ro1$ and $Ro2$ represent the characteristic impedances of the signal and return conductors, respectively, while $Ro3$ represents the characteristic impedance of the virtual conductor.

As in the earlier model, Ina, Ini, and Inr represent the absorbed, incident, and reflected currents at the near end, while Ifa, Ifi, and Ifr represent currents at the far end. The current Ine is that which departs into the environment in the form of electromagnetic radiation. As far as the transmission line is concerned, this represents current which has disappeared from the system. Consequently, there is less current available to lay down antenna-mode charge on the conductors. This current is identified in the diagram as Inf. This means that there are three components to the current reflected from the near end:

- that which is converted into electromagnetic radiation, Ine,
- that which is stored on the outer surface of the cable, Ins, and
- that which is actually transmitted to the far end, Int.

Of course, there are other losses, due to

- voltage drop due to current in the resistance of the two conductors, and
- voltage drop due to current in the dielectric material of the insulation.

These latter two effects are due to the conversion of electrical energy into heat energy.

Losses due to differential mode radiation are indistinguishable from those due to losses in the dielectric material, and could be simulated by a resistor connected between the two cable

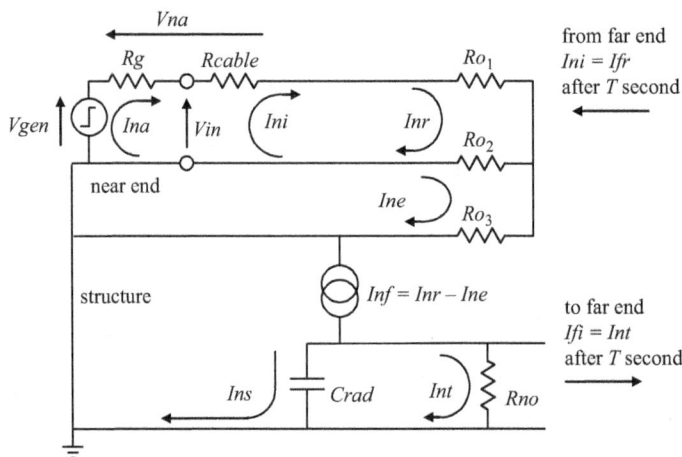

Figure 6.6.1   General circuit model for transient emissions.

conductors. In the model of Figure 6.4.2, all the losses are simulated by a single resistor, *Rcable*.

A comparison between Figures 6.4.2 and 6.6.1 reveals that the significant modification is the inclusion of a new circuit loop carrying the current *Ine*. This is the current which flows into the environment in the form of electromagnetic radiation.

The impedance presented by the cable to the output terminals of the source generator is:

$$Ro = Ro1 + \frac{Ro2 \cdot Ro3}{Ro2 + Ro3} \qquad (6.6.1)$$

This is also the impedance presented to the terminals at the far end.

From inspection of Figure 6.6.1, the equation for the antenna-mode loop is:

$$0 = -Ro2 \cdot (Ini + Inr) + (Ro2 + Ro3) \cdot Ine$$

Hence:

$$Ine = \frac{Ro2}{Ro2 + Ro3} \cdot (Ini + Inr)$$

So:

$$Ine = \text{Loss} \cdot Ina \qquad (6.6.2)$$

where:

$$\text{Loss} = \frac{Ro2}{Ro2 + Ro3} \qquad (6.6.3)$$

The current flowing from the near end toward the far end is now:

$$Inf = Inr - Ine \qquad (6.6.4)$$

The current which stores antenna-mode charge on the cable is similar to (6.4.9):

$$Ins = Inf - \frac{Qns}{Rno \cdot Crad} \qquad (6.6.5)$$

And the current which is actually transmitted to the far-end terminals is:

$$Int = Inf - Ins \qquad (6.6.6)$$

In section 6.5, it was reasoned that the current *Ins* used to create antenna-mode charge on the cable must flow through the source resistance *Rg*. This means that (6.4.5) needs to be modified. The modified relationship becomes:

$$Vna = (Rg + Rcable) \cdot Ina + Rg \cdot Ins \qquad (6.6.7)$$

Since the current used to deliver stored charge varies along the length of the cable, it is assumed that no voltage drop is incurred along the length of the cable by Ins. Taking this new relationship into account, (6.4.6) changes to:

$$Ina = \frac{2 \cdot Ro \cdot Ini + Vgen - Rg \cdot Ins}{Ro + Rg + Rcable} \tag{6.6.8}$$

There is now sufficient information available to modify the worksheet of section 6.4 to analyze the response of the model of Figure 6.6.1. However, several more relationships are needed before the response of the model can be correlated with that of the setup of Figure 6.4.1.

The differential-mode current delivered by the source generator to the cable is:

$$Idiff = Inr + Ini - Ine$$

That is:

$$Idiff = Ina - Ine \tag{6.6.9}$$

The antenna-mode current is, by definition, Ine.

Since the only source of energy is the generator $Vgen$, then the current Ins must flow through the source resistor Rg. Hence, the voltage actually applied to the terminals at the near end must be:

$$Vin = Vgen - Rg \cdot (Ina + Ins) \tag{6.6.10}$$

It can be assumed that the relationship between the voltage monitored by channel 1 of the oscilloscope and the voltage Vin is:

$$K = \frac{Vch1}{Vin} \tag{6.6.11}$$

Similarly, it can be assumed that the relationship between current monitored by the current transformer, $Imon$, and the voltage displayed by channel 2 of the scope is:

$$ZT = \frac{Vch2}{Imon} \tag{6.6.12}$$

Having established all the necessary relationships, the next step is to assemble them into a program that simulates the transient response of the configuration. The first page of the worksheet is devoted to the definition of the input variables associated with the model of Figure 6.6.1 and with the test setup illustrated by Figure 6.4.1. This page is reproduced by Figure 6.6.2.

As with the frequency-response models, this time-domain response model can also be used to correlate the theoretical response with that of the actual hardware. Relevant variables are identified by the phrase 'adjust to suit'.

The second page of the worksheet, Figure 6.6.3, defines the two subroutines used to calculate the responses at the near and far ends of the line after each time step. It is a modified version of the subroutines illustrated by Figure 6.4.4.

Worksheet 6.6, page 1

$Rg := 4.7\ \Omega$ $RL := 10^{7}\ \Omega$ Impedances of source and load

$Vmeas := -0.41\ V$ Measured voltage between horizontal sections of square wave

$Crad := 220 \cdot 10^{-12}F$ Value assigned to radiation capacitor. **Adjust to suit**

$Rcable := 1\ \Omega$ Resistor which simulates cable losses. **Adjust to suit**

$Ro1 := 50\ \Omega$ $Ro2 := Ro1$ $Ro3 := 600\ \Omega$ See Figure 6.6.1 **Adjust to suit**

$Loss := \dfrac{Ro2}{Ro2 + Ro3}$ See (6.6.3)

$Ro := Ro1 + \dfrac{Ro2 \cdot Ro3}{Ro2 + Ro3} = 96.154$ See (6.6.1)

$ZT := 2.27\ \Omega$ See (7.2.6)

$K := \dfrac{50}{96.2}$ See (7.6.1)

$Vg := \dfrac{Vmeas}{K}$ $Vg = -0.789$ See (6.6.11)

$T := 83 \cdot 10^{-9}\ s$ Measured transit time. **Adjust to suit**

$N := 100$ Number of time steps per transit

$dt := \dfrac{T}{N}$ Time of each step

$T1 := 150 \cdot 10^{-9}\ s$ Time of leading edge of square wave, as displayed on oscilloscope

$T2 := 4.1 \cdot 10^{-6}\ s$ Time of trailing edge of square wave, as displayed on oscilloscope

$T3 := 5 \cdot 10^{-6}\ s$ Sweep time of oscilloscope

$N1 := \dfrac{T1}{dt}$ $N2 := \dfrac{T2}{dt}$ $N3 := \dfrac{T3}{dt}$ Number of time steps at which each event occurs.

$n := 1..N3$ $t_n := (n-1) \cdot dt$ Definition of horizontal axis of display

Figure 6.6.2 Definition of input variables.

The third page of the worksheet, illustrated by Figure 6.6.4, defines the main program used to calculate the response of the selected variable over the defined time.

The output variable can be the following:

• *Vch1*: The input voltage waveform, as observed on channel 1 of the oscilloscope. The leading edge is defined by *T1*, the trailing edge by *T2*.

Worksheet 6.6, page 2

$$send(near, INPUT, Vgen) := \begin{vmatrix} Ini \leftarrow INPUT_2 \\ \begin{pmatrix} Ina \\ Int \\ Ine \\ Ins \\ Qns \end{pmatrix} \leftarrow near \\ Ina \leftarrow \dfrac{2 \cdot Ro \cdot Ini + Vgen - Rg \cdot Ins}{Ro + Rg + Rcable} \\ Inr \leftarrow Ina - Ini \\ Ine \leftarrow Loss \cdot Ina \\ Inf \leftarrow Inr - Ine \\ Ins \leftarrow Inf - \dfrac{Qns}{Ro \cdot Crad} \\ Qns \leftarrow Qns + Ins \cdot dt \\ Int \leftarrow Inf - Ins \\ \begin{pmatrix} Ina \\ Int \\ Ine \\ Ins \\ Qns \end{pmatrix} \end{vmatrix}$$

See (6.6.8)

See (6.4.7)

See (6.6.2)

See (6.6.4)

See (6.6.5)

See (6.4.10)

See (6.6.6)

$$recv(INPUT) := \begin{vmatrix} Ifi \leftarrow INPUT_1 \\ Ifa \leftarrow \dfrac{2 \cdot Ro \cdot Ifi}{RL + Ro} \\ Ifr \leftarrow Ifa - Ifi \\ \begin{pmatrix} Ifa \\ Ifr \end{pmatrix} \end{vmatrix}$$

Copy of function defined in Figure 6.4.4

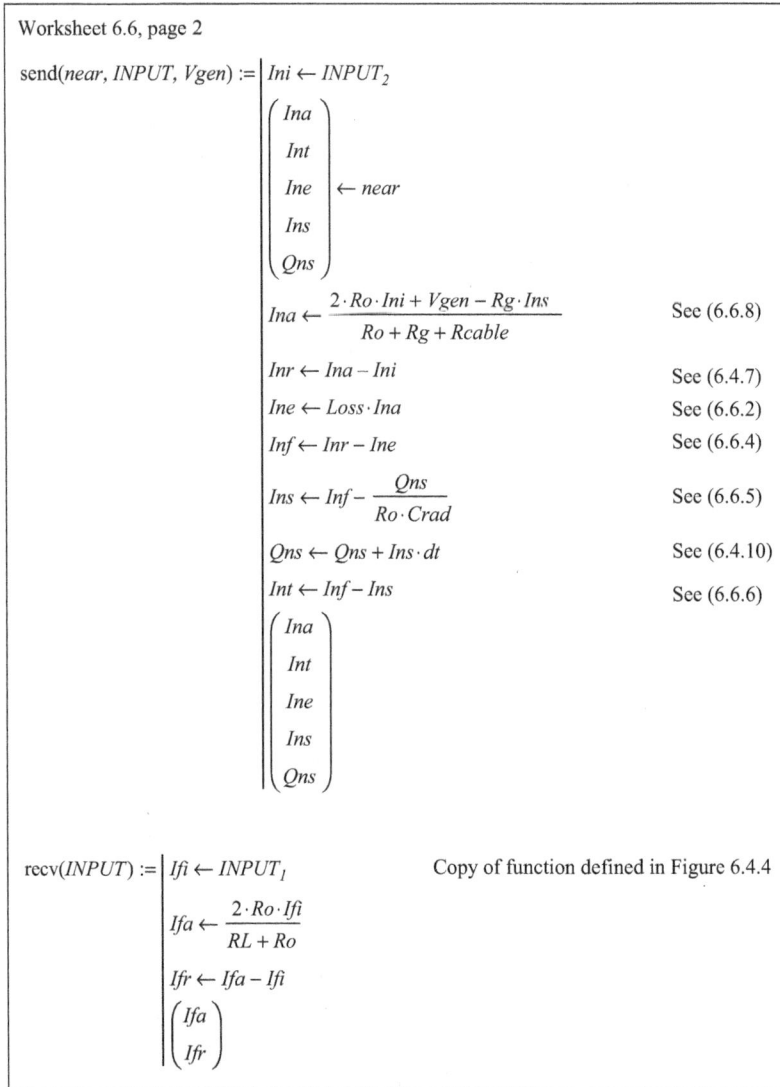

Figure 6.6.3   Subroutines for near-end and far-end calculations.

- *Vdiff*: The differential-mode current waveform, as observed on channel 2 of the oscilloscope.
- *Vrad*: The antenna-mode current waveform, as observed on channel 2 of the oscilloscope.

In the particular case of the routine illustrated by Figure 6.6.4, the simulation would be of *Vch1*.

The input variables have been adjusted to simulate the responses of the three tests described in section 7.6. Several iterations were carried out before a fair correlation was achieved between test results and model. When the final iteration had occurred, no further

Worksheet 6.6, page 3

$$point(n) := \left| \begin{array}{l} m \leftarrow mod(n, N) \\ m \leftarrow N \text{ if } m = 0 \end{array} \right.$$

$$Vch1 := \left| \begin{array}{l} data_{2, N} \leftarrow 0 \\ near_5 \leftarrow 0 \\ \quad \text{for } i \in 1..N3 \\ \qquad \left| \begin{array}{l} Vgen \leftarrow Vg \text{ if } i > N1 \\ Vgen \leftarrow 0 \text{ if } i > N2 \\ p \leftarrow point(i) \\ INPUT \leftarrow data^{\langle p \rangle} \\ near \leftarrow send(near, INPUT, Vgen) \\ \begin{pmatrix} Ina \\ Int \\ Ine \\ Ins \\ Qns \end{pmatrix} \leftarrow near \\ \begin{pmatrix} Ifa \\ Ifr \end{pmatrix} \leftarrow recv(INPUT) \\ OUTPUT \leftarrow \begin{pmatrix} Int \\ Ifr \end{pmatrix} \\ data^{\langle p \rangle} \leftarrow OUTPUT \\ Idiff \leftarrow Ina - Ine \\ Vin \leftarrow Vgen - Rg \cdot (Ina + Ins) \\ Vch1 \leftarrow K \cdot Vin \\ Vdiff \leftarrow ZT \cdot Idiff \\ Vrad \leftarrow ZT \cdot Ine \\ V_i \leftarrow Vch1 \end{array} \right. \\ V \end{array} \right.$$

$Idiff \leftarrow Ina - Ine$	See (6.6.9)
$Vin \leftarrow Vgen - Rg \cdot (Ina + Ins)$	See (6.6.10)
$Vch1 \leftarrow K \cdot Vin$	See (6.6.11)
$Vdiff \leftarrow ZT \cdot Idiff$	See (6.6.12)
$Vrad \leftarrow ZT \cdot Ine$	See (6.6.12)
$V_i \leftarrow Vch1$	Channel 1 selected as output variable

Figure 6.6.4 Main routine used to analyze transient response.

alteration was made to the input variables. The program was run three more times, each time with a different output variable selected.

Then the input variables were assigned to the general circuit model of Figure 6.6.1 to create the representative circuit model of the assembly-under-test (Figure 6.6.5).

Section 7.6 demonstrates the fact that the same circuit model can simulate three different responses of the cable-under-test. This indicates that it is a useful predictor of transient radiation. Worksheet 7.6.9 demonstrates that there is a clear correlation between

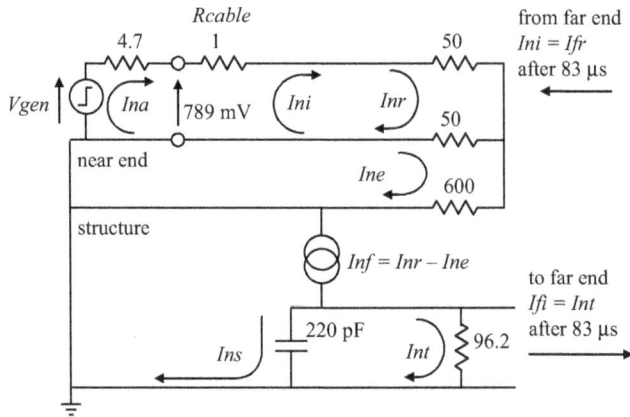

Figure 6.6.5   Representative circuit model of transient emission from cable.

the frequency response model of Figure 7.5.12 and the transient response model of Figure 6.6.5.

Both of these representative models have been related to electromagnetic phenomena and both replicate the actual response of the hardware. So it is fair to claim that the general circuit models on which they are based can be used with a high degree of confidence. Further testing and modeling should confirm their reliability. Sufficient information has been provided for this to be done.

CHAPTER 7

Bench testing

It is normal practice to carry out bench tests of the functional behavior of prototype circuitry during the development of any new product. This identifies problems that had not been predicted during the feasibility study and provides an opportunity to rectify those faults. It also provides an early opportunity to check that the design requirements are being met.

Since EMC is also a functional requirement, it is logical to expect that this aspect of design should be checked at the prototype stage.

Two items of essential equipment are the voltage transformer and the current transformer. Several manufacturers produce such items, but they are highly expensive, and not really suitable for general-purpose use. So the transducers described in this section were assembled from components obtained from a supplier of electronic components.

Section 7.1 describes the construction of a low-cost voltage transformer. This consists of ten turns of wire wound on a spit-core ferrite assembly. A low-value resistor in parallel with the primary winding ensures that the transducer is a wide-band device. The secondary winding is the loop-under-test. A monitor winding enables the amplitude of the injected voltage to be measured.

The frequency response of the transformer was checked by applying a known voltage to the primary winding and measuring the output of the secondary. A circuit model was subjected to analysis using a Mathcad program. Only a few iterations of this program were needed to establish a one-to-one correlation between the test results and the response of the model. By this means, a model was created that could be used to simulate the response of the voltage transformer. This model can be used for analysis in the frequency domain or the time domain.

A current transformer was assembled in much the same way. This time, the primary was the loop-under-test and the secondary provided an output voltage proportional to the input current. A circuit model was created to simulate the device characteristics. This circuit model provides the calibration curve for the device. The process is described in section 7.2.

Section 7.3 describes the construction of a triaxial cable that can be used to minimize interference between cables of the test equipment.

Section 7.4 describes a test on an isolated conductor, where a known voltage is injected into the center of the conductor using the voltage transformer and the current monitored with the current transformer. Data on the length and diameter of the conductor was used to assign value to the components of the dipole model of section 5.2. A close match was obtained between the response of the conductor and the model. However, the peak of the emitted current was significantly higher than that of a dipole antenna. This is probably because there was no resistance in the center of the cable to damp down the oscillations (as there would be in a dipole antenna).

Section 7.5 describes similar test on a twin-conductor cable, where the terminations are open-circuit. A circuit model was developed, which simulates both the differential-mode current and the antenna-mode current.

Section 7.6 provides details of the transient test on the 15 m cable described in sections 6.4 and 6.5. Photographs of the actual waveforms are compared with graphs of the simulated waveforms, providing persuasive evidence of the validity of the model. It is not claimed that this model is correct in every detail. However,

- the one single model simulates the waveforms of three separate signals reasonably accurately,
- every parameter is shown to be related to electromagnetic phenomena,
- it reveals aspects of transmission-line behavior that are not identified in textbooks on electromagnetic theory,
- sufficient information is provided to allow any electronic designer to replicate the tests and perform the analyses,
- clear correlation can be established between the results obtained from the frequency response tests of section 7.5 and those obtained from the transient tests of section 7.6.

Although the test equipment and tests described in sections 7.1–7.5 are limited to a band-width of 20 kHz to 20 MHz, the technique and approach are applicable to a much wider range of frequencies. Each designer will have test equipment which is suitable for use with the equipment under development. It is simply a matter of adapting the available equipment to carry out the type of testing described in this chapter.

The technique of circuit modeling can also be applied to the high-frequency character-ization of components such as capacitors, inductors, and filters. Section 7.7 provides an example of the characterization of a capacitor over the range 200 kHz to 1 GHz.

7.1 Voltage transformer

This section defines the requirement of a voltage transformer intended for use with bench test equipment, describes a particular design, and records details of a set of characterization tests.

Basic requirements for a voltage transformer for use with EMC testing are:

- It should enable a known voltage to be induced in series with the loop-under-test.
- It should have a wide bandwidth.
- It should interface with 50 Ω co-axial cable.

If it is used as an item of general-purpose test equipment and is available for bench testing of equipment under development, then there are two more requirements:

- It should be a split-core transformer, capable of clamping round the cable-under-test.
- It should be a low-cost item.

Figure 7.1.1 illustrates the basic design of such a device and indicates how it can be con-nected to other items of bench test equipment and to the loop-under-test. In order to meet the

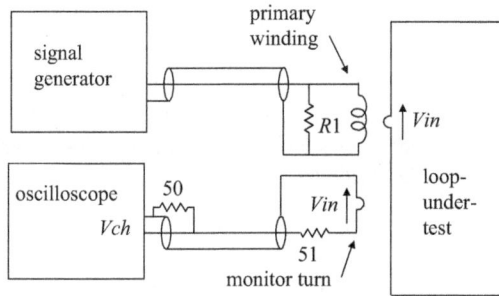

Figure 7.1.1   Use of voltage transformer.

Figure 7.1.2   Voltage transformer.

requirement for a wide bandwidth, the impedance of the primary winding should be higher than $R1$ at the lowest operating frequency.

The requirement to measure the voltage induced in the loop-under-test is met by the monitor turn. If the loop-under-test is a short-circuit, then a high current will be induced, and the applied voltage will be reduced due to the output impedance of the transformer. The voltage detected by the monitor turn will also drop. This means that the monitor turn will measure the actual voltage applied to the loop-under-test.

From the circuitry of Figure 7.1.1, the relationship between the voltage Vin applied to the loop-under-test and the voltage monitored at the appropriate channel of the oscilloscope is:

$$Vin = \frac{50 + 51}{50} \cdot Vch \tag{7.1.1}$$

The 50 Ω input impedance at the oscilloscope is provided by a BNC adaptor and a BNC terminator.

Such a transformer was constructed from parts purchased from suppliers of electronic components, and is illustrated by Figure 7.1.2. The core itself is a 'cable suppression core assembly' part number 04 31 173 551, supplied by Fair Rite Products Corp.

The primary winding was 10 turns of 22 SWG enameled copper wire wound round one of the split cores. The monitor winding was a single turn of enameled copper wire wound round the other core. There were two reasons for this:

- It ensures that the magnetic field being coupled was the same as that which was coupling the loop-under-test.
- It minimizes capacitive coupling with the primary winding, since it is wound on the other half of the split-core ferrite. Figure 7.1.2 illustrates this separation.

The windings were terminated in a standard terminal block, which provided a mount for the other components. A 68 Ω, 2 W resistor was connected in parallel with a 240 Ω, 0.6 W resistor and the primary winding. This provided a 53 Ω resistor in parallel with the primary.

Ideally, the co-axial cable should have been terminated by a 50 Ω resistor to match its characteristic impedance. However, it is inevitable that reflected impedance of the loop-under-test will act as a load in parallel with the termination resistor. To compensate for this loading effect, a value of 53 Ω was fitted.

Two BNC connectors were connected to the terminal block to provide an interface with 50 Ω co-axial cable. A section of a plastic box was used to provide more rigidity to the assembly.

Although this assembly does not meet the quality standard of a professionally engineered item of equipment, it does the job intended and has proved to be reliable over a period of more than 5 years.

Testing was carried out using the setup of Figure 7.1.3. A splitter box was used to provide an input to channel 1 of the oscilloscope. This allowed the input to the voltage transformer to be monitored. The output of the monitor turn was monitored by channel 2. Each 16.5 Ω resistor is a parallel combination of two 33 Ω resistors.

The signal generator was set to provide a sinusoidal output at a particular frequency, and the amplitudes of the peak-to-peak signals were measured on the oscilloscope. This process was carried out over a number of defined frequencies and the results were tabulated. Column 1 of the table of Figure 7.4 gives the frequency in megahertz, column 2 gives the amplitude of the channel 1 in volts, and column 3 gives the amplitude of the signal at channel 2 in millivolt.

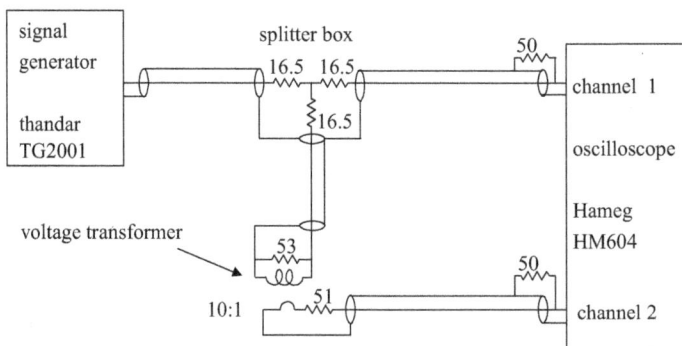

Figure 7.1.3 Characterization of voltage transformer.

Worksheet 7.1, page 1

$$data := \begin{pmatrix} 0.01 & 2.2 & 50 \\ 0.02 & 2.6 & 90 \\ 0.05 & 3.4 & 146 \\ 0.1 & 3.8 & 170 \\ 0.2 & 3.95 & 177 \\ 0.5 & 3.95 & 180 \\ 1 & 4 & 182 \\ 2 & 4 & 180 \\ 5 & 3.9 & 172 \\ 10 & 3.8 & 158 \\ 15 & 3.6 & 140 \\ 19 & 3.45 & 124 \end{pmatrix}$$

Column 1: frequency, mhz
Column 2: peak-to-peak voltage on channel 1, V
Column 3: peak-to-peak voltage on channel 2, mV

$s := 1..\text{rows}(data)$

$f_s := data_{s,1} \cdot 10^6$

$Vch1_s := data_{s,2}$ $Vch2_s := data_{s,3} \cdot 10^{-3}$

$TFt_s := \dfrac{Vch2_s}{Vch1_s}$ $Ref_s := \dfrac{\max(TFt)}{\sqrt{2}}$

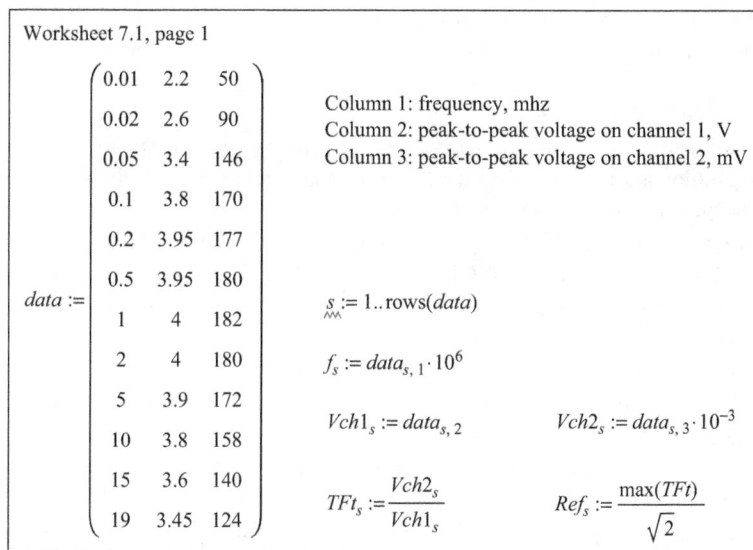

Figure 7.1.4 Using test results to calculate the transfer function.

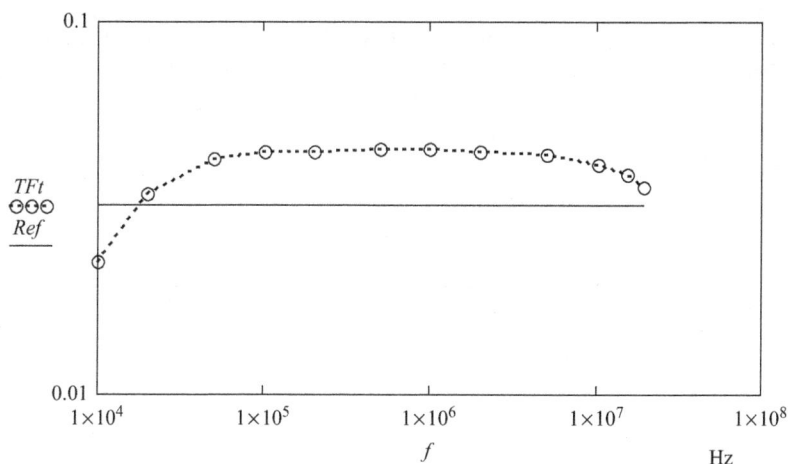

Figure 7.1.5 Transfer function of voltage transformer.

For each spot frequency, the ratio of output voltage $Vch2$ to input voltage $Vch1$ was calculated and this ratio gave the value of the transfer function at that frequency. Plotting this parameter against the frequency, as shown in Figure 7.1.5, gives an indication of the bandwidth of the voltage transformer.

The solid line on the graph defines the 3 dB margin. Comparing the dotted curve with the solid curve gives a clear indication of the bandwidth. In this case, it is from 20 kHz to 20 MHz.

At this point, the characteristics of the device have been defined in terms of a specified test and a table of results. A more compact and informative characterization is to define the transformer in terms of a circuit model. Figure 7.1.6 is such a model. It is based on the textbook treatment of transformers.

Mutual inductance (the inductance shared by primary and secondary) is represented by $L2$. Primary inductance (the inductance due to magnetic flux which links only with the primary winding) is represented by $L1$.

Since the turns-ratio is 10 to 1, the impedance ratio of the secondary reflected in the primary is 100 to 1. Resistance $R3$ is the reflected value of the series resistance of the secondary circuit, and resistance $R4$ is the reflected value of the load resistance at channel 2 input terminals.

Since the simulation is only concerned with amplitudes of the signals, there is no need to know the phase relationship between currents and voltages. So, as far as this model of the setup is concerned, the co-axial cable is transparent.

Resistance $R2$ is the parallel combination of the 68 and 240 Ω resistors. The 16.5 Ω resistors represent the splitter-box components, while the 50 Ω resistors represent the output impedance of the signal generator and the input impedance of channel 2 of the oscilloscope. The capacitance $C1$ represents the capacitance of the primary winding.

The voltage $V1$ represents the voltage as monitored by channel 1 of the scope, $V2$ is the voltage at the junction of the 16.5 Ω resistors, $V3$ is the voltage applied to the input terminals of the transformer, $V4$ is the voltage across the mutual inductance, and $V5$ is 10 times the voltage monitored by channel 2.

Simulating the frequency response of this model over the range 10 kHz to 20 MHz requires that the frequency steps are spaced logarithmically. So a vector needs to be created that defines the set of frequencies to be used. A set of 100 frequencies provides enough data points to ensure a smooth curve on the graph. Figure 7.1.7 shows the section of the Mathcad worksheet which performs this function.

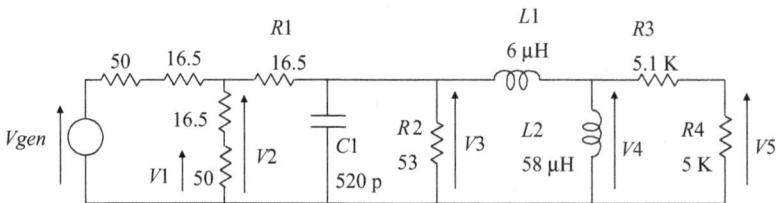

Figure 7.1.6   Representative circuit model of voltage transformer.

Worksheet 7.1, page 2

$$y1 := \log(10 \cdot 10^3) \qquad y2 := \log(20 \cdot 10^6) \qquad m := \frac{y2 - y1}{100}$$

$$i := 1..101 \qquad F_i := \begin{vmatrix} y \leftarrow m \cdot (i - 1) + y1 \\ \\ 10^y \end{vmatrix}$$

Figure 7.1.7   Calculating a set of equally spaced frequencies over a logarithmic scale.

Calculating the transfer function of the circuit model *TFm* is performed by the section of the worksheet illustrated in Figure 7.1.8. The objective is to determine the ratio of *V5* to *V1*, since the former simulates the voltage at channel 2, while the latter simulates the voltage at channel 1. So the amplitude of *V1* is set at unity. This means that the transfer function of the model at any frequency is the amplitude of *V5* divided by the turns-ratio.

Equations used in the subroutine illustrated in Figure 7.1.8 are derived from inspection of the circuit model, using the technique of 'equivalent circuits'.

It is now possible to compare the transfer function derived from test results, *TFt*, with that derived from the circuit model, *TFm*, and this is done in the graph of Figure 7.1.9.

This graph illustrates the fact that the curve produced by the circuit model intersects all the data points derived from testing the assembly. Since the component parts of the model

---

Worksheet 7.1, page 3

$R1 := 16.5 \ \Omega$

$R2 := \dfrac{68 \cdot 240}{68 + 240} = 52.987 \ \Omega$

$R3 := 5100 \ \Omega$

$R4 := 5000 \ \Omega$

$Lp := 64 \cdot 10^{-6} \ \text{H}$

$L1 := 6.0 \cdot 10^{-6} \ \text{H}$

$L2 := Lp - L1 = 5.8 \times 10^{-5} \ \text{H}$

$V1 := 1 \ \text{V}$

$V2 := \dfrac{50 + R1}{50} \cdot V1 = 1.33$

$Turns := 10$

$C1 := 520 \cdot 10^{-12} \ \text{F}$

$$TFm_i := \left| \begin{array}{l} \omega \leftarrow 2 \cdot \pi \cdot F_i \\[1em] Y2 \leftarrow \dfrac{1}{j \cdot \omega \cdot L2} + \dfrac{1}{R3 + R4} \\[1em] Z2 \leftarrow \dfrac{1}{Y2} \\[1em] Y1 \leftarrow \dfrac{1}{R2} + \dfrac{1}{Z2 + j \cdot \omega \cdot L2} + j \cdot \omega \cdot C1 \\[1em] Z1 \leftarrow \dfrac{1}{Y1} \\[1em] V3 \leftarrow \dfrac{Z1}{Z1 + R1} \cdot V2 \\[1em] V4 \leftarrow \dfrac{Z2}{Z2 + j \cdot \omega \cdot L1} \cdot V3 \\[1em] V5 \leftarrow \dfrac{R4}{R3 + R4} \cdot V4 \\[1em] \dfrac{|V5|}{Turns} \end{array} \right.$$

Figure 7.1.8 Calculating the transfer function of the model.

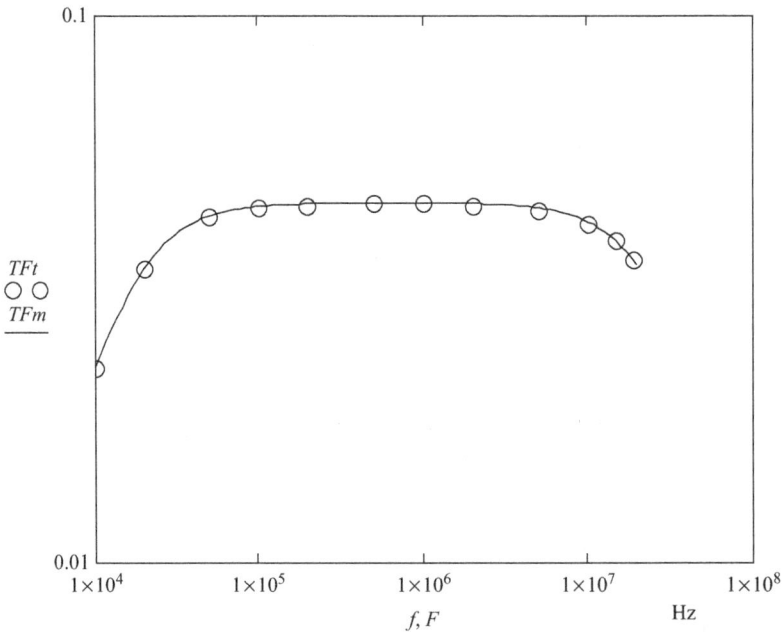

Figure 7.1.9 Transfer functions derived from testing and modelling.

simulating the voltage transformer can be identified, it means that those parts can be used to define its characteristics.

It is useful to describe the process by which the two curves of Figure 7.1.9 were brought into co-incidence. It was basically an iterative process.

Initially, guess values were assigned to Lp, $L1$, and $C1$. Values of other circuit components were fixed. Then the value of one of the parameters at the top of the worksheet was altered slightly and the page was scrolled up to display the final graph. The curve on the graph did not change until the mouse was clicked. This allowed the change in the position of the curve to be noted. If the curves moved closer together, then the same incremental change was made to the parameter.

Initially, it was assumed that capacitor $C1$ was very small, and attention was focused on the low-frequency response. The value of Lp (see third line of Figure 7.1.8) was adjusted to 'line up' the two curves at the low end of the frequency range, and $L1$ was adjusted to 'line up' the mid-frequency response.

It is useful to note that $L2$ is a dependent variable. This means that the total inductance of the primary winding of the transformer model can be altered with a single parameter change.

Surprisingly little iteration was needed to achieve coincidence over the lower and mid-frequency ranges. Finally, the value of $C1$ was adjusted to 'line up' the high-frequency roll-off.

The useable range of this transformer is from 20 kHz to 20 MHz. It is reasonable to expect that transformers can be constructed to match the operating frequency range of any particular system-under-review, for example, by using a smaller transformer to extend the frequency range upwards or by adding more turns on the secondary to extend the range downwards.

7.2 Current transformer

The purpose of the current transformer is to enable measurements to be made of the current in system cables. In configurations where the loop is completed by conductive components of the structure the measurement would be of the common-mode current. Where the far end of the cable is isolated (for example, with loudspeaker or microphone cables), the measurement would be of antenna-mode current.

If the core is clamped round the conductor carrying the signal current, then it would be differential-mode current that is being measured.

In the analytical treatment of the preceding chapters, a clear distinction is made between these current modes. For example, with a three-core mains conductor, the live/neutral loop would carry differential-mode current, the neutral/earth loop would carry common-mode current, and the antenna-mode current would flow in loop formed by the earth conductor and the virtual conductor representing the environment.

Figure 7.2.1 illustrates the method of coupling the test equipment to the loop-under-test. With such a configuration, the load reflected into the primary circuit (the loop-under-test) is the load presented to the secondary divided by the square of the number of turns. Hence, the greater the number of turns, the less the value of the reflected load, and the less will be the effect of the test equipment on the system-under-review.

The load 'seen' by the secondary winding is the 51 Ω resistor $R1$ in parallel with the 50 Ω impedance of the screened cable. Figure 7.2.2 illustrates the relationship.

As far as this load is concerned, it 'sees' a current source $Isec$, where

$$Isec = \frac{Iprim}{Turns} \tag{7.2.1}$$

and $Iprim$ is the current in the loop-under-test. The variable $Turns$ is the number of turns on the primary winding.

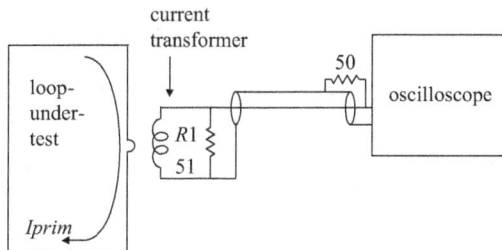

Figure 7.2.1 Use of current transformer.

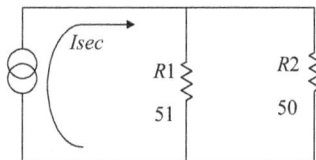

Figure 7.2.2 Transformer as a current source.

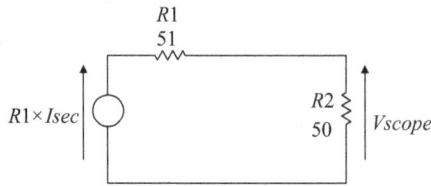

Figure 7.2.3   Transformer as a voltage source.

Figure 7.2.4   Current transformer assembly.

From the viewpoint of the input terminals of the oscilloscope, the signal is generated by a voltage source, as illustrated by Figure 7.2.3. In all the tests described in this chapter, a 50 Ω resistor is inserted in parallel with each scope input.

In the particular transformer described here, the core is exactly the same as that used for the voltage transformer, and the secondary winding comprises 10 turns of 22 SWG enamelled copper wire. Figure 7.2.4 illustrates the assembly. The tie wrap is removable, and is used to ensure that the two halves of the core are tightly clamped together.

To characterize this transformer, a simple coupling jig was assembled. This delivered a primary current of known amplitude and known frequency, and was designed to ensure that the loop-under-test was tightly coupled to the transformer.

The test setup is shown in Figure 7.2.5. Channel 1 of the oscilloscope was used to measure the current delivered to the transformer, that is, the current in the 50 Ω resistor placed at the scope input connector. Channel 2 was used to monitor the output of the transformer assembly.

As with the test on the voltage transformer, measurements were taken of the peak-to-peak amplitude of the sine waves displayed on the screen of the scope.

Figure 7.2.6 tabulates the results and illustrates how the transfer impedance ZTt is calculated, by dividing the voltage at channel 2 input by the current in the primary winding.

It is worth noting that the parameter resulting from these calculations is derived from the ratio of two measurements. It can be assumed that the amplifiers for channels 1 and 2 in the

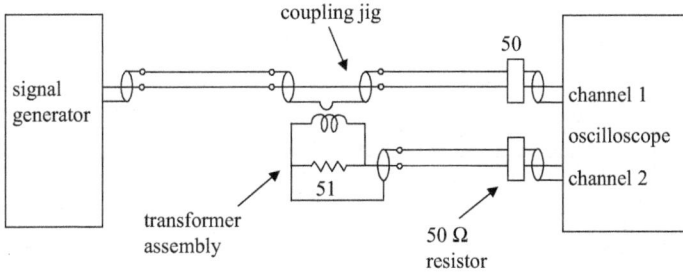

Figure 7.2.5 Setup for calibration of current transformer.

Worksheet 7.2, page 1

$$data := \begin{pmatrix} 0.005 & 8 & 98 \\ 0.01 & 8 & 175 \\ 0.02 & 8 & 270 \\ 0.05 & 8 & 340 \\ 0.1 & 8 & 360 \\ 0.2 & 8 & 365 \\ 0.5 & 8 & 375 \\ 1 & 8 & 370 \\ 2 & 8 & 370 \\ 5 & 7.9 & 365 \\ 10 & 7.8 & 360 \\ 15 & 7.4 & 335 \\ 19 & 6.9 & 330 \end{pmatrix}$$

column 1: frequency, MHz
column 2: channel 1 voltage, V
column 3: channel 2 voltage, mV

$s := 1..\text{rows}(data)$ $f_s = data_{s,1} \cdot 10^6$

$$ZTt_s := \left| \begin{array}{l} Vch1 \leftarrow data_{s,2} \\ Vch2 \leftarrow data_{s,3} \cdot 10^{-3} \\ Iprim \leftarrow \dfrac{Vch1}{50} \\ \dfrac{Vch2}{Iprim} \end{array} \right.$$

Figure 7.2.6 Using the test results to calculate the transfer impedance.

oscilloscope are identical. This being so, most of the errors in the absolute measurements are cancelled out.

Mathcad software was then used to display the results. As with the response of the voltage transformer, this was a set of points that could be joined up to create a frequency response characteristic.

Again, the use of circuit theory enables the creation of circuit model of the link between the loop-under-test and the oscilloscope input. Figure 7.2.7 is a development of the simple model of Figure 7.2.2. *R*2 and *R*3 represent the 51 Ω resistor in the transformer assembly and the resistor at channel 2 input connector, respectively.

*L*1 represents the inductance of the transformer winding, while *R*1 represents transformer losses. These losses could be due to the magnetic field from the loop-under-test which does not link with the transformer core. Another cause of losses is the eddy current in the core.

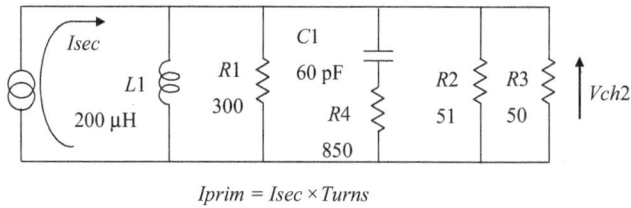

$$Iprim = Isec \times Turns$$

Figure 7.2.7 Representative circuit model of current transformer.

Worksheet 7.2, page 2

$y1 := \log(5 \cdot 10^3)$ $\qquad$ $y2 := \log(20 \cdot 10^6)$ $\qquad$ $m := \dfrac{y2 - y1}{100}$

$i := 1..101$

$F_i := \begin{vmatrix} y \leftarrow m \cdot (i - 1) + y1 \\ 10^y \end{vmatrix}$ $\qquad\qquad$ Calculting a set of 100 frequencies, equally spaced between 5 and 20 MHz.

$R1 := 300 \ \Omega$ $\qquad\qquad$ $R2 := 51 \ \Omega$ $\qquad\qquad$ $R3 := 50 \ \Omega$

$R4 := 850 \ \Omega$ $\qquad\qquad$ $L1 := 200 \cdot 10^{-6} \ H$ $\qquad\qquad$ $C1 := 60 \cdot 10^{-12} \ F$

$Turns := 10$

$ZTm_i := \begin{vmatrix} \omega \leftarrow 2 \cdot \pi \cdot F_i \\[6pt] Z1 \leftarrow R4 + \dfrac{1}{j \cdot \omega \cdot C1} \\[6pt] Y2 \leftarrow \dfrac{1}{R1} + \dfrac{1}{R2} + \dfrac{1}{R3} + \dfrac{1}{j \cdot \omega \cdot L1} + \dfrac{1}{Z1} \\[6pt] Z2 \leftarrow \dfrac{1}{Y2} \\[6pt] ZT \leftarrow \dfrac{|Z2|}{Turns} \end{vmatrix}$

Figure 7.2.8 Calculating the transfer impedance of the circuit model.

Capacitance $C1$ and resistor $R4$ were added to the model to simulate additional losses at frequencies over 2 MHz. This model is amenable to the use of Simulation Programs with Integrated Circuit Emphasis (SPICE) software. It would have been possible to use such software to produce a frequency response curve similar to that of ZTt. Achieving close correlation between theoretical results and test results would have involved downloading the results of the analysis to a computer file, picking up that file with Mathcad software, and then using Mathcad to compare the two curves.

Such a process was avoided by carrying out the frequency response analysis in the same worksheet that was used to display the test results. Figure 7.2.8 illustrates the program involved.

This first calculates a set of 101 frequencies, on a logarithmic scale, between 5 KHz and 20 MHz, and stores them in the vector F. It then defines the component values of the circuit model as well as the number of turns on the secondary.

The equations used in the function ZTm are derived from an inspection of the circuit model. The impedance $Z2$ is the impedance as 'seen' by the current generator. It defines the ratio of $Vch2$ to $Isec$. Dividing $Z2$ by the number of turns gives the ratio of $Vch2$ to $Iprim$, that is, the transfer impedance of the circuit model.

Figure 7.2.9 illustrates the correlation between the test results and the response of the model. Although there was an initial discrepancy between the two curves, a few adjustments of $L1$ and $R1$ led to a curve which intersected the data points at the low frequencies. Varying the values of $C1$ and $R4$ led to a curve which intersected the data point above 2 MHz as well.

The existence of this model makes it possible to deduce the amplitude of the current in the loop-under-test by noting the amplitude and frequency of the signal observed on the oscilloscope. That is, the response of the model of Figure 7.2.8 provides the calibration curve for the current transformer.

The fact that the frequency response of the transformer assembly is flat over a wide bandwidth means that it can also be used to monitor the amplitude and waveform of transient currents. There is one important proviso: the bandwidth of the waveform being monitored must lie within the bandwidth of the device.

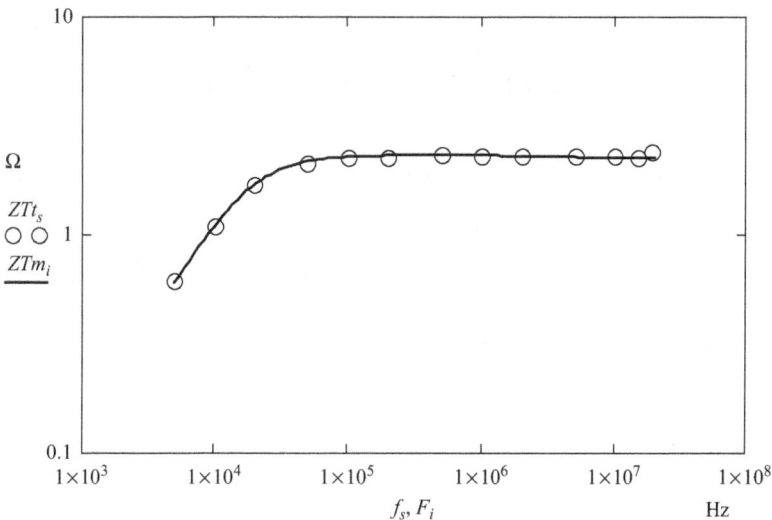

Figure 7.2.9 Transfer impedance from test results, ZTt, and from circuit model, ZTm.

A first approximation to the transient response characteristic can be made by assuming that the inductor $L1$ is open-circuit, while the capacitor $C1$ is replaced by a short-circuit. The admittance of the circuit becomes $Ysec$, where:

$$Ysec = \frac{1}{R1} + \frac{1}{R2} + \frac{1}{R3} + \frac{1}{R4} \qquad (7.2.2)$$

and

$$Rsec = \frac{1}{Ysec} \qquad (7.2.3)$$

This gives:

$$Vch = Rsec \cdot Isec \qquad (7.2.4)$$

Voltage at the oscilloscope input can be related to the current in the primary loop by invoking (7.2.1):

$$Vch = \frac{Rsec}{Turns} \cdot Isec \qquad (7.2.5)$$

For this particular device, the transfer impedance to use with transient test analysis is:

$$RT = \frac{Rsec}{Turns} = 2.27 \qquad (7.2.6)$$

7.3 Triaxial cable

In any test setup, there will be cables which connect the test equipment to the equipment-under-test (EUT). It is normal practice for these cables to be co-axial, because such cables are the most practical and efficient means of transmitting signals from one location to another, as well as being the least likely to cause interference problems.

Even so, stray coupling does exist between co-axial cables. Practical experience has shown that this coupling can create unwanted signals which completely obscure the signals from the EUT.

One way of reducing this coupling is to use triaxial cables as illustrated by Figure 7.3.1. It was not particularly difficult to manufacture such an assembly. The protective boots were removed from the BNC connectors at each end of an RG58 cable assembly and one end of an 18 Ω resistor connected to each shell. A roll of wire braid was cut to length and fitted over the assembly to form an outer sheath. The 18 Ω resistors were then connected to the outer sheath, care being taken to ensure the outer braid did not touch the shells of the connectors. Finally, the assembly was sheathed in an insulating braid.

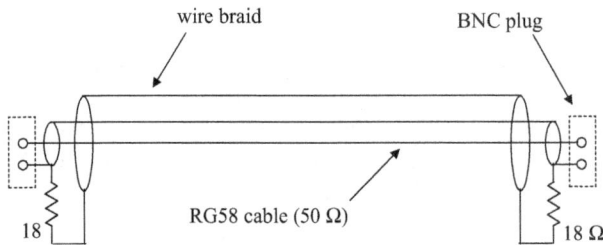

Figure 7.3.1   Triaxial cable.

The inner and outer braids act as a transmission line, and the characteristic impedance can be calculated using:

$$Ro = \frac{1}{2 \cdot \pi} \cdot \sqrt{\frac{\mu_o \cdot \mu_r}{\varepsilon_o \cdot \varepsilon_r}} \cdot \ln\left(\frac{r_{3,3}}{r_{2,2}}\right) \tag{7.3.1}$$

where $r_{2,2}$ and $r_{3,3}$ are the radii of the inner and outer braids, respectively.

For RG58 cable, the diameter of the screen is 3.3 mm and that of the outer sheath is 5 mm. Allowing for a loose fitting braid, it can be assumed that the braid diameter is 6 mm. If it is also assumed that the relative permittivity is 4, then invocation of (7.3.1) gives the value of Ro as 18 Ω.

In the presence of an external field, antenna-mode current will flow in the outer skin of the outer braid. Voltage developed along the length of the braid will cause common-mode current to flow in the loop formed by the inner and outer braid. Since these two conductors act as a transmission line, any signal arriving at either end is absorbed in an 18 Ω resistor. Since these resistors are the same as the characteristic impedance of the line, there will be little or no reflection. This means that most of the energy delivered to the cable in the form of electromagnetic radiation will be dissipated in a resistive load. To achieve good performance at very high frequencies, several resistors would need to be assembled in an annular ring at each termination to minimize the inductance of the resistive elements.

As far as radiation from the differential-mode signal is concerned, any voltage developed along the length of the inner braid will cause a circulating current in the outer transmission line. Again, the energy of the unwanted signal will be absorbed by the two end resistors.

The net result is a significant improvement in the shielding effectiveness of the RG58 cable.

7.4 The isolated conductor

The voltage transformer provides a means of injecting a defined voltage into any conducting loop without making physical contact between the conductors of the EUT and the test equipment. The current transformer provides the means of measuring the amplitude of the induced current. This being so, it is possible to carry out tests on the simplest configuration which can be envisaged (the isolated conductor). So this is the first test recorded here.

Since the behavior of an isolated conductor was expected to approximate to that of a half-wave dipole, and since the estimated value of the half-wave frequency of a 15 m length is 10 MHz, the selection of such a length brings the first resonant frequency within the operating range of the test equipment.

Figure 7.4.1 illustrates the setup. As with the calibration process for the current transformer, the output of the signal generator was varied over a range of spot frequencies. Note was taken of the peak-to-peak amplitude of the signal displayed on both channels of the oscilloscope and the results tabulated.

Although such a procedure appears more cumbersome than one which uses a programmable signal generator and spectrum analyzer, the process has five distinct advantages:

● Using the same item of equipment to compare input and output signals eliminates many problems of calibration.

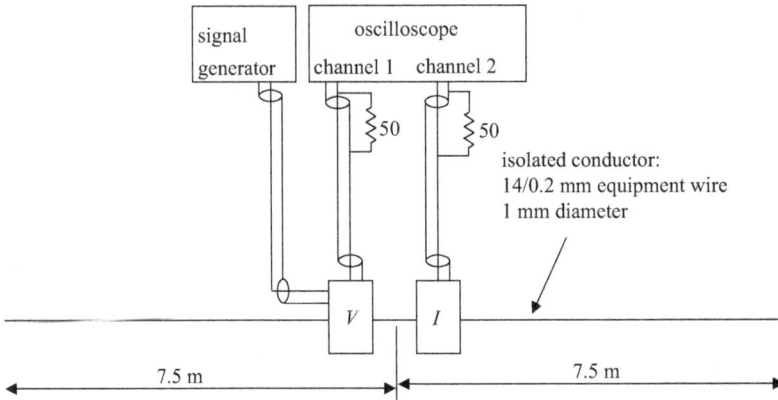

Figure 7.4.1   Testing the response of an isolated conductor.

- Monitoring the waveforms provides a visible indication of any distortion which might be present.
- The user is not inundated with vast quantities of data to be processed and interpreted.
- Oscilloscopes and signal generators are ubiquitous items of equipment in any electronics laboratory.
- All the items of test equipment can be provided by designers with a limited budget.

Figure 7.4.2 displays the recorded data in the form of a three-column array and defines a function *Ytest(s)* which can be used to calculate the admittance characteristic. This constitutes the first page of the Mathcad worksheet which analyzes the results.

The range variable *s* is used to identify a particular row in the array. The relationship between the voltage *Vin* induced in the conductor and the voltage at channel 2 of the oscilloscope *Vch2* is defined by equation (7.1.1). The admittance of the secondary winding of the current transformer is calculated by using the set of equations defined in the subroutine of Figure 7.2.8. Multiplying this value by the channel 2 voltage *Vch2* gives the value of the current *Isec* in the secondary winding. Multiplying *Isec* by the number of turns gives the current in the conductor, *Iprim*.

The ratio of the current in the conductor to the applied voltage *Vin* gives a value for the admittance at the spot frequency f_s. All the values are recorded in the vector *Yt*. Figure 7.4.3 illustrates the frequency response of this parameter over the range 1–20 MHz.

As far as the predicted response was concerned, the shape of this curve was extremely encouraging. There is a single peak at just below 10 MHz and an indication that there could be a second peak at just below 30 MHz. This would relate to resonances at the quarter-wave and three-quarter-wave frequency.

It could reasonably be expected that the dipole model of Figure 5.2.2 would be capable of simulating the response. Such a task would be just a matter of assigning numerical values to the components, defining the relevant equations, and setting these out in the worksheet. The general circuit model is reproduced in Figure 7.4.4.

A brief examination of the tabulated data reveals that the quarter-wave frequency f_q of the conductor-under-review is 7.83 MHz. Given knowledge of his frequency and the measured length of the cable, it is possible to use (2.3.10) to calculate the propagation velocity.

Worksheet 7.4, page 1

$$data := \begin{pmatrix} 1 & 400 & 0.4 \\ 2 & 400 & 0.4 \\ 3 & 400 & 1 \\ 4 & 395 & 1.9 \\ 5 & 390 & 3.4 \\ 6 & 390 & 5.9 \\ 6.5 & 385 & 8.4 \\ 7 & 385 & 14.5 \\ 7.5 & 380 & 26 \\ 7.83 & 360 & 36 \\ 8 & 350 & 31.5 \\ 8.5 & 345 & 16.5 \\ 9 & 345 & 10.5 \\ 10 & 345 & 6.2 \\ 11 & 345 & 6 \\ 12 & 345 & 4.4 \\ 13 & 340 & 3.4 \\ 14 & 340 & 3 \\ 15 & 330 & 3 \\ 16 & 330 & 2.4 \\ 17 & 320 & 1.4 \\ 18 & 315 & 2.2 \\ 19 & 310 & 3.4 \\ 19.6 & 305 & 4.2 \end{pmatrix}$$

column 1: frequency, MHz
column 2: channel 1 voltage, mV
column 3: channel 2 voltage, mV

Yt is the admittance derived from test results

From Figure 7.2.7:

$R1 := 300\ \Omega$ $\qquad$ $R2 := 51\ \Omega$

$R3 := 50\ \Omega$ $\qquad$ $R4 := 850\ \Omega$

$L1 := 200 \cdot 10^{-6}\ H$ $\qquad$ $C1 := 60 \cdot 10^{-12}\ F$

$Turns := 10$

$s := 1..\text{rows}(data)$ $\qquad$ $f_s := data_{s,1} \cdot 10^6$

$$\text{Ytest}(s) := \begin{vmatrix} Vch1 \leftarrow data_{s,2} \cdot 10^{-3} \\[4pt] Vin \leftarrow \dfrac{51 + 50}{50} \cdot Vch1 \\[4pt] Vch2 \leftarrow data_{s,3} \cdot 10^{-3} \\[4pt] \omega \leftarrow 2 \cdot \pi \cdot f_s \\[4pt] Z1 \leftarrow R4 + \dfrac{1}{j \cdot \omega \cdot C1} \\[4pt] Y2 \leftarrow \dfrac{1}{R1} + \dfrac{1}{R2} + \dfrac{1}{R3} + \dfrac{1}{j \cdot \omega \cdot L1} + \dfrac{1}{Z1} \\[4pt] Isec \leftarrow |Y2| \cdot Vch2 \\[4pt] Iprim \leftarrow Isec \cdot Turns \\[4pt] \dfrac{Iprim}{Vin} \end{vmatrix}$$

$Yt_s = \text{Ytest}(s)$

Figure 7.4.2 Using test data to calculate the admittance characteristic.

Since there was no magnetic material in the cable, the value of μ_r can be assumed to be unity. So (2.3.11) can be used to derive a value for the relative permittivity.

Equations (2.3.1) and (2.3.2) can then be used to calculate theoretical values for the primitive capacitance Cp and primitive inductance Lp. Since (5.1.3) gives the theoretical value for the radiation resistance $Rrad$, there is now enough information to assign values to the components of the circuit model of Figure 7.4.4.

A/V

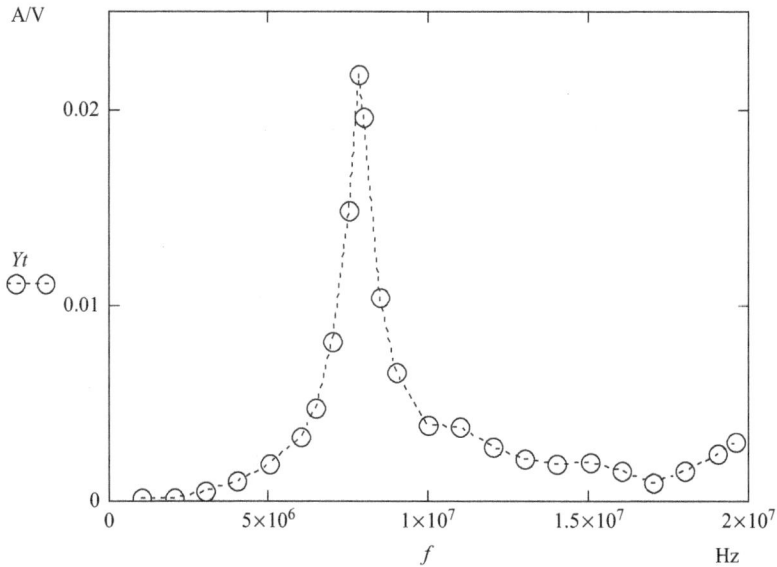

Figure 7.4.3 Admittance of conductor, derived from test results.

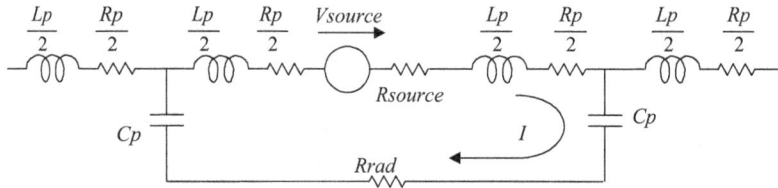

Figure 7.4.4 General circuit model of a single isolated conductor.

Figure 7.4.5 is a copy of the page of the worksheet that calculates the values of the reactive and resistive parameters of the conductor. The frequency at which skin effect comes into play is calculated to be 6.89 kHz. The only other parameters to be defined are the radiation resistance and the amplitude of the voltage source. The former is set at the same value as a conventional dipole, 73 Ω. The latter is set at unity to allow the response to be defined as amperes per volt.

Assigning these values to the general circuit model leads to the model of Figure 7.4.6.

The Mathcad function to calculate the value of the admittance at each frequency is illustrated in Figure 7.4.7. This is an adaptation of the function Zbranch*(f)* depicted in Figure 4.3.4.

Distributed parameters Z1 and Z2 are determined, and then used to calculate the loop impedance Z3. Dividing the amplitude of the voltage source *Vsource* by the magnitude of the loop impedance gives a value for the current which will be flowing in the loop.

Since the magnitude of *Vsource* is unity, the output variable of the function Ymodel*(i, Vsource)* is the admittance of the loop shown, at the frequency F_i. The input variable *i* is an integer which points to the relevant row of the frequency vector *F*. The other input variable

Worksheet 7.4, page 3. (Worksheet 7.4 page 2 gives figure 7.4.3)

$\varepsilon_o := 8.854 \cdot 10^{-12}$ F/m $\mu_o := 4 \cdot \pi \cdot 10^{-7}$ H/m $\mu_r := 1$ $c := 3 \cdot 10^8$ m/s

$\rho := 1.7 \cdot 10^{-8}$ Ω m   resistivity of copper

$l := 7.5$ m   length of monopole

$r := 0.5 \cdot 10^{-3}$ m   from micrometer measurement of conductor diameter

$Fx := \dfrac{4 \cdot \rho}{\mu_o \cdot \pi \cdot r^2} = 6.89 \times 10^4$ crossover frequency
see (2.5.14)

$Lp := \dfrac{\mu_o \cdot \mu_r \cdot l}{2 \cdot \pi} \cdot \ln\left(\dfrac{l}{r}\right) = 1.442 \times 10^{-5}$ see (2.3.2)

$\dfrac{Lp}{2} = 7.212 \times 10^{-6}$ inductive component value
for model

$f_q := 7.83 \times 10^6$ Hz   from tabulated data

$v := 4 \cdot l \cdot f_q$ $v = 2.349 \times 10^8$ see (2.3.10)

$\varepsilon_r := \left(\dfrac{c}{v}\right)^2 = 1.631$ see (2.3.11)

$Cp := \dfrac{2 \cdot \pi \cdot \varepsilon_o \cdot \varepsilon_r \cdot l}{\ln\left(\dfrac{l}{r}\right)}$ $Cp = 7.077 \times 10^{-11}$ see (2.3.1)

$Rrad := 73$ Ω   see (5.1.3)

$Vsource := 1$ V   allows admittance value to be calculated

$Rss := \dfrac{\rho \cdot l}{\pi \cdot r^2} = 0.162$ steady-state resistance;
see (2.5.11)

$Gp := 0$ S   good quality insulation assumed

$Rsource := 1$ Ω   guess value for impedance of current transformer

Figure 7.4.5 Calculating values for components of circuit model.

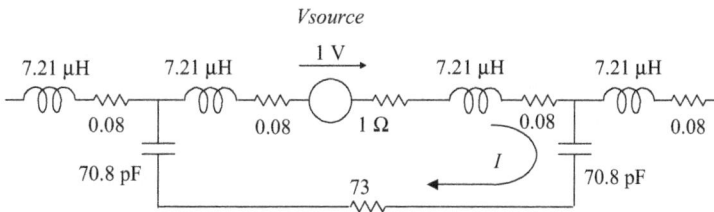

Figure 7.4.6 Initial circuit model of conductor-under-review.

Worksheet 7.4, page 4

$i := 1..200$ $\qquad F_i := i \cdot 10^5$

$$\text{Ymodel}(i, \textit{Vsource}) := \left| \begin{array}{l} Rp \leftarrow Rss \cdot \sqrt{1 + \dfrac{F_i}{F_x}} \\[2ex] \omega \leftarrow 2 \cdot \pi \cdot F_i \\[1ex] \theta \leftarrow \sqrt{(Rp + j \cdot \omega \cdot Lp) \cdot (Gp + j \cdot \omega \cdot Cp)} \\[2ex] Zo \leftarrow \sqrt{\dfrac{Rp + j \cdot \omega \cdot Lp}{Gp + j \cdot \omega \cdot Cp}} \\[2ex] Z1 \leftarrow Zo \cdot \tanh\left(\dfrac{\theta}{2}\right) \\[2ex] Z2 \leftarrow Zo \cdot \operatorname{csch}(\theta) \\[1ex] Z3 \leftarrow 2 \cdot Z1 + 2 \cdot Z2 + Rrad + Rsource \\[1ex] \dfrac{\textit{Vsource}}{|Z3|} \end{array} \right.$$

$Ym_i := \text{Ymodel}\,(i, \textit{Vsource})$

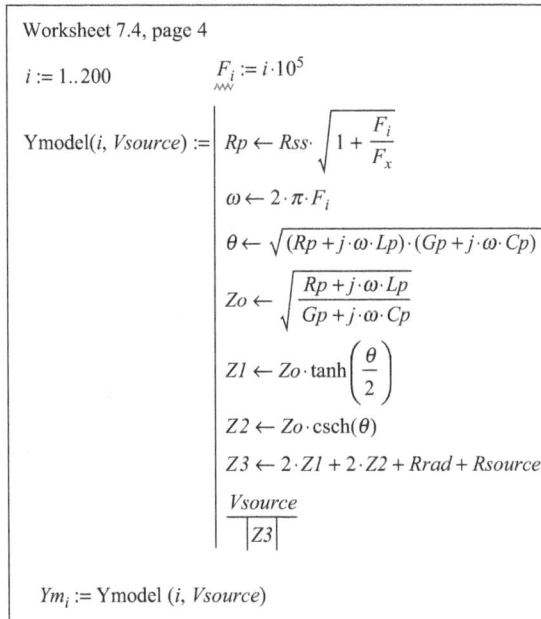

Figure 7.4.7 Calculating the admittance of the circuit model over a range of frequencies.

Vsource allows the response to be determined when the source voltage itself is a function of frequency. (Figure 5.3.7 illustrates how the threat voltage of the incoming wave can vary.)

The vector *Ym* stores the results of the calculations over the same range of frequencies that were used in the test procedure. Displaying the response of this model and those derived from test results gives the two curves of Figure 7.4.8.

The close correspondence in the region between 1 and 5 MHz indicates that the value calculated for the primitive capacitance *Cp* was fairly accurate. The fact that the peaks of both models occur at the same frequency indicates that the value of *Lp* was also accurate. However, there is an obvious discrepancy between the magnitudes of the two peaks.

Discovering the reason for this discrepancy involved several attempts to vary component values of the model and note the effect this had on its response. At this point in the assessment, it was assumed that the value of 73 Ω for *Rrad* was sacrosanct.

Eventually, it was found that increasing the amplitude of the voltage source to 1.65 V brought the two curves into near-coincidence over the critical range between 5 and 10 MHz. Figure 7.4.9 is a copy of the final page of the worksheet, and illustrates the resultant two curves.

In hindsight, the explanation as to why the source voltage needed to be increased is fairly simple. Unlike the conventional dipole antenna which is connected to the source termination resistance of a transmitter or the load termination of a receiver, an isolated length of wire contains no resistors to absorb the electromagnetic energy. So energy which is not radiated out into the environment is stored as a voltage across the capacitors. This immediately creates a reflected current which flows back through the transformers, and this reflected current is a source of further radiation.

Worksheet 7.4, page 5

$Cp = 7.077 \cdot 10^{-11}$ F $Lp = 1.442 \cdot 10^{-5}$ H $Fx = 6.89 \cdot 10^{4}$ Hz

$Rrad = 73\ \Omega$ $Rsource = 1\ \Omega$ $Rss = 0.162\ \Omega$

$Vsource = 1\ \Omega$

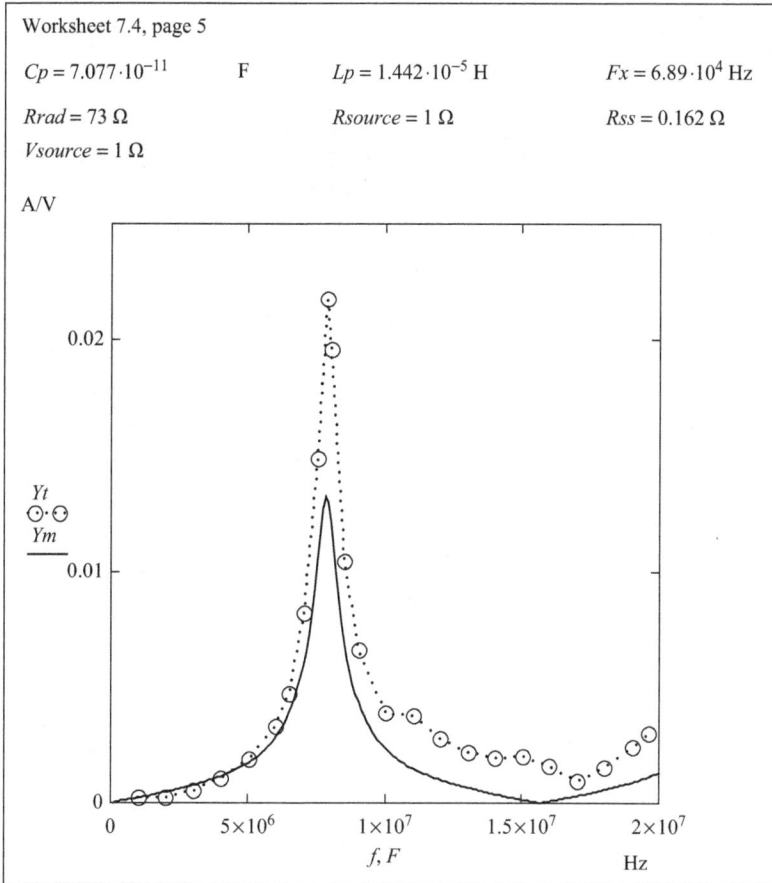

Figure 7.4.8 Comparing response of the test with that of the initial circuit model.

Hence, the final circuit model turns out to be very similar to that of the initial model. Figure 7.4.10 shows that the only change is the addition of a second voltage source to represent the effect of stored energy.

Since the total energy is shared equally between current and voltage and since the current has departed from the conductor, it follows that half the energy is stored in the capacitors. Since voltage is proportional to the square root of power, it is reasonable to assume that the maximum value for the voltage developed across the capacitors is:

$$Vstored = \sqrt{2} \cdot Vsource \approx 0.71 \cdot Vsource \qquad (7.4.1)$$

This value of 0.71 represents a system in which there are no other losses. So the observed value of 0.65 is entirely plausible.

However, the addition of the source *Vstored* has caused the theoretical curve to be higher than the actual response over the range 0–5 MHz. This deviation was corrected by removing the extra voltage source and reducing the value of the radiation resistance to 42 Ω. Figure 7.4.11 shows the resultant curve. This correlates well with the actual response in the

Worksheet 7.4, page 6

$Vsource := 1.65 \text{ V}$ $Ym_i := Ymodel(i, Vsource)$

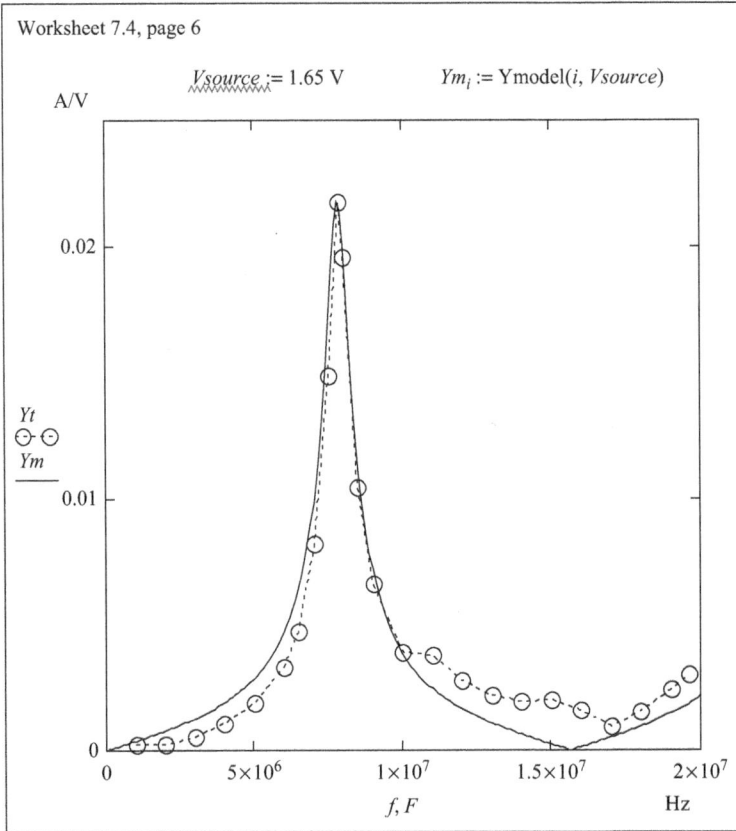

Figure 7.4.9   Response of model, taking into account stored energy.

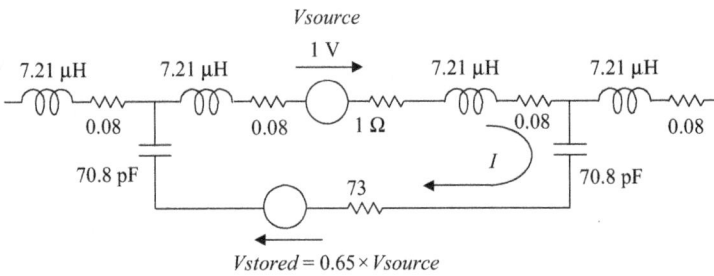

$Vstored = 0.65 \times Vsource$

Figure 7.4.10   Introducing a voltage source to simulate stored energy.

range 0–5 MHz, as well as replicating the amplitude of the peak response. This test can be used to measure the radiation resistance of the conductor.

Deviations in the region between 10 and 20 MHz can be explained by the fact that, above the quarter-wave frequency, current is flowing backwards and forwards along the cable, rather like waves in a harbor. Simulating this effect would require extra complexity to be introduced into the model.

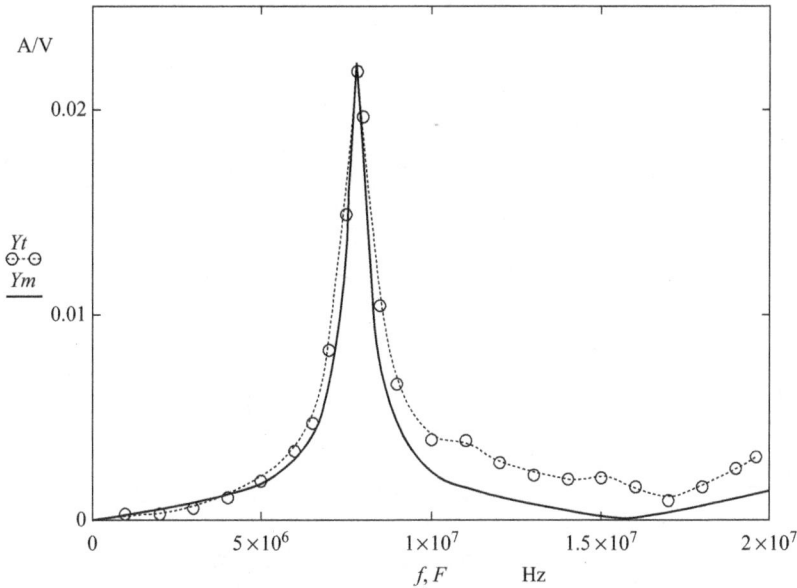

Figure 7.4.11   Response of model with value of radiation resistance set at 42 Ω.

However, as far as EMC is concerned, the critical frequencies are when the peaks are high. So minor deviations between reality and model at other frequencies are of little concern when assessing the probability that the equipment will pass the formal EMC tests.

The close correlation between the response of the actual system and the theoretical model effectively validates the concepts introduced in the previous chapters. Specifically, these are:

● the derivation of the formulae for the primitive inductance Lp in section 2.2,
● the derivation of the formulae for the primitive capacitance Cp in section 2.1,
● the distributed impedance circuit model, derived in section 4.1,
● the circuit model for the dipole, postulated in section 5.2.

The close correlation also demonstrates that simple test equipment is capable of providing extremely accurate measurements. For example, the fact that the response peaks at 7.83 MHz means that, for this particular setup, the propagation velocity of antenna-mode current is 235 m/μs. (See the value for v in Figure 7.4.5.)

7.5 Cable characterization

The method described in the previous section can be developed to characterize specific cables, that is, to determine the representative circuit model for a particular cable. By carrying out tests on a cable that is completely isolated, the effects of external components can be eliminated. Measurements which result are specific to the cable.

In the example described here, a circuit model is derived for a 15 m length of two-core power cable. Two setups are involved, the first to measure the emitted radiation, and the second to measure the wire-to-wire coupling.

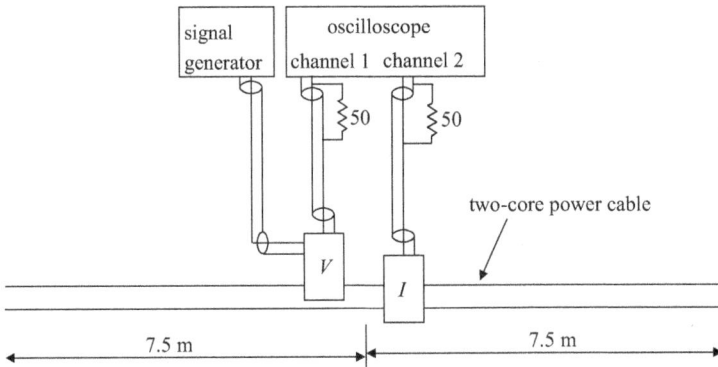

Figure 7.5.1 Test to measure emitted radiation characteristics of cable.

Figure 7.5.1 illustrates the first setup. This configuration is essentially a practical implementation of the general circuit model of an isolated cable, derived from theoretical considerations in the section on the virtual conductor. Figure 5.2.8 illustrates this model and (5.2.9) and (5.2.10) define formulae for component values. Since each half of the 15 m cable is represented by a triple-T network, the length l of each network is 7.5 m.

A voltage transformer is used to inject a voltage in series with one conductor of the cable. Both ends of the cable are open-circuit. If these terminations had both been short-circuited, then it would be clear that the transformer was inducing a voltage in the loop formed by the signal and return conductors (a differential-mode voltage). The existence of open-circuit terminations does not change the fact that a differential-mode voltage is being injected.

This injected voltage creates a current that flows along the signal conductor. Electromagnetic coupling between the two conductors creates a current that flows in the opposite direction along the return conductor (the differential-mode current). The injected voltage also creates an antenna-mode current that flows along the cable and converts into an electromagnetic wave.

A current transformer clamped round both conductors was used to measure the amplitude of the antenna-mode current. The range of frequencies was from 1 to 20 MHz and the output displayed in Figure 7.5.2 is the transfer admittance YT, in amperes per volt. A worksheet similar to that illustrated by Figure 7.4.2 was used to create this graph.

Figure 7.5.3 illustrates the second setup. Again, the voltage transformer injects a differential-mode voltage into the cable. But this time the monitored current is the differential-mode current. In Figure 7.5.4 there are two peaks, at about 5.5 MHz and at about 16.8 MHz.

The general circuit model for antenna-mode coupling of a transmission line is derived in the section on the virtual conductor, and illustrated by Figure 5.2.8. Rearranging this in the form of a bridge circuit leads to Figure 7.5.5. This layout simplifies the definition of the two current loops. In the configuration-under-review the voltage source is in series with conductor 1; so this fact is reflected in the location of *Vsource* in the diagram.

Transforming the circuit components into distributed impedances leads to Figure 7.5.6.

Component values can be assigned by following a systematic process, starting with the measurement of the conductor radii r_{11} and r_{22}, the spacing between conductors, r_{12}, and the length l of the cable. Using equations available in previous chapters, initial values can be

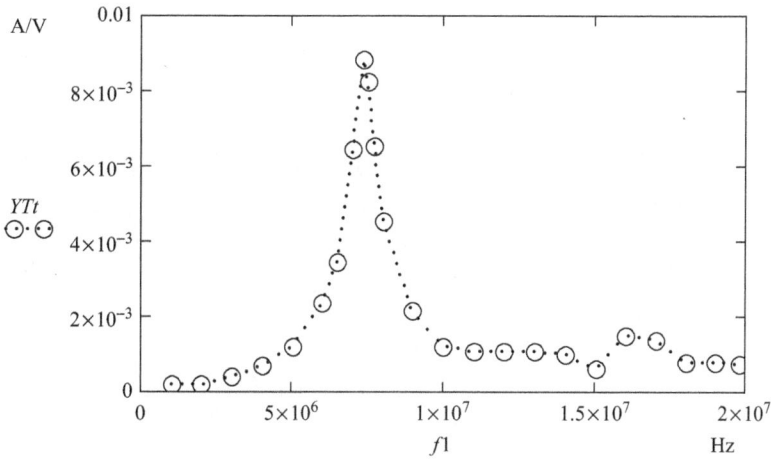

Figure 7.5.2   Results of radiated emission test.

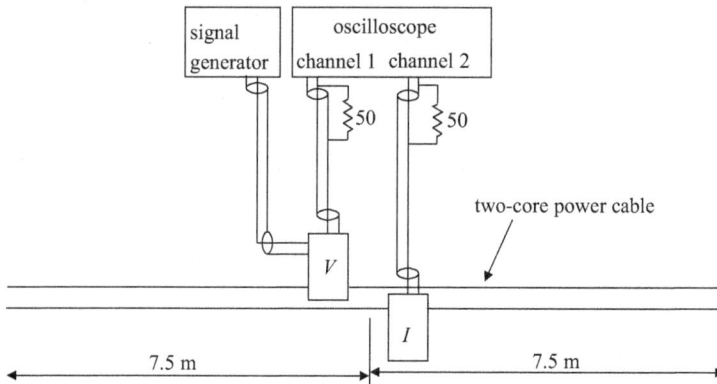

Figure 7.5.3   Test to measure transmission-line characteristics of cable.

assigned to the resistive and inductive components. This is carried out by the page of the worksheet illustrated by Figure 7.5.7.

Since the conductors of all cabling in electrical systems are sheathed or supported by insulating material, and since this material has a significant effect on the propagation velocities of the signals, it is necessary to determine the value of the relative permittivity of the dielectric before values for the capacitors can be defined. This is done on the page of the worksheet illustrated on Figure 7.5.8.

Having provided initial values for all the components of the model, the next step is to simulate the responses of both tests. This is carried out by the main program, illustrated by Figure 7.5.9. There are two outputs: the vector YTm, which simulates the response of the radiated emission test, and Ym, which simulates wire-to-wire coupling.

Response characteristics of the model are compared with those of the actual tests in the final page of the worksheet. These are displayed here as Figures 7.5.10 and 7.5.11.

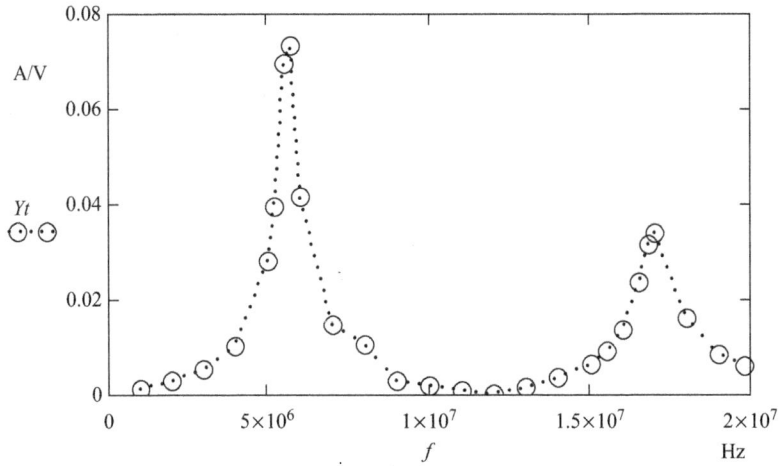

Figure 7.5.4   Results of wire-to-wire coupling test.

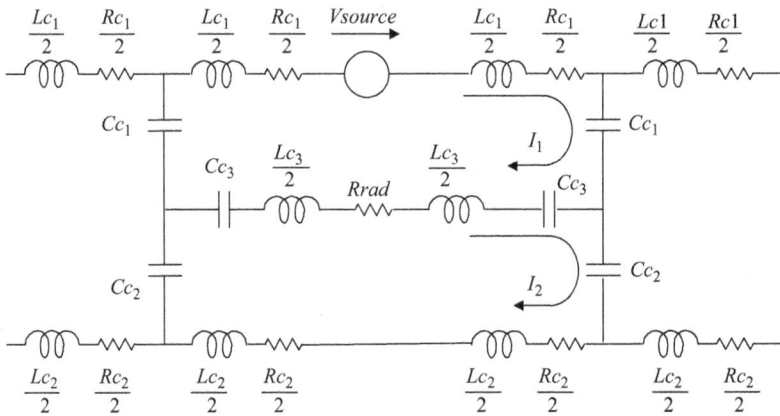

Figure 7.5.5   General circuit model of an isolated cable.

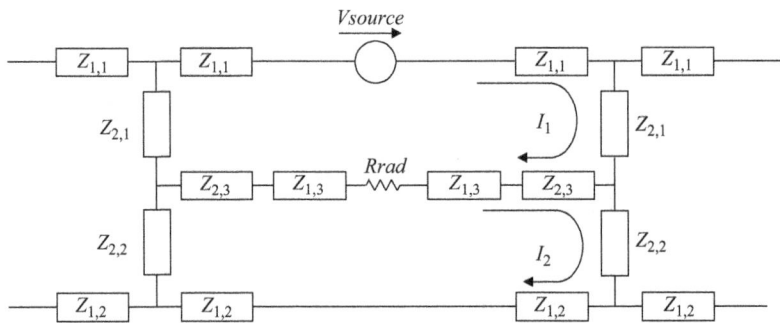

Figure 7.5.6   Distributed parameter circuit model.

Worksheet 7.5, page 4

$\rho := 1.7 \cdot 10^{-8} \, \Omega \; m$ $\mu_o := 4 \cdot \pi \cdot 10^{-7} \, H/m$ $\mu_r := 1$

$r11 := 0.48 \cdot 10^{-3} \, m$ $r22 := r11$ Radii of conductors

$r12 := 1.95 \cdot 10^{-3} \, m$ Spacing between conductors. See Figure 2.7.4

$l := 7.5 \, m$ Half the length of the cable

$Ra := 0.75 \, \Omega$ Steady-state resistance of length l of the cable. **Value selected during analysis**

$Rss := \begin{pmatrix} Ra \\ Ra \\ 0 \end{pmatrix}$ Placing steady-state resitance values in the vector Rss. Conductor 3 represents the environment

$Fx := \dfrac{4 \cdot \rho}{\mu_o \cdot \pi \cdot r11^2} = 7.476 \times 10^4$ Frequency at which skin effect starts to modify resistance see (2.5.14)

$Lc_1 := \dfrac{\mu_o \cdot \mu_r \cdot l}{2 \cdot \pi} \cdot \ln\left(\dfrac{r12}{r11}\right)$

$Lc_2 := \dfrac{\mu_o \cdot \mu_r \cdot l}{2 \cdot \pi} \cdot \ln\left(\dfrac{r12}{r22}\right)$ Calculating inductance values for the circuit model and placing these values in a three-element vector see (5.25)

$Lc_3 := \dfrac{\mu_o \cdot \mu_r \cdot l}{2 \cdot \pi} \cdot \ln\left(\dfrac{l}{r12}\right)$

$\dfrac{Lc}{2} = \begin{pmatrix} 1.051 \times 10^{-6} \\ 1.051 \times 10^{-6} \\ 6.191 \times 10^{-6} \end{pmatrix} \; H$ Values of inductors of circuit model

$\dfrac{Rss}{2} = \begin{pmatrix} 0.375 \\ 0.375 \\ 0 \end{pmatrix} \; \Omega$ Values of resistors of circuit model

Radiation resistance
Value selected during analysis

$Rrad := 50 \, \Omega$ Initially set at 73 Ω

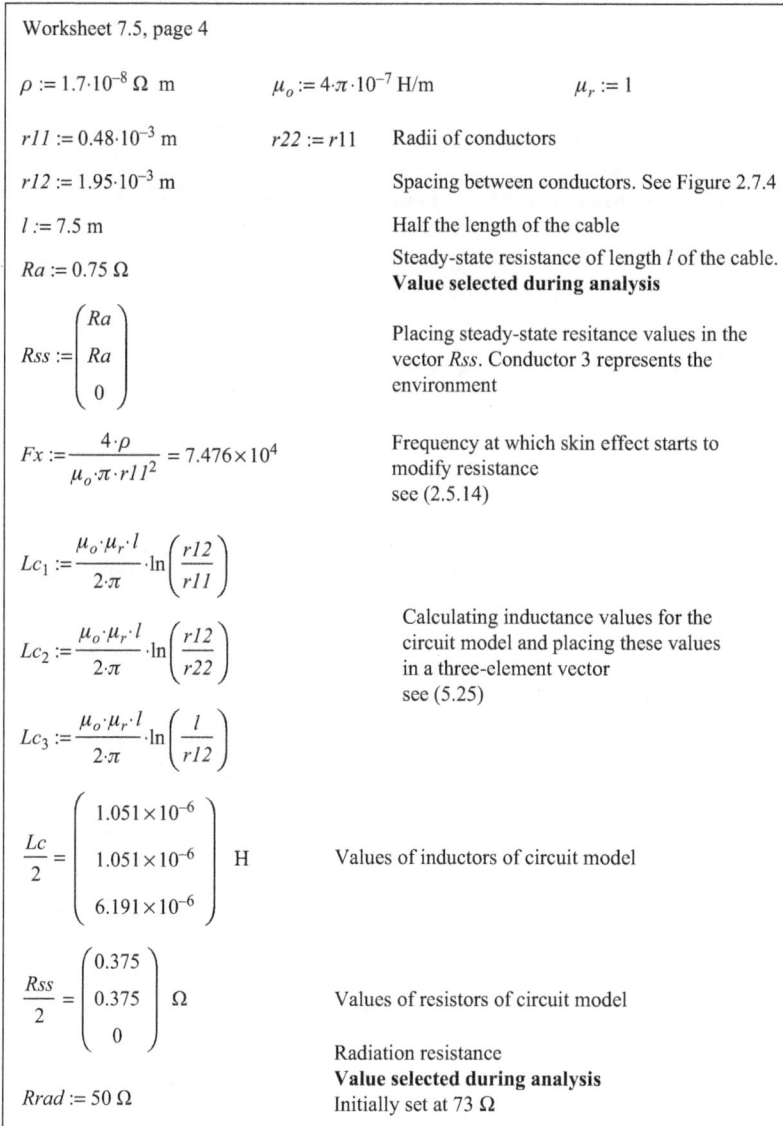

Figure 7.5.7   Calculation of values for resistors and inductors.

On the first run of the program, there were noticeable errors in the response of the model. However,

- varying the value of the conductor resistance Ra (see Figure 7.5.7) brought the amplitude of Ym closer to that of Yt,
- varying the value of the radiation resistor $Rrad$ brought the amplitude of the peak of YTm closer to that of YTt,

Worksheet 7.5, page 5

$\varepsilon_o := 8.854 \cdot 10^{-12}$ $c := 3 \cdot 10^8$

$fqa := 5.66 \cdot 10^6$ — The frequency at which the peak occurs in test of wire-to-wire coupling

$va := 4 \cdot l \cdot fqa = 1.698 \times 10^8$ — Velocity of propagation of electromagnetic wave along transmission line. see (2.3.10)

$\varepsilon_{ra} := \left(\dfrac{c}{va} \right)^2$ — Relative permittivity of dielectric of cable, acting as a transmission line. see (2.3.11)

$\varepsilon_{ra} = 3.122$ — Value of relative permittivity to be used to define capacitor values for conductors

$fqb := 7.55 \cdot 10^6$ — The frequency at which the peak occurs in radiated emission test

$vb := 4 \cdot l \cdot fqb = 2.265 \times 10^8$ — Velocity of propagation of electromagnetic wave along cable. see (2.3.10)

$\varepsilon_{rb} := \left(\dfrac{c}{vb} \right)^2$ — Relative permittivity of dielectric of cable when it is acting as an aerial. see (2.3.11)

$\varepsilon_{rb} = 1.754$ — Value of relative permittivity to be used to define capacitor value for monopole

$Cc_1 := \dfrac{2 \cdot \pi \cdot \varepsilon_o \cdot \varepsilon_{ra} \cdot l}{\ln\left(\dfrac{r12}{r11} \right)}$

$Cc_2 := Cc_1$ — Calculating capacitance values for the circuit model and placing these values in a three-element vector. see (5.2.10)

$Cc_3 := \dfrac{2 \cdot \pi \cdot \varepsilon_o \cdot \varepsilon_{rb} \cdot l}{\ln\left(\dfrac{l}{r12} \right)}$

$Cc = \begin{pmatrix} 9.291 \times 10^{-10} \\ 9.291 \times 10^{-10} \\ 8.867 \times 10^{-11} \end{pmatrix}$ — Values of capacitors of circuit model

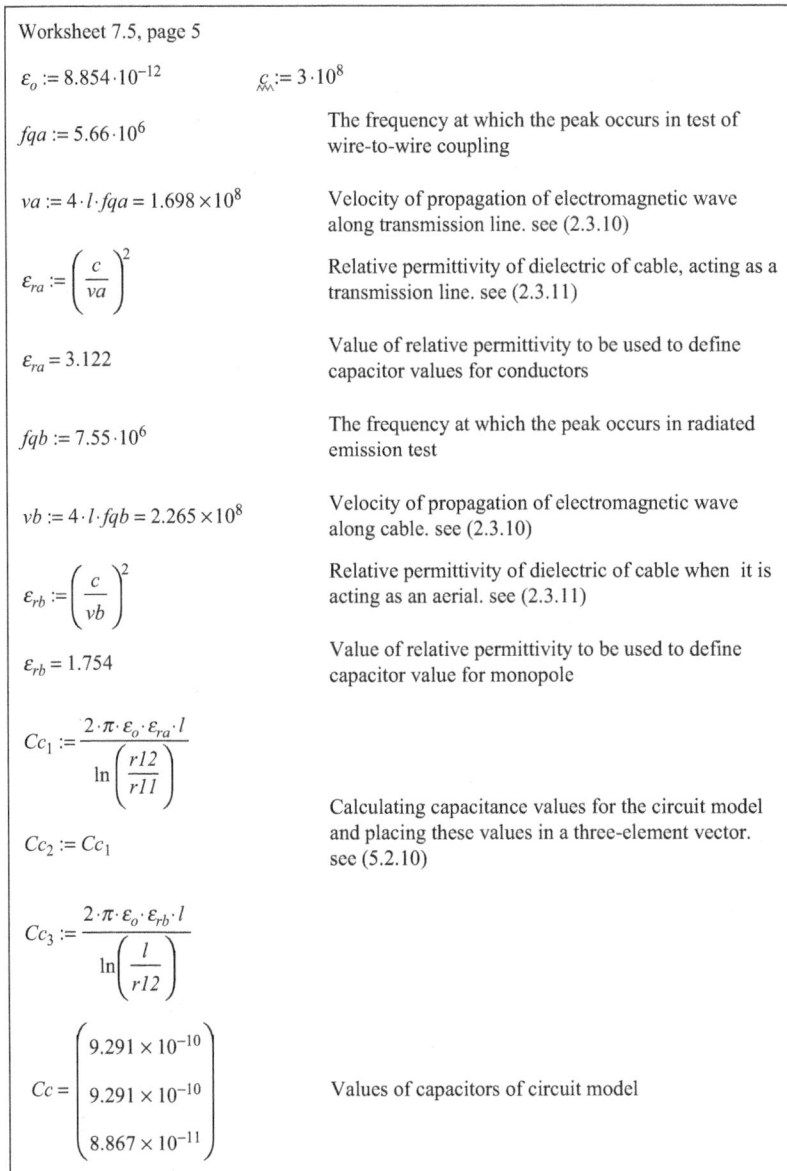

Figure 7.5.8 Calculating values for the capacitors.

- varying the value of the frequency *fqa* (see Figure 7.5.8) brought the peak of *Ym* closer to that of *Yt*,
- varying the value of *fqb* brought the peak of *YTm* closer to that of *YTt*.

Of course, varying any of the above parameters modified the responses of both curves. Even so, these side effects were minimal and very few iterations of the program were needed to achieve the correlation depicted in these two final graphs.

Worksheet 7.5, page 6

$i := 1..200$ $F_i := i \cdot 10^5 \text{ Hz}$ Defining the frequency range for the model

$\text{Zbranch}(s) := \bigg| \omega \leftarrow 2 \cdot \pi \cdot F_s$

$\text{for } k \in 1..3$

$\bigg| Rc_k \leftarrow Rss_k \cdot \sqrt{1 + \dfrac{F_s}{Fx}}$

$\theta \leftarrow \sqrt{(Rc_k + j \cdot \omega \cdot Lc_k) \cdot j \cdot \omega \cdot Cc_k}$

$Zo \leftarrow \sqrt{\dfrac{Rc_k + j \cdot \omega \cdot Lc_k}{j \cdot \omega \cdot Cc_k}}$

$z_{1,k} \leftarrow Zo \cdot \tanh\left(\dfrac{\theta}{2}\right)$

$z_{2,k} \leftarrow Zo \cdot \text{csch}(\theta)$

z

Copy of function introduced in
see Figure 4.3.4

Equations for loop impedances, derived from inspection of see Figure 7.5.6

$\text{Zloop}(s) := \bigg| Z \leftarrow \text{Zbranch}(s)$

$Z11 \leftarrow 2 \cdot (Z_{1,1} + Z_{2,1} + Z_{1,3} + Z_{2,3}) + Rrad$

$Z12 \leftarrow -2 \cdot (Z_{1,3} + Z_{2,3}) - Rrad$

$Z22 \leftarrow 2 \cdot (Z_{1,2} + Z_{2,2} + Z_{1,3} + Z_{2,3}) + Rrad$

$\begin{pmatrix} Z11 & Z12 \\ Z12 & Z22 \end{pmatrix}$

$V := \begin{pmatrix} 1 \\ 0 \end{pmatrix} V$ $YTm_i := \bigg| Z \leftarrow \text{Zloop}(i)$ Calculating response of transfer
                                            admittance for radiated emission
$\bigg| I \leftarrow \text{lsolve}(Z, V)$ from the cable
$\big| I_1 - I_2 \big|$

$Ym_i := \bigg| Z \leftarrow \text{Zloop}(i)$ Calculating response of admittance of
                                            open-circuit transmission line
$\bigg| I \leftarrow \text{lsolve}(Z, V)$
$\big| I_2 \big|$

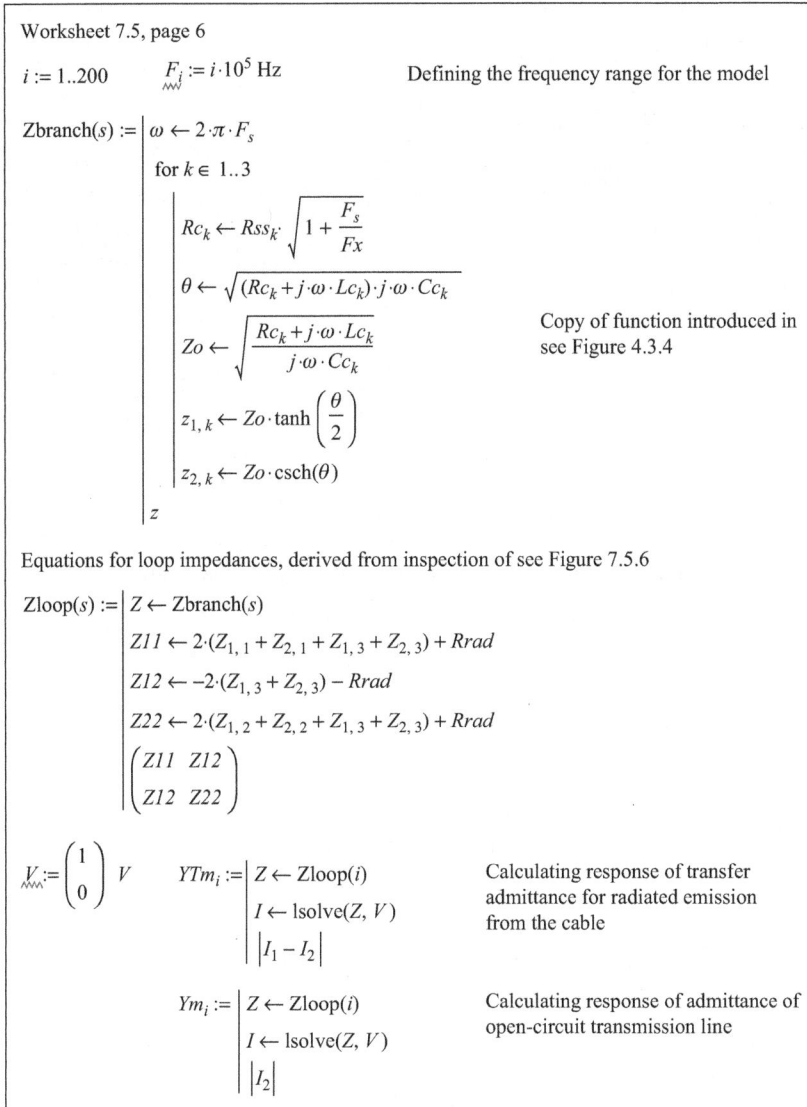

Figure 7.5.9   The main program: calculating the response of the circuit model.

It is worth noting that although there are 25 different components in the circuit model, only four independent variables were required to correlate the curves in both graphs. The component values used in the final run of the program are recorded at the bottom of Figures 7.5.7 and 7.5.8. Assigning these values to the variables of the general circuit model of Figure 7.5.5 leads to the representative circuit model for the configuration-under-review (Figure 7.5.12).

There was no need to include a second voltage source in the model as was done for the model of the single wire in section 7.4. All that needed to be done to line up the related peaks was to reduce the value of the radiation resistance from 73 Ω to 50 Ω.

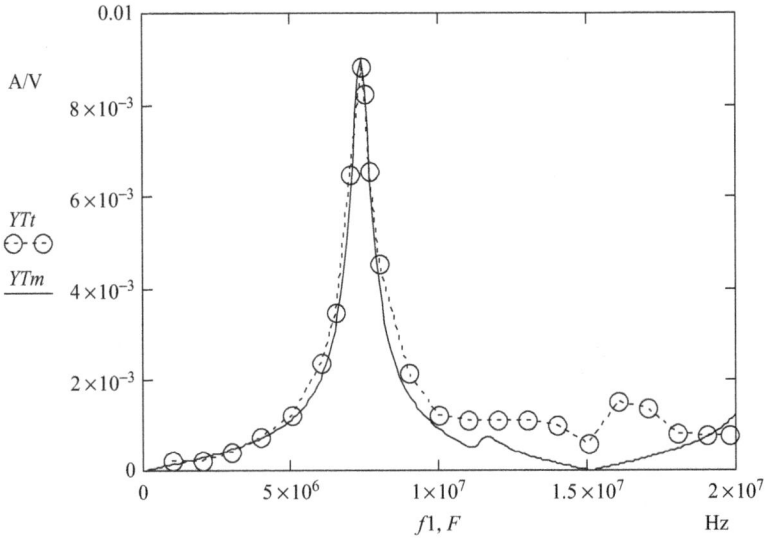

Figure 7.5.10 Radiated emission of cable and model.

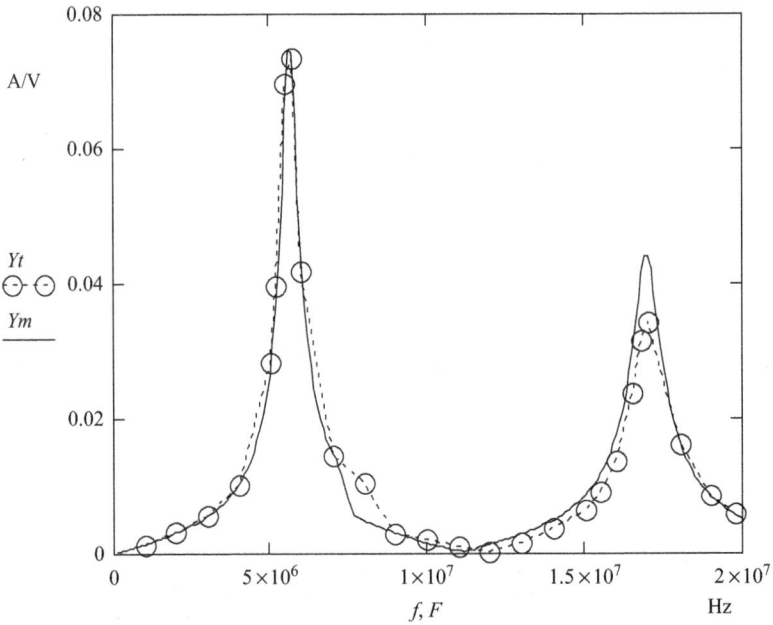

Figure 7.5.11 Transmission-line responses of cable and model.

The ability to create a circuit model of such a cable leads to several advantages:

• The basic feature of a circuit model of a length of cable is that component values are proportional to length of the cable. Per-unit-length parameters can be derived and later

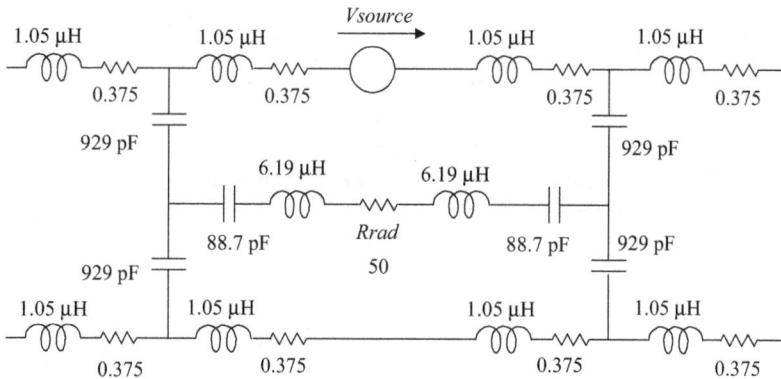

Figure 7.5.12 Representative circuit model of cable under test.

used to define models for any length of cable used in the system-under-review. (It is important that the cross section of the cable is uniform.)

- Measurements can be carried out on long cables at relatively low frequencies and used to predict the performance of shorter cable at higher frequencies.
- Having determined a circuit model for a particular cable, the file can be stored in a library, similar to that used by SPICE software for devices such as operational amplifiers or Schmitt triggers.
- Combining the models of the interface equipment with that of the cable will allow the radiated emission of the system-under-review to be simulated.
- Since the same model is valid no matter where the location of the voltage source, it is also possible to simulate the response of the system to external interference.

It can be reasoned that the field pattern of antenna-mode propagation is almost totally external to the cable and so includes more air than it does plastic insulation. The differential-mode propagation is more confined to the plastic insulation, so travels slower.

However, this set of tests does more than provide a reason for the difference. It defines the propagation velocities and provides an actual measurement of the relative permittivities. From Figure 7.5.8:

- Propagation of antenna-mode current, $vb = 227$ m/μs
- Propagation velocity of differential-mode current, $va = 170$ m/μs
- Relative permittivity of cable, acting as an antenna, $\varepsilon_{rb} = 1.76$
- Relative permittivity of transmission line, $\varepsilon_{ra} = 3.12$

It is also worth re-emphasising the fact that the value of the radiation resistor of Figure 7.5.12 was obtained by varying the value of *Rrad* in the model to line up the amplitude of the peak values of *Yt* and *Ym* in Figure 7.5.11. Effectively, the value of *Rrad* was derived from the test results. For any particular cable, the radiation resistance at the half-wave frequency is assumed to be a constant.

From the worksheet of Figure 7.5.7:

measured value of radiation resistance, *Rrad* 50 Ω

This particular model was derived from tests of a particular cable. Tests on another cable would result in a different set of values to assign to the general model of Figure 5.2.8. The significant feature of this approach is the existence of a general circuit model that will reduce the time it takes for a representative model to be created for any particular assembly.

Since the cross section of the cable is constant, the capacitors, inductors, and resistors are all proportional to length of the cable. Since the terminals are open-circuit, there are no components at the interfaces to upset the measurements. This means that the representative model of Figure 7.5.12 can be extrapolated to allow the performance of a longer or shorter cable to be assessed.

The theory of antennae suggests that a value of 73 Ω should be assigned to the radiation resistance. This particular test suggests that a value of 50 Ω would be more appropriate when carrying out a worst-case analysis of radiation susceptibility. Sections 5.6 and 5.7 give more information on this aspect. In the end, the choice of the value to use for *Rrad* in any system analysis is a matter of engineering judgment.

7.6 Cable transients

Since the signal generator used with the setup of Figure 7.5.1 was capable of generating square waves as well as sine waves, it seemed logical to check the response of the cable to transient signals. However, the waveform supplied by the output of the voltage transformer deviated significantly from the ideal square wave. So it was decided to deliver the input signal to the cable via a resistor network. The setup is illustrated by Figure 7.6.1.

This configuration allowed the cable to be analyzed as a transmission line. Since the far end was open-circuit and the resistance at the near end was very low, it was to be expected that multiple reflections would take place and that the monitored current would provide some insight into what was happening along the cable.

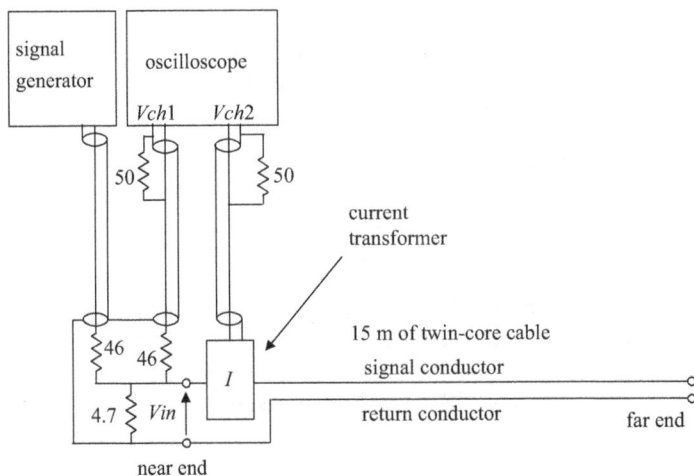

Figure 7.6.1 Setup for radiated transient test.

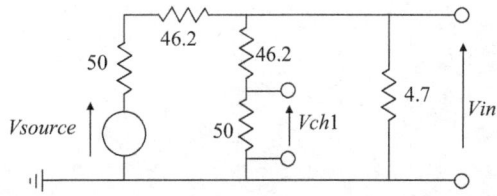

Figure 7.6.2 Circuit model of test setup.

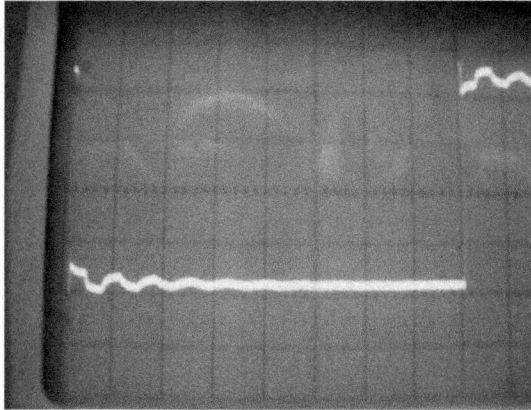

Figure 7.6.3 Waveform of signal at channel 1 of oscilloscope (500ns/div, 0.1V/div).

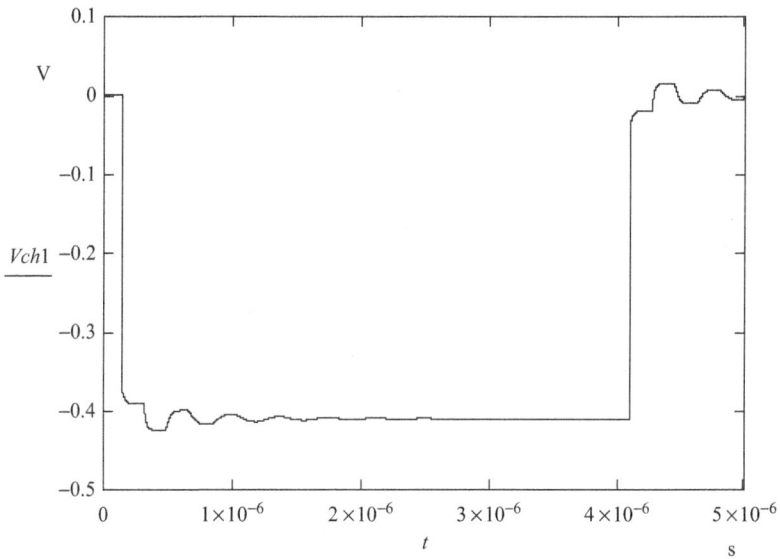

Figure 7.6.4 Simulated waveform of channel 1 input voltage. Setup as Figure 7.6.1.

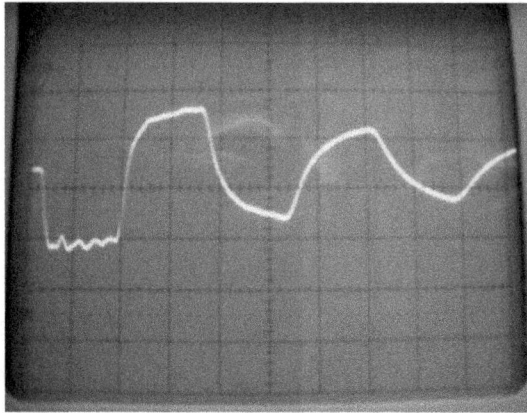

Figure 7.6.5   Voltage at channel 2 (100ns/div, 10mV/div).

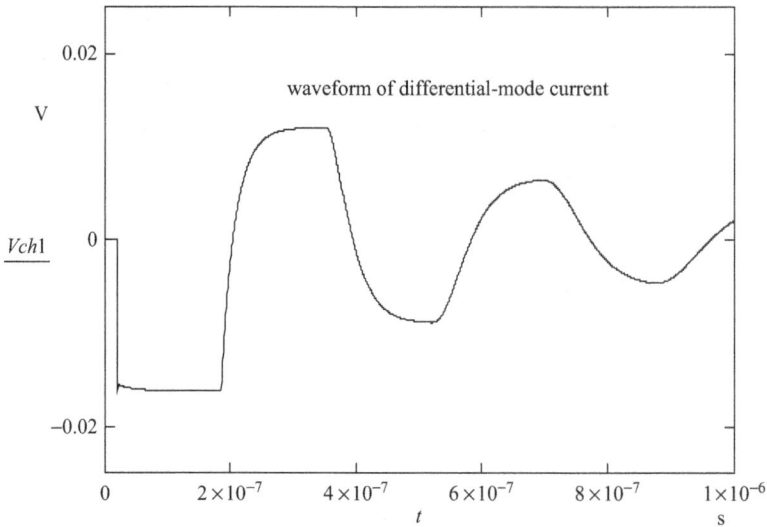

Figure 7.6.6   Simulated voltage at channel 2 input. Setup as Figure 7.6.1.

Figure 7.6.2 illustrates how the voltage monitored at channel 1 input terminals is related to the voltage delivered to the cable. Inspection of this model gives:

$$Vch1 = \frac{50}{96.2} \cdot Vin \tag{7.6.1}$$

Figure 7.6.3 is a photograph of the actual waveform monitored by channel 1. The current delivered to the line causes minor perturbations to the input voltage for about 2 μs after each step change, after which *Vin* reaches its open-circuit value. The period for the square wave was about 8 μs. That is, the oscillator frequency was set at about 125 kHz.

The program described in section 6.6 was used to create the waveform of Figure 7.6.4. Input variables were set to simulate an initial falling edge $T1$ at 150 ns and a rising edge $T2$ at 4.1 μs (see Figure 6.6.2). Output variable was selected to be $Vch1$ (see Figure 6.6.4).

Figure 7.6.5 is a photograph of the waveform displayed by channel 2 of the oscilloscope. The oscillations during the first 200 ns could be due to reflections emanating from the ground conductors of the test equipment. They could also be due to cable or connector discontinuities.

The program of section 6.6 was run again, with input variables set to simulate an initial falling edge $T1$ at 20 ns and a sweep time $T3$ of 1 μs, and the output variable was selected to be $Vdiff$. Figure 7.6.6 was the result. The period between the first and second edges of the waveform is twice the differential-mode propagation delay of the 15 m cable – the time taken for the reflected wave to return to its source.

The setup of Figure 7.6.1 was modified to route both conductors of the cable through the core of the current transformer. Antenna-mode current was now being monitored.

Running the program of worksheet 6.6 a third time, with the time $T1$ of the initial rising edge set at 120 ns, the time $T2$ of the falling edge set at 1.65 μs, and with $Vrad$ selected as the output variable, gave Figure 7.6.8. The rise and fall times of the simulation were set by the value of dt. From the worksheet of Figure 6.6.2:

$$dt = \frac{T}{N} = \frac{83 \times 10^{-9}}{100} = 830 \text{ ps}$$

Comparing Figure 7.6.5 with Figure 7.6.6 reveals that the simulated waveform of the differential-mode current $Idiff$ is very similar to the actual waveform. The most striking aspect is the fact that the waveform gradually changes from a square wave to a sine wave.

There are some deviations, however:

- The leading edge of the step on the observed signal is not quite as fast as the theoretical step. This is almost certainly due to the finite response times of the signal generator and oscilloscope.
- There is a ripple on the actual waveform which lasts for about 200 ns. This is probably due to reflected current returning from the conductors of the structure.

The similarity between Figure 7.6.4 and Figure 7.6.3 is just as significant. These compare the monitored voltage at channel 1 with the simulated voltage $Vch1$. The ripple during the first 2 μs following each step change is due to voltage drop in the source resistance Rg caused by current delivered to the cable.

Most notable is the fact that the corners of the leading and trailing edges are rounded. If the loading effect had been due solely to the current waveform displayed on Figure 7.6.6, then the corners would have been much sharper. The fact that they are rounded is a clear demonstration that the source resistance is carrying antenna charging current as well as differential-mode current.

There is one significant difference between Figure 7.6.3 and Figure 7.6.4. There is a sharp 'spike' after each step change of the monitored waveform. This is due to capacitive coupling between the output of the signal generator and the input to channel 1 of the scope. The two terminals connecting co-axial inner conductors to the 46 Ω resistors shown in Figure 7.6.1 were very close to each other. However, the existence of this spike does not obscure the fact that the ends of each step change are rounded.

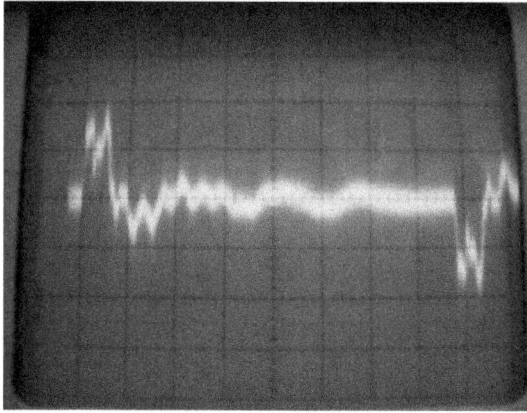

Figure 7.6.7   Antenna-mode current, monitored on scope (200ns/div, 2mV/div).

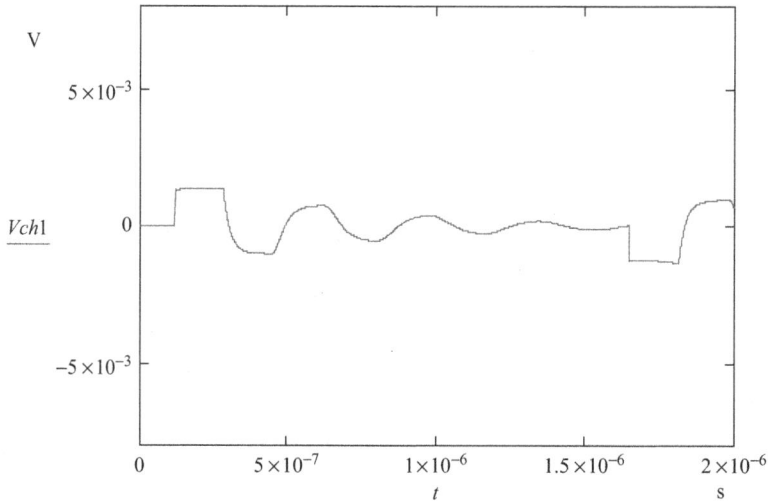

Figure 7.6.8   Simulated voltage at channel 2 input. Antenna current monitored.

Figures 7.6.7 and 7.6.8 also display significant similarities. These compare the waveform of the antenna-mode current carried by the cable with the waveform generated by the simulation. It is quite clear that the waveform resulting from the simulation follows the same basic pattern as that observed on the oscilloscope, that is, a damped oscillation that is a miniature version of *Idiff*.

However, there is also a higher frequency component superimposed on top of the damped oscillation. This lasts for many more cycles than the underlying waveform of Figure 7.6.8, indicating that the damping it experiences is insignificant. The probable cause is a current component that is oscillating backwards and forwards between cable and structure. This component is detectable as a ripple superimposed on the differential-mode waveform of Figure 7.6.5.

The fact that the test results and the simulated waveforms are closely correlated provides a fair degree of confidence in the reliability of the model. The waveform of figure 7.6.6 is certainly more representative of reality than that of Figure 6.2.6. A point worth remembering is that, although the capacitors and inductors of that simple model are replaced by the parameters of propagation time and characteristic impedance, these properties continue to exist. (Section 6.3 covers this aspect in more detail.)

Since the delay-line model has introduced the parameter of time into the computations, it becomes possible to think in terms of the energy stored in the line. That is, the static energy of electric charge and the kinetic energy of moving charges. In fact, (6.1.5) for an RLC circuit is very similar to the equation of motion of a mass on a spring.

The inclusion of the network of *Crad*, *Ro*, and *Inf* of Figure 6.6.1 enables the effect of energy storage caused by antenna-mode current to be separated from the effect of storage of differential-mode energy in the delay line. When a step voltage is applied to the open-circuit line, the differential-mode current delivered to that line is constant for a period equal to twice the propagation time, and so is the antenna-mode current. After that period, the capacitor *Crad* has been charged up and contains a lot of potential energy. The initial flow of current into *Crad* represents the kinetic energy required to deliver that charge.

The fact that the current waveform reverts rapidly from a square wave to a sinusoidal wave at the natural frequency of the assembly shows that the transient energy of the initial step has been converted into the dynamic energy of an L-C circuit. From Figure 7.6.5, the time for one cycle of this waveform is about 0.34 μs, corresponding to about 2.9 MHz. That is, a frequency that is close to the quarter-wave frequency of a 15 m cable acting as a monopole antenna. Figure 7.5.2 shows that, at the frequency of resonance (of a 15 m dipole), the radiation emitted by the assembly reaches a peak value. Since the amplitude of this sinusoidal wave of Figure 7.6.5 decays rapidly, it is reasonable to assume that this loss of energy is mostly due to radiated emission.

The inclusion of *Ro*3 in the model of Figure 6.6.1 allows the amplitude of the antenna-mode current itself to be evaluated.

The setup of Figure 7.6.1 offers another way of deriving values for cable parameters, simply by using data from the transient response. Since the voltage and current are almost constant during the first 166 μs, the value of the characteristic impedance *Ro* can be calculated quite quickly. Since the transit time is also known, the loop inductance *La* and loop capacitance *Ca* of the 15 m cable can also be derived, as illustrated in the worksheet of Figure 7.6.9. The worksheet also derives values for these reactive components, using data from the frequency response tests on the same cable.

	Transient analysis	**Frequency response analysis**
Capacitance between conductors	780 pF	929 pF
Loop inductance	8.8 μH	8.4 μH

Reasonably close correlation between these results shows that it is possible to obtain a first estimate of the values of both the inductance and the capacitance of a cable, just by using data from a couple of transient waveforms.

Worksheet 7.6.4

From the waveform of Figure 7.6.2, the amplitude of the step voltage measured by channel 1, at the end of the first exponential rise, is:

$$Vch1 := 0.39 \text{ V}$$

The voltage applied to the input terminals of the line is related to $Vch1$ by Equation (7.6.1)

$$Vin = \frac{96.2}{50} \cdot Vch1 = 0.75 \text{ V}$$

From the waveform of Figure 7.47, the voltage measured by channel 2 is:

$$Vch2 := 0.016 \text{ V}$$

The ratio of the amplitude of $Vch2$ to the amplitude of the current monitored by the current transformer is given by Equation (7.2.6)

$$RT := 2.27 \text{ }\Omega$$

Invoking Equation (7.2.5) $\quad Iout = \dfrac{Vch2}{RT} = 7.048 \times 10^{-3} \text{ A}$

The ratio of Vin to $Iout$ gives a measure of the characteristic impedance of the cable:

$$Ro := \frac{Vin}{Iout} = 106 \text{ }\Omega$$

The waveform of Figure 7.6.5 also gives a measure of the time taken for a current step to propagate to the far end and return to the near end. This is the time difference between the first and second edges of the waveform. This gives:

$$T := \frac{166 \times 10^{-9}}{2} \cdot \sec$$

The capacitance and inductance of the cable can be determined by invoking Equations (6.3.7) and (6.3.6)

$$Ca := \frac{T}{Ro} = 7.8 \times 10^{-10} \text{ F}$$

$$La := T \cdot Ro = 8.8 \times 10^{-6} \text{ H}$$

These results can be compared with those obtained from cable characterisation tests. From Figure 7.5.12

$$Ca\_ := 2 \cdot \frac{1}{2} \cdot 929 \cdot pF = 9.29 \times 10^{-10} \text{ F}$$

$$La\_ := 8 \cdot 1.05 \text{ }\mu H = 8.4 \times 10^{-6} \text{ H}$$

Figure 7.6.9 Calculating values of components derived for 15 m cable, using data from transient tests and frequency response tests.

It is useful to summarize the reasoning involved in the transient analysis and to identify its significance in the design of electronic equipment.

Any step voltage between two terminals of a cable will cause a transient wavefront to propagate along the cable at a velocity somewhat less than the speed of light. During the transit time, some of the energy will be radiated into the environment and some will be stored temporarily in the form of static charge.

The static charge between cable and environment will disappear in the form of a high-frequency signal which rapidly decays in amplitude (see Figure 7.6.7). This means that both the static energy in the charge and the dynamic energy used to create that charge are converted into radiated interference.

The charge between the conductors will eventually settle down to a steady-state value, but not before some of the energy is converted into differential-mode radiation (see Figure 7.6.5).

This will happen with every step change in voltage. Since any signal is effectively a continuous series of step changes, it is inevitable that a significant portion of the signal on a twin-conductor cable will radiate away in the form of EMI.

If the separation between send and return conductors is uncontrolled, then most of the transient energy will be converted into unwanted radiation. This is very useful for covert surveillance, but not much else.

The only way of minimizing the amount of electromagnetic energy in a transmission line which is lost to the environment is to use a co-axial cable, a screened pair, or a waveguide. If an unscreened twin-conductor cable is used, the only way of reducing the level of radiated energy is to absorb it in a resistive component. Sections 8.5 and 8.6 describe a range of design techniques that enable this objective to be achieved.

7.7 Capacitor characterization

It is entirely possible to use the technique of circuit modeling to create representative circuit models of components such as inductors, capacitors, and filters. The setup would be as illustrated in Figure 7.7.1.

In this instance, the component-under-test is a capacitor, mounted in the test jig. The jig is coupled to a signal generator using a 50 Ω co-axial cable. A similar cable is used to route the output signal to a signal monitor. A transfer function analyzer was used to provide the input signal and to monitor the output. Alternatively, the source could be a signal generator and the monitor could be an oscilloscope or a spectrum analyzer.

Figure 7.7.2 shows the general circuit model for this setup. Since the output impedance of the generator is 50 Ω and the characteristic impedance of the co-axial cable is 50 Ω, the cable

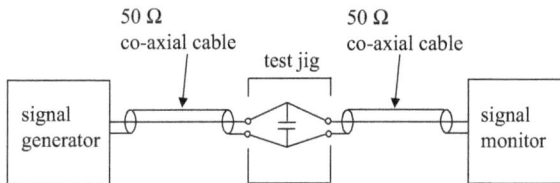

Figure 7.7.1   Test setup for characterizing components.

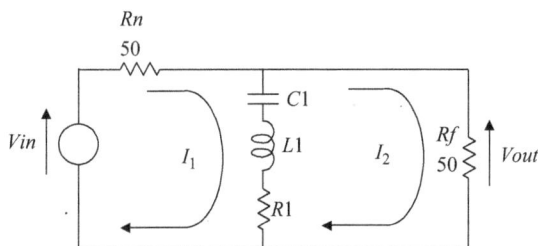

Figure 7.7.2   General circuit model for capacitor characterization.

between source and jig is effectively transparent. The same applies to the output link, that between jig and signal monitor.

Since all capacitors possess inductive and resistive properties, the simplest model for this particular component-under-test is the series LCR circuit. For this type of test, the relevant parameter is the transfer impedance.

$$ZT = \frac{Vout}{I_1} \qquad (7.7.1)$$

So the test procedure would be to apply a constant-voltage variable frequency to the test jig and note the amplitude of the output signal at a set of spot frequencies. This gives the frequency response of the transfer impedance ZTt, that is, the transfer impedance derived from test results.

It is a simple matter to write a program to calculate the frequency response of the transfer impedance ZTm of the circuit model. Both responses can then be plotted on the same graph. Initial guess values are assigned to $C1$, $L1$, and $R1$, and these parameters are then treated as input variables. Adjusting $C1$ will bring the negative slopes of both curves into close cor-

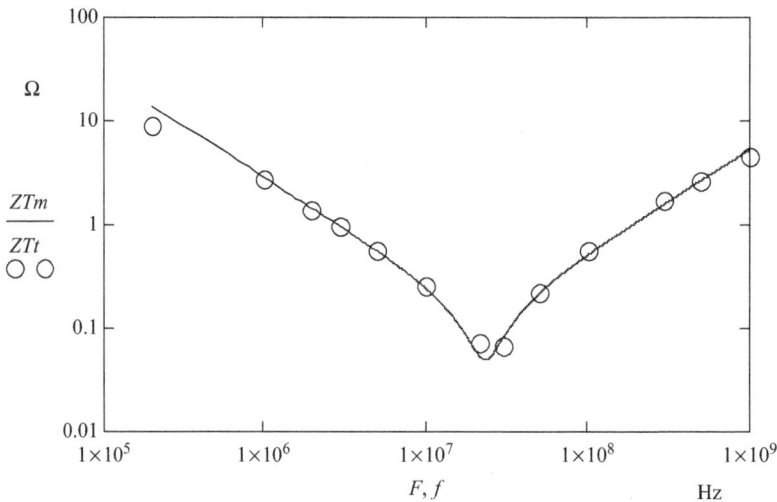

Figure 7.7.3   Transfer impedance of test setup and circuit model.

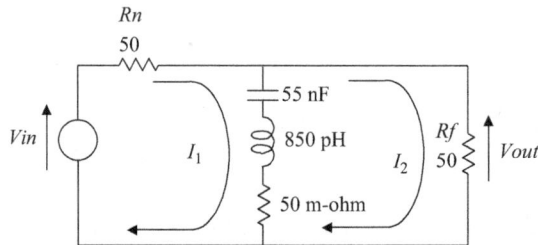

Figure 7.7.4   Representative circuit model of Murata 100 nF capacitor-under-test.

Worksheet 7.7, page 1

$$D = \begin{pmatrix} 0.2 & 430 & -9020 & \text{"58nF"} \\ 1 & 190 & -2700 & \text{"53nF"} \\ 2 & 93 & -1400 & \text{"53nF"} \\ 3 & 71 & -975 & \text{"54nF"} \\ 5 & 30 & -554 & \text{"54nF"} \\ 10 & 30 & -251 & \text{"62nF"} \\ 21.6 & 49 & -53 & \text{"0"} \\ 30 & 30 & 60 & \text{"396pH"} \\ 50 & 27 & 221 & \text{"696pH"} \\ 100 & 79 & 553 & \text{"877pH"} \\ 300 & 400 & 1670 & \text{"881pH"} \\ 500 & 944 & 2500 & \text{"793pH"} \\ 1000 & 2500 & 3700 & \text{"586pH"} \end{pmatrix}$$

column 1: frequency, MHz
column 2: resistance, mΩ
column 3: reactance, mΩ
column 4: text

$s := 1..\text{rows}(D)$

$fs := D_{s,1} \cdot 10^6$

$R_s := D_{s,2} \cdot 10^{-3}$

$X_s := D_{s,3} \cdot 10^{-3}$

$ZTt_s := \left| R_s + j \cdot X_s \right|$

Data copied from:
www.ediss-electric.com/bypass_caps_pdf/1_100n0201X5R_murata.pdf

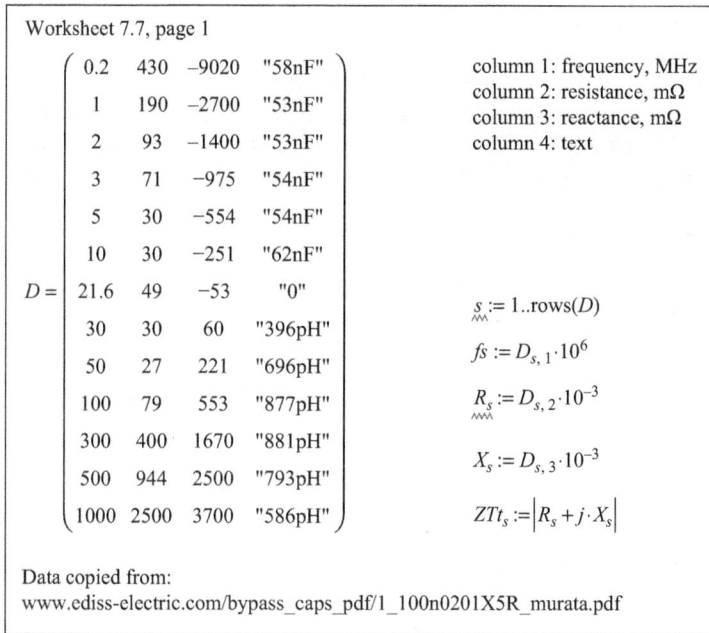

Figure 7.7.5 Calculating frequency response of transfer impedance from test data.

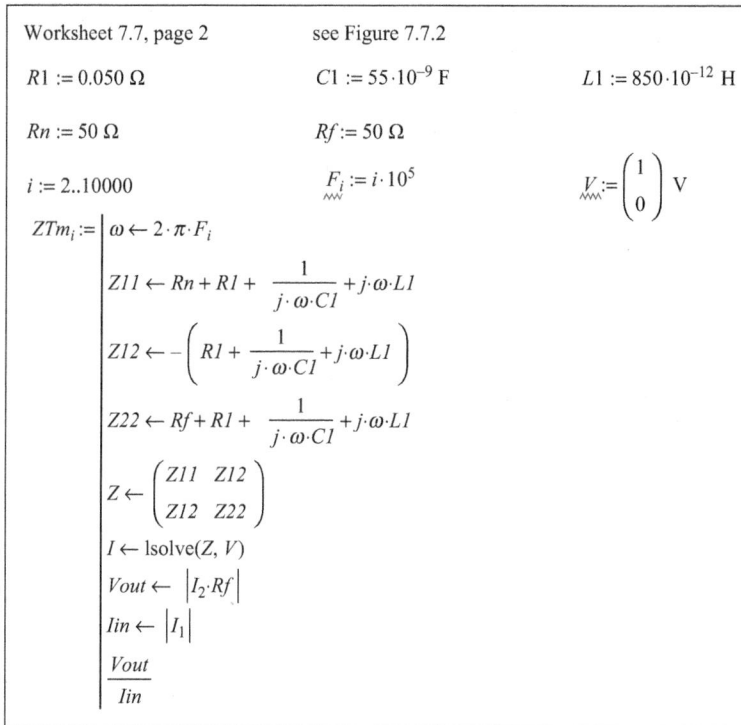

Worksheet 7.7, page 2   see Figure 7.7.2

$R1 := 0.050\ \Omega$ $C1 := 55 \cdot 10^{-9}\ \text{F}$ $L1 := 850 \cdot 10^{-12}\ \text{H}$

$Rn := 50\ \Omega$ $Rf := 50\ \Omega$

$i := 2..10000$ $F_i := i \cdot 10^5$ $V := \begin{pmatrix} 1 \\ 0 \end{pmatrix}\ \text{V}$

$ZTm_i := \Bigg|$ $\omega \leftarrow 2 \cdot \pi \cdot F_i$

$Z11 \leftarrow Rn + R1 + \dfrac{1}{j \cdot \omega \cdot C1} + j \cdot \omega \cdot L1$

$Z12 \leftarrow -\left(R1 + \dfrac{1}{j \cdot \omega \cdot C1} + j \cdot \omega \cdot L1 \right)$

$Z22 \leftarrow Rf + R1 + \dfrac{1}{j \cdot \omega \cdot C1} + j \cdot \omega \cdot L1$

$Z \leftarrow \begin{pmatrix} Z11 & Z12 \\ Z12 & Z22 \end{pmatrix}$

$I \leftarrow \text{lsolve}(Z, V)$

$Vout \leftarrow \left| I_2 \cdot Rf \right|$

$Iin \leftarrow \left| I_1 \right|$

$\dfrac{Vout}{Iin}$

Figure 7.7.6 Calculating frequency response of transfer impedance of circuit model.

relation, adjusting $L1$ will do the same for the positive slopes, and adjusting $R1$ will align the minimum values.

When the two curves are as closely correlated as possible, the new values of $C1$, $L1$, and $R1$ can then be assigned to the relevant component in the general circuit model. This gives the representative circuit model for the capacitor-under-test. This analytical process was carried out using data from a test on a 60 nF capacitor, one of many tests carried out by Roy Ediss [7.1].

Figure 7.7.3 illustrates the final graph and Figure 7.7.4 shows the representative circuit model. A copy of the Mathcad worksheet used to create the graph is provided by Figures 7.7.5 and 7.7.6.

CHAPTER 8

Practical design

There are a number of design concepts and techniques in existence which are aimed at improving EMC. Many of them are supported by the analytical approach. But some are not.

A concept that can be traced back for many decades is that of the 'equipotential ground plane'. There is no such thing. Since the 'method of images' of electromagnetic theory is often quoted as the basis for this concept, it is useful to identify the fallacies underlying the correlation. Section 8.1 does just that. Designers should resist the temptation to treat the structure as a convenient return path for signals and power. The most effective use of the conducting structure is as a shield.

One deduction which is continually reinforced during the process of creating circuit models is the fact that the signal and return conductors should be as close together as possible.

Section 8.2 shows that this requirement is implemented to good effect in the design of printed circuit boards by routing the signal tracks as close to the ground plane as possible. It also shows that, for signals between boards and between equipment, this requirement is best achieved by allocating a return conductor to every signal or power conductor.

In terms of EMC, one of the most counterproductive concepts is that of the 'single-point ground'. Section 8.3 illustrates how this technique guarantees the creation of intractable problems. In contrast, the existence of a multitude of 'ground loops' causes a significant reduction in the level of interference.

Improvements achieved by the inclusion of dedicated return conductors can be nullified by inappropriate design of the interface circuitry of equipment units. Section 8.4 identifies the need for current balance in the interconnecting cables, and illustrates a few methods of achieving this objective.

Section 8.5 focuses on the fact that optimum transmission is achieved by terminating transmission lines with a resistance equal in value to the characteristic impedance. If the signal is transmitted efficiently, it follows that emissions due to interference are minimal. Circuit modeling confirms that a configuration which exhibits low emission is less susceptible to external interference.

Whether the common-mode loop is open-circuit at one end or short-circuited at both ends, it is inevitable that there will be reflections. Electrical energy will be stored temporarily in this loop and will quickly depart into the environment in the form of interference. It is possible to absorb this energy and damp down the oscillations. Section 8.6 describes ways of doing this.

Section 8.7 deals briefly with the subject of equipment shielding and identifies the basic requirements. Measures used to protect buildings and equipment from the indirect effects of lightning are also described.

Every technique described in this chapter defines the interface circuitry at both ends of the signal link. This provides visibility of both the signal loop and the common-mode loop. Without such visibility, there is no point in even trying to assess the interference coupling characteristics, let alone analyze them.

By defining the interface circuits at both ends of the link, an important feature of printed circuit board design is identified; the interface circuits provide a buffer between the internal wiring of each board and the cabling used to carry signals between boards. This means that the interface circuits can be designed to handle the higher levels of interference which are bound to exist on the longer conductors.

The length of the signal link is an important parameter. The longer the link, the lower is the frequency at which interference reaches its highest peak. Since the power available is inversely proportional to the square of the frequency, the longest links deliver the highest level of interference energy. The buffer circuits on the printed circuit boards should be able to absorb whatever unwanted energy arrives via the signal link.

8.1 Grounding

Reliance on the use of the conducting structure as the universal return path for all signals and all supplies is probably the most prevalent cause of EMC problems. This could be due to the widespread belief in the existence of the equipotential ground plane. There is no such thing.

Ground planes are an extremely useful design feature of printed circuit boards and integrated circuits. But this does not mean that they are equipotential surfaces. Nor does it mean that a conductor designated as 'ground' or 'earth' is automatically a zero voltage reference point for all signals in the system.

The concept of a surface at which all points are at zero voltage probably arises from a misunderstanding of the method of images used in electromagnetic field theory. So it is a useful exercise to review the reasoning behind the technique.

The image solution method is usually illustrated by considering an infinite length of line charge parallel to a grounded conductor of infinite length [3.1]. It is possible to simulate the electric field distribution by replacing the ground plane with an insulating surface and placing an image conductor the same distance below the plane as the actual conductor is above the surface.

Figure 8.1.1 illustrates a configuration where two conductors are routed over a plane surface. A steady voltage is applied between the near end of conductor 1 and the near end of the ground plane (conductor 3). The far end of conductor 1 is open-circuit. Conductor 2 is short-circuited to ground at both ends. Contour lines, depicting intermediate voltages, are also shown on the figure.

Since the voltage applied between conductor 1 and the ground plane is constant, no current is flowing. Under these circumstances, the ground plane is indeed an equipotential surface. But with no current flowing there can be no signal and no interference. It is also useful to note that the contour lines do not intersect conductor 2. The field pattern is asymmetrical.

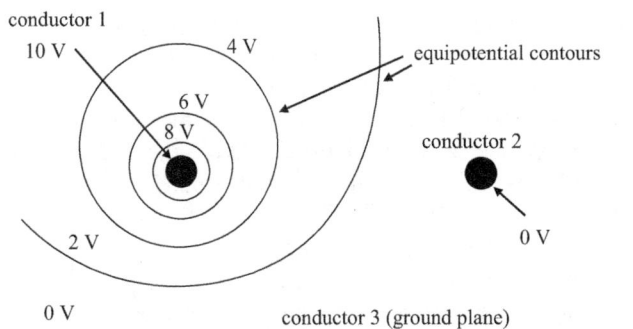

Figure 8.1.1 Two conductors over a ground plane.

If, now, the far end of conductor 1 is short-circuited to the ground plane and a current at constant rate of change is caused to flow in that conductor, returning via the ground plane, then, although the actual voltages would be different, the pattern of contour lines would remain exactly the same. However, there would be voltage difference between each cross section.

If the method of images had been used to create this field pattern, it would have been assumed that the image of conductor 1 had been carrying exactly the same current. This means that the voltage developed along conductor 1 would be balanced by a voltage along the image conductor. Since this image conductor represents the effect of the plane, it follows that the voltage developed along the ground would be the same magnitude as the voltage along conductor 1.

The coupling mechanism can be analyzed by assuming that a low-frequency sinusoidal waveform is applied to the near-end terminals of conductors 1 and 3, and that the short-circuit link between the near end of conductor 2 and ground is replaced by a voltmeter. Figure 8.1.2 is a circuit diagram depicting the inductive properties of the three conductors. A sinusoidal voltage $V2$ would be developed between the near-end terminals of conductor 2 and ground. Since the impedance of a voltmeter is high, then minimal current would flow in conductor 2, and $V2$ would be a measure of the voltage along $L3$, the inductance of the ground plane.

If a voltage is developed along the ground plane, then that conductor cannot possibly be regarded as an equipotential surface.

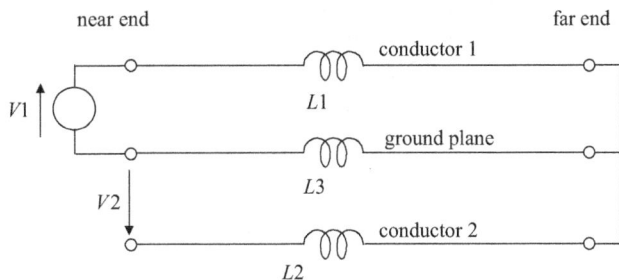

Figure 8.1.2 Inductive properties of two conductors over ground plane.

If the frequency of the voltage source is increased, capacitive coupling becomes more significant. Also, proximity effects will cause the return current in conductor 3 to flow through a restricted cross section. The resistance presented by the ground will also increase. The end result is that the simple model of Figure 8.1.2 can best be modeled by the triple-T network of Figure 2.7.6. The ground plane can be a source of both electric and magnetic field coupling.

Acceptance of the concept of the equipotential ground has inevitably led to further misconceptions. Notable among these is the belief that all connections made to the 'ground' conductor are automatically held at zero voltage.

In circuit diagrams, it is often desired to define that a connection exists between two separate points without the need to draw a line between those points. Normal convention is to identify those points with a unique symbol. This avoids the use of 'tramlines' crisscrossing the diagram, and allows the reader to focus attention on the functional purpose of the circuit. The most frequently used symbols are for 'ground' or 'earth'. From a functional point of view, it is assumed that any voltage difference between such points can only have a marginal effect on circuit function. It is then assumed that, for all practical purposes, there is no voltage difference.

However, such an assumption is only valid if the area enclosed by the relevant circuit loop is extremely small. This means that there are few problems when the convention is used with dual-layer or multilayer boards. In these assemblies, the traces used to link components are very close to the ground plane. Hence, the area enclosed by the differential-mode loop is always very small. The ground plane provides a return path for the current, just where it is needed. Section 8.3 provides more information on this aspect of design.

But that is not the case with single-sided boards. When such a board is used, great care is needed to ensure that the loop area involved is as small as possible. To do this, it is necessary to use solid lines to depict conductors on the board. Symbols indicating the presence of hidden connecting links have no place on such a diagram.

If it transpires that the conductive trace linking two points on a single-sided board follows a long circuitous route, then there is an increased probability that the signal in question will be prone to interference problems. If the designer does not have visibility of this track, then the existence of an interference problem will come as a surprise.

At the time when electronic circuitry was based on the use of valves, the components were usually mounted on an aluminum chassis and the circuit diagram depicted this conductor as a solid line at the bottom of the page. The power supply rail was also included at the top of the page, identifying the R-C networks used to minimize the interaction between each stage of the supply from the next. Visibility of every circuit path was maintained. This allowed the designer to minimize the area of every circuit loop, whether that loop carried power supply current, signal current, or both. Care was taken to ensure that high current in the output stages did not affect the sensitive circuitry at the front end.

If this disciplined approach was utilized during the design of single-sided printed circuit boards, then a significant improvement in the EMC of that particular circuit function could be achieved. In the case of single-sided boards, such an approach might be desirable. In cases where the connecting links are between boards of an equipment unit, this approach would be extremely important. In cases where the link is between equipment units, such an approach is essential.

Assumption of the existence of an equipotential conductor is evident in every circuit diagram which includes the ground symbol or any of its variants. This may be reasonable if the purpose is to describe the function of the circuit, but totally counterproductive if the objective is to analyze EMC. The existence of these symbols means that the return path for every signal current is undefined. The same applies to power supply currents.

If half the circuitry is undefined at the initial design stage, then subsequent stages such as detailed drawing, component manufacture, wiring, and assembly are left to the discretion of whoever is carrying out the task in question. The EMC of the system is effectively out of control. By the time the completed system is submitted for EMC testing, the only thing to do is cross one's fingers and hope for the best.

If the physical relationships between the signal and return conductors are defined at the initial stage of the project, then the designer is able to use all the analytical tools described in the preceding chapters and all the techniques described in the following sections. Visibility of this relationship can be maintained at all subsequent stages of development.

It can be concluded that the conductors of the structure should not be used as a convenient return path for signals and power. Any current flowing along the structure will create a voltage along that length and any voltage between different locations on the structure is a source of interference.

Up to now, the focus of this section has been on what the structure should not be used for. To identify just what it should be used for, it is necessary to go back to Figure 8.1.1. Here, it shows that neither the equipotential contours of an electric field nor the magnetic lines of force of a magnetic field penetrate the ground plane. Any circuitry on the underside of this barrier is shielded from the worst effects of the electromagnetic field above. From the point of view of EMC, this is a desirable design feature.

Another significant aspect of the magnetic field distribution emerges. The lines of force do not link with conductor 2. This is because a loop has been formed between this conductor and ground. Transformer action ensures that the current in this loop creates an opposing magnetic field that precisely balances the field emanating from conductor 1. That is, this secondary loop tends to act as a shield that minimizes the penetration of the magnetic field to the right of conductor 2.

This action is put to good use in the design of Faraday cages and the assembly of conductors used to protect buildings from the worst effects of lightning. So, it can also be reasoned that, even where the conductors of the structure do not present an impenetrable surface, they still perform a very useful shielding function.

If the structure is designed to form a lattice network of conducting loops, then any interference will create a circulating current in each loop, and the field created by these circulating currents will tend to balance the incoming interference. Such a structure is behaving as shield which radiates unwanted electromagnetic fields back into the environment.

8.2 Conductor pairing

As indicated in section 8.1, the concept of the equipotential ground conductor is extremely attractive to those involved in the design of systems which are installed on a conductive framework such as on vehicles, aircraft, or spacecraft. It is reasoned that, since the framework provides a very convenient return path for all currents in the system, there is no need to install unnecessary return conductors which cost money and add mass.

This impression is reinforced by the known effectiveness of the ground plane in multilayer printed circuits, where it does seem to behave as an equipotential surface. Since the reason for this is not immediately apparent, it is useful to review the behavior of the ground plane in a little more detail.

Figure 8.2.1 illustrates part of a printed circuit board assembly where two adjacent tracks are connected to two voltage sources and the far ends are short-circuited to the ground plane.

Using the technique of composite conductors described in Chapter 3, it is possible to determine the distribution of current in a cross section of the printed circuit board.

If it is assumed that the voltage sources apply step voltages of 1 V of opposite polarity simultaneously to both tracks, then the rate of change of current in all three conductors can be calculated using the process described in Chapter 3. The resultant distribution in all three conductors is illustrated in Figure 8.2.2.

It is evident that the current in the left-hand track is positive and that in the right-hand track is negative. Also, the rate of change of current in each track is unevenly distributed, with the greatest rate of change occurring at the outer edges.

However, the most significant feature of the illustration is the fact that most of the current in each track returns via the area of the ground plane that is immediately adjacent. Effectively, the ground plane provides a return path for the current precisely where it is needed.

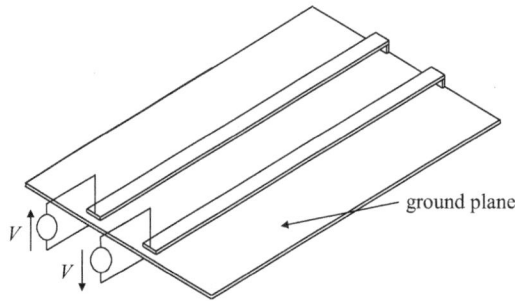

Figure 8.2.1   Two printed circuit tracks over a ground plane.

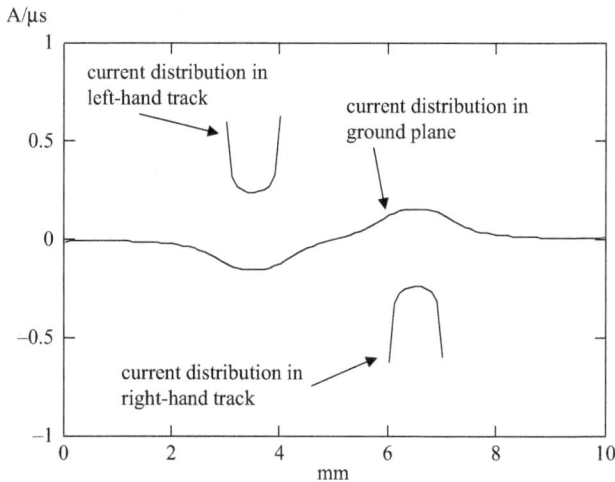

Figure 8.2.2   Distribution of current in cross section of pcb.

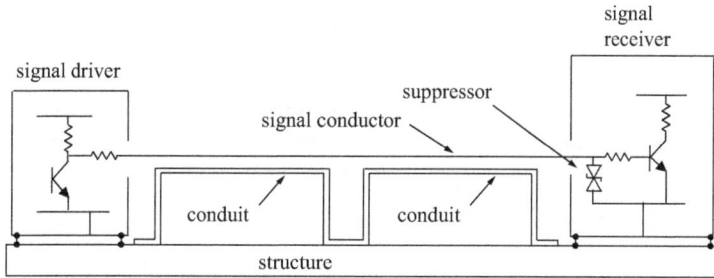

Figure 8.2.3 Using the framework as a return conductor.

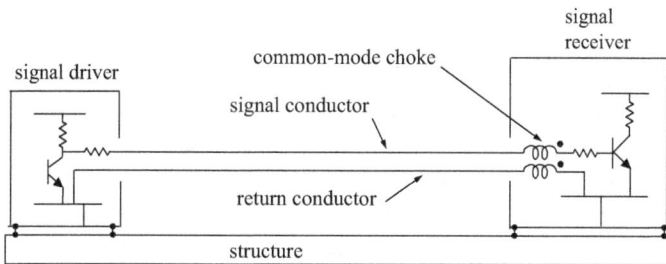

Figure 8.2.4 Allocating a dedicated return conductor.

This effect is also present in systems where the supporting framework is constructed of conducting material. In such a system, it is possible to route interconnecting cables along that framework to utilize the structure as a return conductor. Figure 8.2.3 illustrates such a setup.

However, it is never possible to match the close coupling achieved on a printed circuit board. It is inevitable that wide gaps will occur between cable and structure, and the area enclosed by the current loop can become quite large. Stray coupling is increased. Reducing the level of this unwanted coupling can only be achieved by minimizing the separation between cable and structure. Adding cable trays or cable conduits as shown on Figur 8.2.3 will help toward this objective. Effectively, the structure is being brought closer to the cable.

Even so, the signal link is still quite susceptible to interference transients. The longer the cable, the higher the level of interference. It becomes advisable to include suppressors at the interface circuitry to prevent high-voltage transients from damaging the more sensitive semiconductor devices.

Although suppressors can protect the equipment from damage, they cannot prevent interference pulses from getting through to the processing circuitry. Incorrect messages can be received. Data can be corrupted.

If, at the initial design stage, it is decided to allocate a return conductor to every signal conductor, then this will create two separate loops, the differential-mode loop and the common-mode loop. Analysis of this configuration is provided in section 2.8, where it is concluded that the signal and return conductors should be held as closely together as possible along the entire route from driver to receiver. It is much easier to hold two conductors together than to minimize the spacing between cable and structure.

Given this configuration, there are many ways of minimizing interference coupling. For example, the suppressor can be replaced by a common-mode choke, as shown in Figure 8.2.4.

This significantly reduces the amplitude of interference pulses in the common-mode loop while having minimal effect on performance.

Also, it is very helpful to reduce the area of the common-mode loop by routing the signal/return pair very close to the structure. It generally reduces the antenna efficiency of that loop.

8.3 Ground loops

The single-point ground is another concept that is well past its sell-by date. Such a configuration is shown in Figure 8.3.1, where three different printed circuit boards are used to process the signal. Each board is supplied with a steady-state voltage of 15 V, derived from a stabilized supply module.

The intent of the single-point ground is to prevent the supply return currents from sharing a common conductor. It is reasoned that if there is no common conductor along which a voltage can be developed, then there will be no interference. This could possibly be the case if the current drawn by each printed circuit board is constant in value. However, this never happens in practical situations; current in every conductor is continuously changing.

Loop areas enclosed by the signal currents are extremely large. Moreover, the loops formed by signal and supply currents share a common area. Magnetic flux will link with both loops. Transformer action will ensure that high levels of self-induced interference exist between the signals carried between boards. That is, intra-system interference is likely to cause significant problems. Such a high level of sensitivity does not bode well for the time when the completed equipment is subjected to formal EMC testing.

Figure 8.3.2 illustrates how the performance can be dramatically improved. Changes to the configuration are twofold:

- The supply conductors are kept close together.
- A return conductor is allocated to each signal conductor.

Comparing Figures 8.3.1 and 8.3.2 enables the essential difference between these two configurations to be identified. The latter circuit contains ground loops.

Ground loops can be treated as common-mode loops. Similarly, signal loops and power supply loops can be treated as differential-mode loops. An important feature of this configuration is the fact that it is possible to minimize the loop area of every differential-mode loop.

The way is now open for the circuit analysis techniques described in the previous chapters to be employed. Assessment of the signal links in the light of section 4.3 should allow potential problems to be identified. Common-mode rejection can be included in the design of the interface circuitry.

It is also worth noting that on a multilayer circuit board, the ground plane caries the return current for every signal current carried by the copper tracks on the adjacent surface. This gives rise to a large number of circulating currents. Figure 8.2.2 illustrates the fact that the ground conductor does indeed carry currents which flow in opposite directions, and the only way for this to happen is for a loop current to exist in the flat conducting surface. The ground plane is effectively a large number of ground loops.

The protection afforded by ground loops is even more important at system level, where cables are used to route signals between equipment units. If each unit of equipment is shielded, as illustrated in Figure 8.7.1, if every cable is shielded by a screen braid, and if

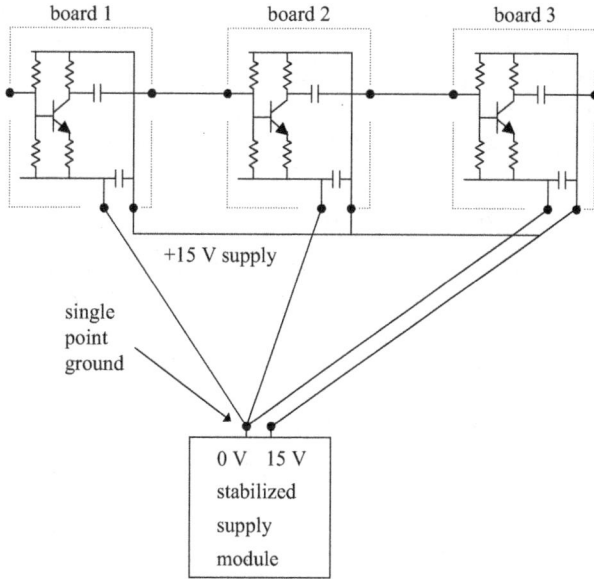

Figure 8.3.1 Single point ground.

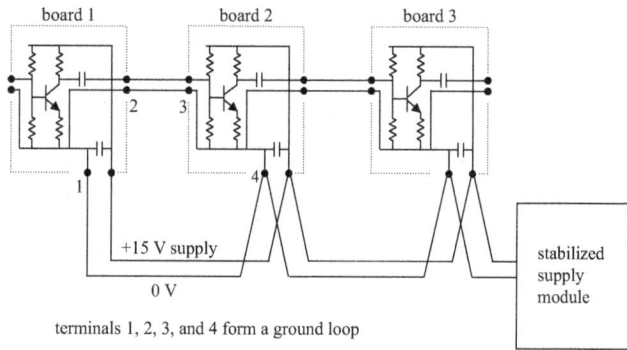

Figure 8.3.2 Introduction of ground loops.

those braids are bonded at each end to the units they interconnect, then the entire system will be shielded. Circuitry within the system will be protected from the worst effects of external fields, and radiated emission will be attenuated by the shield barrier. In such a configuration, there will be a multitude of ground loops.

It does not matter whether the link-under-review is on a printed circuit board, interconnects two boards, or interconnects two units of equipment, the existence of the ground loop is an essential part of good EMC design.

In large systems, it is also good practice to route the cables as close to the conducting structure as possible. Any high-amplitude transient in a conducting member of the structure, such as those due to a lightning strike, will create an electromagnetic field that will couple with every conductor routed alongside that member. This will induce essentially the same voltage in series with every conductor of the cable. Differential voltages will be minimal, and interference will be reduced.

8.4 Common-mode rejection

The configuration illustrated in Figure 8.3.2 is perfectly adequate for the vast majority of signal links inside units of equipment. However, there could be significant interference problems when the links are between widely separated units of equipment. There are also situations where unwanted transient signals appear inside a single unit. Such interference manifests itself as large spikes on an oscilloscope screen. The amplitude of these spikes can usually be reduced by the introduction of common-mode rejection circuits.

8.4.1 Differential amplifier

Figure 8.4.1 illustrates one of the simplest methods, the differential amplifier. This is an operational amplifier with the feedback and input networks organized such that:

$$V3 = V1 - V2 \tag{8.4.1}$$

That is, the output voltage of the amplifier in the signal receiver is a function of the voltage difference between the two input terminals. Equally important is the relationship between the signal and return current. Ideally, this is:

$$Isignal = Ireturn \tag{8.4.2}$$

If this condition is achieved, then any voltage developed along the signal conductor is precisely balanced by an equal and opposite voltage developed along the return conductor.

In theory, this means that the signal passed between the transmitter and the receiver is immune to any interference signal appearing in the common-mode loop. That is, the source Vcm has no effect on the transmitted signal.

In practice, this is not the case. Several factors are involved. All four resistors should be the same value, within a close tolerance. To minimize reflections, the value chosen for $R1$

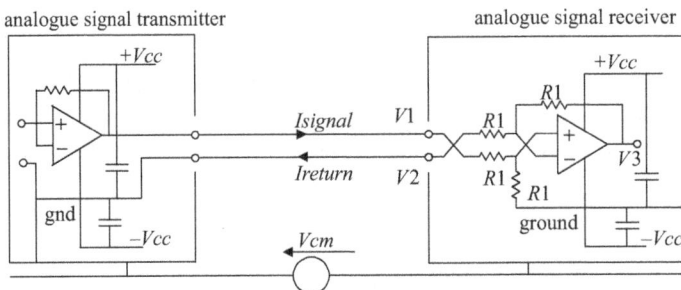

Figure 8.4.1 Differential amplifier.

should be the same as the characteristic impedance of each conductor. At high frequencies, the level of rejection also depends on the length of the two cable conductors being identical. If the amplitude of the interference exceeds the linear range of the amplifier, signal distortion will occur. Common-mode rejection is also limited by the bandwidth of the amplifier.

However, these considerations do not really diminish the usefulness of the differential amplifier. Protection from common-mode interference is provided over a wide range of frequencies. The useful bandwidth is from zero frequency to the upper frequency limit of the amplifier, and some amplifiers have a very fast response indeed.

Such a configuration is amenable to analysis using the techniques described in the previous chapters. Figure 2.7.6 illustrates how the connections between the two units can be modeled.

8.4.2 Differential logic driver

Figure 8.4.2 illustrates one way of transmitting logic signals between widely separated units of equipment. A screened pair is used to shield the signal and return conductors.

The amplifiers in the differential logic driver are fed by complementary inputs from a bi-stable circuit. When the non-inverted signal is 'high', the inverted signal is 'low', and vice versa. These two inputs are fed to the comparator at the differential logic receiver.

In the absence of any connection to the receiver, the two input terminals of the comparator would be held at half the supply voltage, Vcc, by the two potential dividers. (This means that some hysteresis needs to be built into the amplifier.) When connection is made to the transmitter, there will be current flowing from the transmitter to the receiver down one conductor, and an equal amount flowing from receiver to transmitter down the other, assuming that the supply voltage is the same in both items of equipment.

Electromagnetic coupling between the conductors will maintain this balance during transient conditions. As with analogue signal transmission, the response will be limited only by the capabilities of the active devices.

Since both the 'inverted' and 'non-inverted' signal voltages are referenced to 'ground', a connection is needed between the units to carry any spurious transient current during a change in the state of the logic signal. A three-conductor cable is needed. This requirement is best met by the installation of a screened twisted pair.

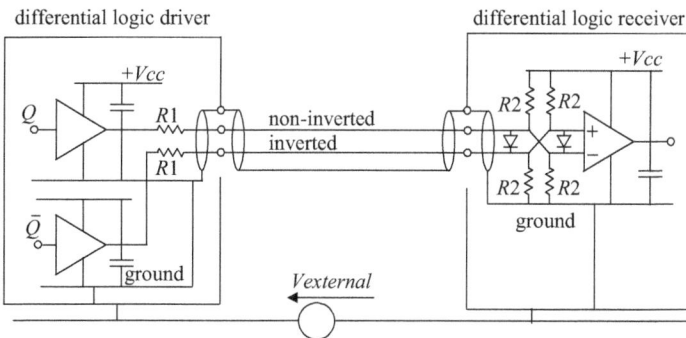

Figure 8.4.2 Differential logic signal transmission.

It will be noted that the screen is connected to the shield of both units of equipment as well as to the grounded conductors. Since this balance depends on the cable conductors being of equal length and the values of the resistors matching precisely, there will always be some coupling of common-mode voltage into the output signal.

8.4.3 Differential analogue driver

The screened pair can also be used to carry an analogue signal. In this case, the driver would be as shown in Figure 8.4.3. The advantage of using this configuration is the fact that the voltage output of the driver is balanced with respect to local ground, as is the input current to the receiver. At high frequencies there will be significant common-mode coupling with the screen. Since both the differential-mode voltage and differential-mode current are balanced with respect to the screen, the voltages due to common-mode coupling will tend to balance each other out. Since perfect balance is unachievable, there will always be some common-mode current in the screen.

If the units are not shielded, then the cable screen will still connect the power supply returns.

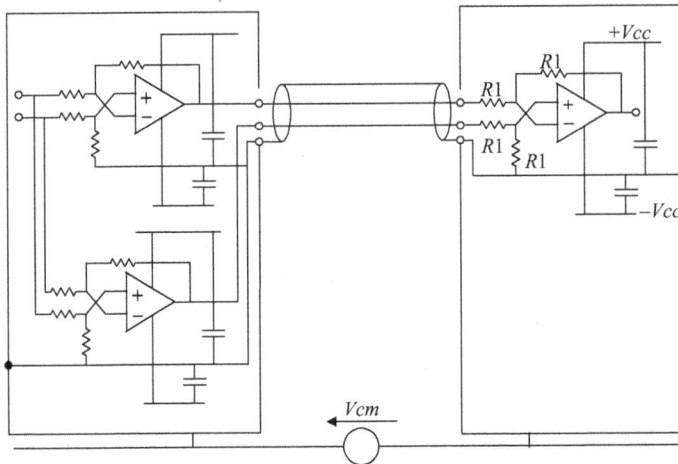

Figure 8.4.3   Differential analogue driver.

8.4.4 Common-mode choke

Figure 8.4.4 illustrates the use of the common-mode choke. Transformer T1 consists of a number of turns of twin-conductor cable, bifilar wound onto a ferrite core. As far as the common-mode current is concerned, the transformer behaves as an inductor.

With a differential-mode signal, the currents in the two conductors are equal in value but opposite in direction. Since the inductances of the two conductors are almost identical, the voltages induced along the cable will tend to cancel out. As far as the differential-mode signal is concerned, the impedance presented by the transformer is negligible. However, some minimal coupling between common-mode and differential-mode signals is bound to exist.

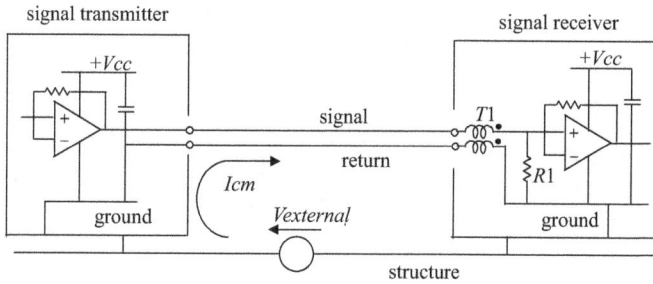

Figure 8.4.4 The common-mode choke.

Since the transformer is transparent to differential-mode signals, its effect on the common-mode loop can be modeled by inserting an inductance in series with the structure.

There will be a small amount of flux linking each conductor which does not link the other. In the classical model of a transformer there would be a primary inductance and a secondary inductance. It is important that a bifilar winding is used because this will minimize the primary and secondary inductances. Also, the balancing action is enhanced by the capacitive coupling.

Since the transformer behaves as an inductor in the common-mode loop, its effectiveness will increase with frequency. However, this also means that there will be little or no common-mode rejection at the low-frequency end of the response curve.

At the upper end of the response, capacitive effects come into play, and these will limit the level of rejection offered to unwanted signals. There is also a distinct possibility that resonances will occur; a transformer acts as an open-circuit at high frequencies. 'Soft' ferrites which behave resistively can help to ameliorate this effect. Hence, it is a useful exercise to identify the frequencies at which the common-mode rejection is minimal. Section 4.3 helps with this task.

The common-mode choke can sometimes be useful in minimizing troublesome effects after the equipment has been manufactured. Split-core ferrites are available, and these can be used to clamp round cables.

8.4.5 Transformer coupling

Another way of achieving current balance between the signal and return conductors is to break the connection between return conductor and ground at the receiving end of the line.

With analogue signals, a transformer could be used, as illustrated in Figure 8.4.5.

The problem with transformer coupling is that the bandwidth of the signal is limited. The purpose of the resistor across the primary winding is to increase the useful bandwidth. However, if the purpose of the link is to transmit a narrow-band signal, then the transformer is the obvious choice.

This configuration gives excellent common-mode rejection at low frequencies. However, the immunity to interference reduces with frequency, due mostly to the inter-winding capacitance of the transformer. Resonance can occur at the quarter-wave frequency of the line, and the interference can actually be amplified.

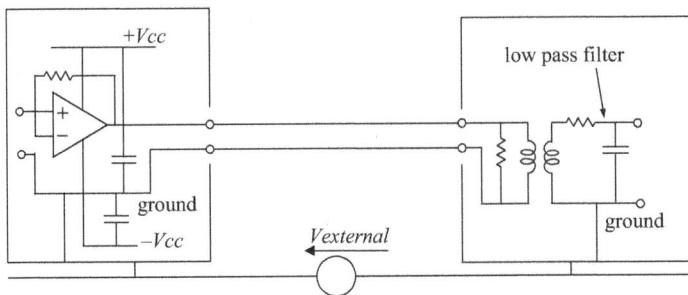

Figure 8.4.5   Floating transformer at receiving end of the line.

The response to interference in the common-mode loop is very similar to the curve of Figure 4.3.6. To avoid problems with high-frequency coupling, it is necessary to ensure that the upper frequency of the signal is significantly less than the quarter-wave resonance of the cable. Also, it is good idea to include a low-pass filter at the receiver to limit its bandwidth to that of the expected signal.

8.4.6 Center-tapped transformer

A transformer can also be used in the signal transmitting circuitry. Figure 8.4.6 illustrates how a center-tapped transformer can be used to create a balanced output. The resistor in parallel with the primary winding is needed to cater for the fact that the input impedance of the transformer is frequency dependent.

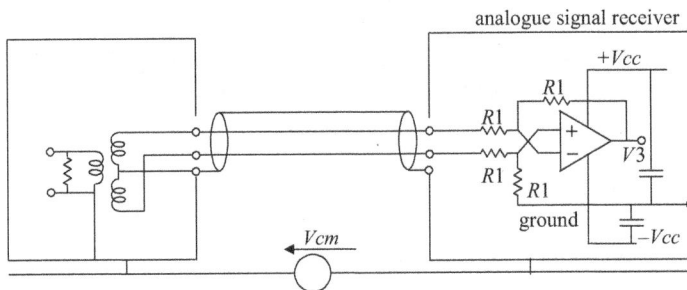

Figure 8.4.6   Transformer driver.

8.4.7 Opto-isolator

With logic signals, isolation between the return conductor and local ground can be achieved by using an opto-coupler, as illustrated in Figure 8.4.7.

Again, it is necessary to guard against high levels of interference at the quarter-wave frequency of the cable.

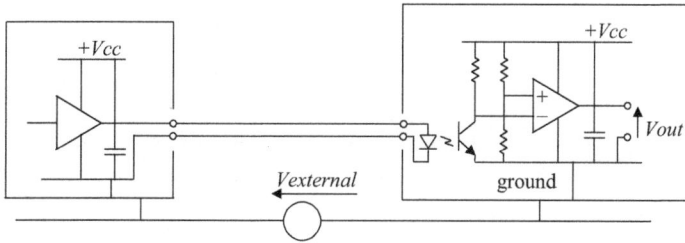

Figure 8.4.7 Opto-isolation for logic signals.

8.5 Differential-mode damping

Although transmission-line equations are not often used in the design of low-frequency circuits, they still apply at those frequencies. There is no cut-off point in the spectrum below which transmission-line theory ceases to be valid.

One important concept is that of characteristic impedance. Optimum transmission is obtained when the input impedance of the receiver is equal to the characteristic impedance of the line. There is no reflection of energy. If the source resistance is also equal to the characteristic impedance, then there are no reflections at either end. This is standard practice when dealing with radio-frequency signals. It is worthwhile considering the use of the characteristic impedance to terminate the lines in the configurations of Figures 8.4.1 and 8.4.2.

If the line is operating at maximum efficiency, then it must also be experiencing minimum loss. Since much of the loss is due to radiation, it follows that by terminating the line correctly, minimum radiation will occur.

It is shown in section 2.9 that the transfer admittance for conducted susceptibility is exactly the same as for conducted emission. Hence, this configuration also experiences minimum susceptibility.

8.5.1 Transient damping

By its very nature, logic circuitry creates a continuous succession of transient steps in the current delivered by the power supply rails on a printed circuit board. These transients have minimal effect on the supply voltage because of the smoothing capacitors positioned at strategic locations on the board.

Even so, there are still transient current demands on the cables delivering power from the stabilized supply module. This causes transient voltages to be developed along the cable conductors.

In Figure 8.5.1, it is assumed that pcb1 handles relatively high current surges compared to pcb2. The resultant voltage transients can be simulated by the source, *Vtransient*, shown on the figure. As far as transients are concerned, the inductor behaves like an open-circuit and a capacitor behaves like a short-circuit. So the voltage transient is delivered to the cable via the 120 Ω resistor. Similarly, the step voltage arriving at pcb2 'sees' a resistive load of 120 Ω. If the characteristic impedance of the supply conductors is 120 Ω, then there will be no reflections at pcb2.

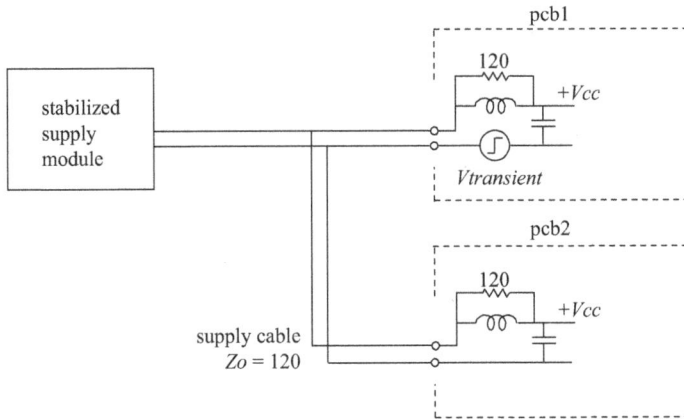

Figure 8.5.1   Transient damping on printed circuit boards.

If there are no reflections, then there will be no 'ringing' transients in the supply cable. Removing these ringing pulses will remove a potential source of interference.

8.5.2 Mains filtering

The fact that a capacitor behaves like a short-circuit to transients means that any filter capacitor between live and neutral of the 230 V, 50 Hz supply will reflect all of the energy of an incident pulse back up the line. When an equipment unit is switched on, all of the transient energy is reflected back into the mains supply.

From the point of view of the equipment-under-review, this is a 'don't care' situation. The capacitor has provided an effective block to the incoming interference. If every equipment unit is protected in this way, then the transient pulse will simply reverberate backwards and forwards along the supply cable. This appears as a ringing transient, which decays to zero eventually. Since the decay mechanism is due to the fact that the energy is converted to electromagnetic radiation, this means that the act of switching any equipment on or off will inevitably create a burst of high energy radiation.

Since the mains supply is routed virtually everywhere in a building, this transient has the capability of disturbing the operation of any other equipment in the vicinity. One way of avoiding this situation is to use the filter network of Figure 8.5.1. Another method is illustrated in Figure 8.5.2. Here, the live and neutral conductors can be viewed as a transmission line supplying differential-mode power. The neutral and earth conductors would then be carrying common-mode loop current.

Any fast transient current arriving from the supply is routed via the capacitor $C1$ into the load resistor $R1$. If $R1$ is equal in value to the characteristic impedance of the supply conductors, then there will be no differential-mode reflection. Similarly, the resistor $R2$ absorbs any transient power in the common-mode loop.

The use of resistors in the mains filters makes it possible to reduce the level of high-frequency interference emanating from the power supply cables, at least in the immediate

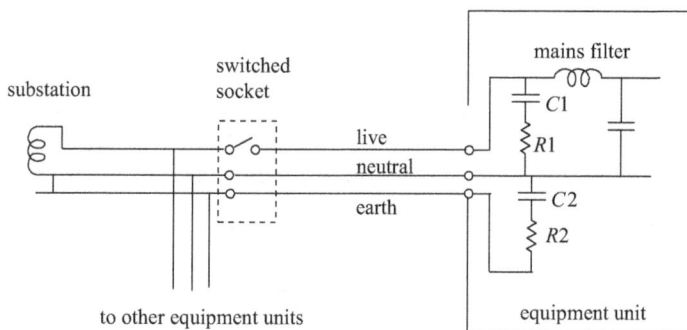

Figure 8.5.2   Mains filter with transient damping.

vicinity of the equipment-under-review. Although perfect impedance matching is unlikely, such terminations offer a dramatic improvement on filters which reflect all the energy.

Although the mains supply conductors are not designed as controlled-impedance transmission lines and they have many stubs and branches along them, the fact remains that there are usually several meters of cable between the equipment and the mains socket, and tens of meters of three-core cable between the mains socket and the local switchboard. To a first approximation, these cables can be represented by a triple-T circuit model such as the one illustrated in the general circuit model of Figure 4.2.4. (If there was no point in trying to simulate the characteristics of the mains supply, then Line Input Simulation Networks (LISNs) would not exist.)

The tests described in section 7.5 show that a two-core mains cable has characteristics which can be closely defined. The same is true of three-core cable. It is entirely possible for designers to obtain a length of such cable and carry out measurements which define the inductance, capacitance, and resistance of each conductor of that cable, and calculate the values of the characteristic impedances.

Given knowledge of the values of the characteristic impedances of the line, it is possible to define the optimum values for the resistors for the filter. If such resistors were to be included, they would damp down the transients in the length of cable which supplies the equipment-under-review. They would also attenuate interference caused by 'broadband over power lines'.

8.5.3 Solenoid switching

Another rich source of transient pulses is the action of solenoids. A fair amount of energy is stored by the current in the winding. When the current source is switched off, the energy in the magnetic field is converted to a high transient voltage, and this acts as a step voltage input to the supply conductors. All of the energy is then converted to a ringing transient in the cable. The amplitude of the transient dies down fairly quickly, because power has radiated away in the form of an electromagnetic field.

Figure 8.5.3 illustrates how this transient energy can be safely absorbed. When the switch $SW1$ closes, a step voltage is propagated down the line formed by the supply and return conductors. If the resistor $R1$ is equal to the characteristic impedance of the line, then there

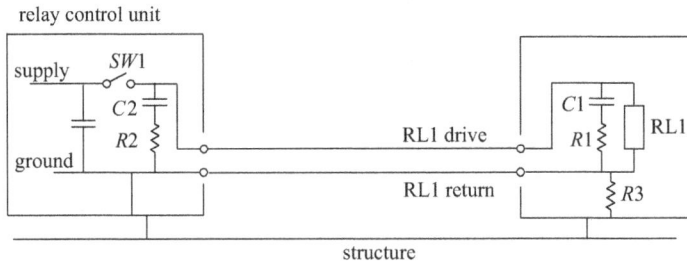

Figure 8.5.3   Relay transient damping.

will be no reflection from the initial edge. If the value of the capacitor is selected such that the current waveform is critically damped, then the waveshape of the current in the cable will be a single pulse, decaying to a constant value. High-frequency ringing should be minimal.

More importantly, when the switch opens, most of the energy stored by current in the inductance of the solenoid will be absorbed by $R1$. Most of the energy stored in the capacitance of the line will be absorbed by $R2$.

It is also a fact that when $SW1$ is closing, the contacts will bounce several times. The input supply to the solenoid is not a clean step function. Rather, it is a series of pulses of varying mark/space ratio. There is no way of avoiding the creation of some level of interference. However, the amplitude of the radiated field can be minimized by using a conductor pair to carry the supply current and by using damping resistors at each end, as Figure 8.5.3 illustrates.

If the structure had been used to carry the return current, then the amplitude of the radiated field would have been much larger.

8.5.4 Commercial filters

It is useful for designers to be aware of the fact that commercial filters can cause unexpected problems. These devices are characterized by subjecting them to a standard test. A signal generator is used to apply a voltage to the input terminals of the device and the output voltage is measured. The output impedance of the signal generator is 50 Ω and the load presented by the voltage measuring equipment is 50 Ω. Maintaining a constant voltage and varying the supply frequency allow a response curve to be derived. This is the definitive curve recorded on manufacturer's data sheets.

These curves usually exhibit a flat response at low frequencies, then form a downward slope that indicates that as the frequency rises, the attenuation increases. It should not be assumed that when the filter is installed in an item of equipment, the attenuation it provides will match that of the manufacturer's specification.

Usually, the filter presents a capacitive load at the input terminals. This capacitor resonates with the inductance of the line to provide a peak in the response. This peak indicates that there is a voltage gain, and sometimes the gain can be as high as 20 dB.

If the purpose of the filter is to remove high-frequency content from the signal arriving at the equipment input terminals (or from the signal delivered to the output terminals), then it is necessary to include the cable parameters in the simulation as well as the circuitry at the far end, in analyzing its performance.

8.5.5 Use of carbon

Carbon is a material worth considering when selecting which components to use, since it has a relatively high resistance that can be used to damp down reflections. For example, carbon-composite resistors have minimal inductance and capacitance, making them very useful as damping resistors.

The high-voltage co-axial cables used on the photon detector assembly, part of the Space Telescope, have a layer of carbon composite material between the inner conductor and the insulation, and a graphite layer between the insulation and the shield. It is possible for current transients to occur in the high-voltage supplies due to switch-on and switch-off conditions as well as to partial discharges. Fast transients are rich in high-frequency components. Skin effect will ensure that these components of the current flow in the adjacent conducting surfaces. Since these surfaces are composed of carbon material, the transient current will flow along a highly resistive path.

Carbon tips on the static dischargers on aircraft present a high resistance in series with the discharge path, minimizing the amplitude of each transient.

8.6 Common-mode damping

As indicated in section 4.4, there are always peaks in the response of the transfer admittance, whether a grounded or a floating configuration is adopted. It is usually possible to select which of these is the best choice for a particular application. Sometimes neither configuration provides the required level of common-mode rejection, and it is necessary to damp down the peaks in the response.

The only way of reducing the level of these peaks is to absorb the unwanted energy. Although it is possible for a soft ferrite to absorb some of the energy in the resistive region of its frequency range, it is much better to be able to insert a pre-defined value of resistance into the common-mode loop over a pre-defined range of frequencies.

8.6.1 Common-mode resistor

One method of including a resistor in the common-mode loop would be to add a third winding to the common-mode choke and connect a low value resistor between the terminals, as illustrated in Figure 8.6.1. Since the bifilar winding carries the common-mode current,

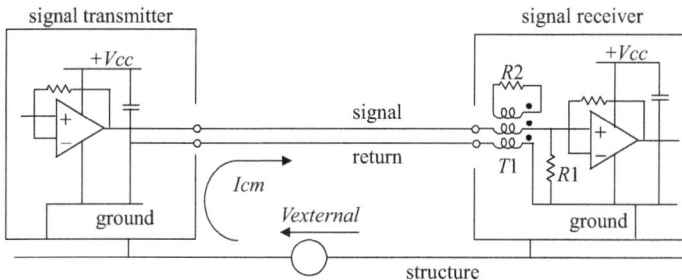

Figure 8.6.1   The common-mode resistor.

there will be a voltage induced in this third winding proportional to the rate of change of this current. The 'choke' now really needs to be viewed as a transformer. Since the third winding is loaded by the resistor R2, the effect is to insert a resistance in series with the common-mode loop. If the third winding has the same number of turns as the bifilar winding, then the value of the resistance inserted into the common-mode loop will be R2.

At all frequencies, this resistor is transparent to the signal current. But at the frequencies for which the transformer is effective, the device acts to absorb the unwanted energy in the common-mode loop. The value of the resistor is determined by the characteristic impedance of that loop. Damping action is effective at all frequencies where the impedance of the winding is greater than the resistance. Figure 7.5 illustrates the fact that the frequency range can be quite wide.

One benefit from this technique is that the value of the common-mode resistor can now be defined by the designer. The other is that it can be used at relatively low frequencies.

8.6.2 Minimizing pickup

Another possible use for the common-mode resistor is to prevent radio-frequency signals being picked up by a microphone lead. It is not at all obvious that this would be a problem with public address systems which restrict their bandwidth to the audio-frequency band. However, common-mode signals arriving at the pre-amplifier can be converted to differential-mode signals by the mechanisms described in previous chapters. Any non-linearity in the circuitry of the mixer can result in demodulation of the radio-frequency signal. The demodulated signal will then be processed in exactly the same way as the intended audio signal.

To deal with this problem, a ferrite core can be clamped round the cable and a short piece of wire can be threaded through the core to act as the third winding and connected to a resistor of defined value (say 100 Ω). This acts to provide a resistive load in series with the antenna-mode current. It is effective over the entire bandwidth at which the toroid acts as a transformer. Figure 8.6.2 illustrates the setup. This may not eliminate the problem of inadvertent demodulation, but it will reduce the interference level to a point where the signal is inaudible.

It is important that the device be fitted as close to the grounded end of the cable as possible, because it is at this end the current will reach a maximum value. Transformer action depends on the rate of change of current. At the microphone end of the cable, there is no

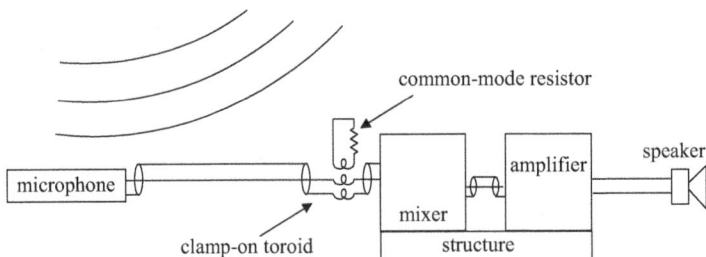

Figure 8.6.2   Antenna-mode damping.

antenna-mode current. If the device were to be located near the microphone it would be totally ineffective.

An alternative solution to such a problem would be to use a shorter length of microphone cable to change its quarter-wave frequency. This was the solution adopted for a public address system in a church, when it had been possible to hear faint voices coming from the speakers during quiet times of an evening service.

8.6.3 Triaxial cable

Another method of absorbing power in the common-mode loop is illustrated in Figure 8.6.3. This method is particularly useful in situations where connections are made between the bench test equipment and the equipment-under-test. In such setups, it is inevitable that the interconnecting cables will be routed in close proximity to each other and that cross-coupling will occur. Even with co-axial cables, the effect of this type of interference can be quite noticeable, particularly under resonance conditions. The need arises for extra shielding to be provided and for this shielding to be free from resonances.

A triaxial cable can be assembled by sliding a braided sheath over an insulated co-axial cable. This assembly then consists of an inner and an outer transmission line, where the outer line is formed by the two braids. The inductance of this outer line can be calculated by setting $r23$ and $r33$ to the same value in (2.10.3). This allows the capacitance to be determined using (2.3.3), enabling the characteristic impedance to be derived from (4.1.7). This sets the value for $R2$.

If the braid is connected to the shell of each co-axial connector via $R2$, then any tendency for current in this inner loop to resonate will be severely damped.

To achieve good performance at very high frequencies, several resistors would need to be assembled in an annular ring at each termination to minimize the inductance of the resistive components.

The setup of Figure 8.6.3 can first be viewed as a source of interference. Any high-frequency signal passing between these two units will flow along the inner conductor of the co-axial cable, returning via the screen. A small voltage will be developed along this screen, causing current to flow in the loop formed by the screen and the outer shield. The resistors $R2$ will reduce the amplitude of this current and will minimize reflections in this loop.

With any setup linking the test equipment to the equipment-under-test, there are bound to be at least three such cables. It is inevitable that these will be routed in close proximity, in

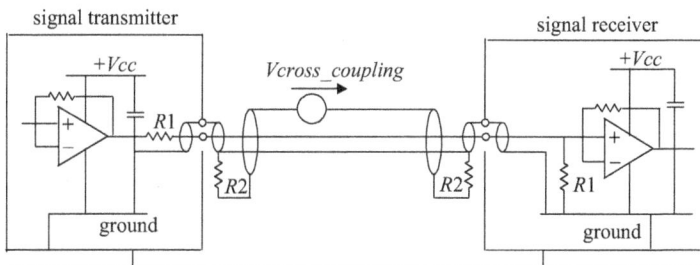

Figure 8.6.3   Triaxial cable.

some cases even touching. Although an outer woven braid of insulating material will prevent direct contact, electromagnetic coupling cannot be avoided.

If the setup of Figure 8.6.3 is now viewed as a victim of such stray coupling, then the effect of this coupling can be represented by a source voltage *Vcross_coupling* in series with the outer shield. This will cause a current to flow in the outer loop. Since the resistors *R2* limit the amplitude of this current, the voltage developed along the length of the inner screen will be extremely small.

This means that the interference caused by the coupling of signals between cables of the test setup will be extremely small. The introduction of damping resistors into the screen/shield loops represents an improvement on the level of shielding that can be provided by a configuration where the outer shields are simply shorted to the connector shells.

8.6.4 Transformer inter-winding

Figure 8.6.4 illustrates another method of absorbing unwanted power.

Some transformers are constructed with an inter-winding between the primary and secondary windings. One end is left open-circuit and the other is usually connected to the structure. Capacitive coupling between the windings allows any transient interference to be short-circuited to earth, and minimizes the amplitude of the transient current between primary and secondary windings.

However, the existence of a capacitor at the end of a transmission line means that it is effectively short-circuited. The line is certain to resonate. Interference current at these frequencies will be amplified, just like in a tuned circuit.

Such a situation can be avoided by inserting a resistor in series with the inter-winding. This will damp down any resonant peaks.

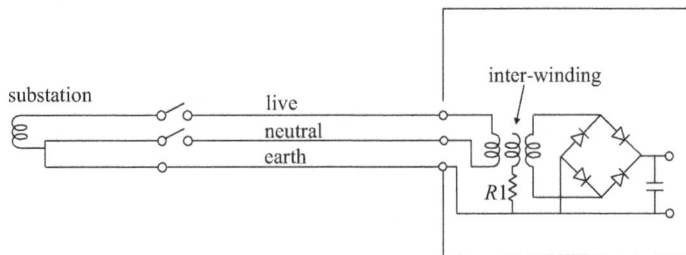

Figure 8.6.4 Absorbing power via screen inter-winding.

8.6.5 Common-mode filter

Figure 8.6.5 illustrates a method of absorbing radio-frequency power picked up by long cables. It uses a firing line for an electro-explosive device (EED) as an example. Under normal operation, a steady-state voltage is applied to the firing line. Since the EED is usually a low-resistance device, a high current will flow and the device will heat up sufficiently to cause ignition.

Under all other conditions, it is essential that a safety margin exists between any interference current and the no-fire current of the device. It could be that the system is intended to

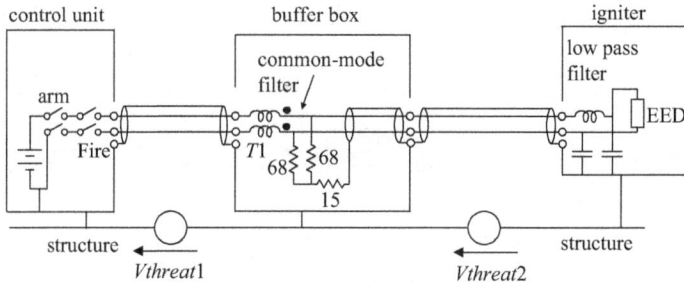

Figure 8.6.5   The common-mode filter.

operate under conditions where the electromagnetic environment is extremely harsh, such as when it is located near a high-power transmitter.

A fair level of protection is provided by the use of a differential-mode supply, the installation of a low-pass filter at the igniter end of the line, and the shielding afforded by the screen of the twisted pair. However, the firing line is still prone to resonate at some frequencies, and at these frequencies, the protection can be compromised.

The purpose of the buffer box is to ensure that the high peaks due to resonance never occur even in the most severe environmental conditions. Under such worst-case conditions, a high threat voltage *Vthreat*1 appears in the common-mode loop between control unit and buffer box. Since transformer *T*1 acts as an inductor to common-mode current, it limits the common-mode current very significantly.

A threat voltage *Vthreat*2 also exists between the buffer box and the igniter. This creates a current in the loop formed by screen and structure. Current in the screen develops a voltage along its resistance, and this small voltage appears as a voltage source in the common-mode loop formed by the screen and the firing line. Since the resistors in the buffer box match the characteristic impedances of the conductors they terminate, there is no resonance. Also, the resistor network acts as a bridge circuit to balance the current in the two conductors of the firing line.

The result is a firing circuit that is safe to use in very hostile environments.

8.6.6 Transformer-coupled resistors

Apart from lightning pulses, the most powerful source of interference is the main power supply. Although the power is delivered at a very low frequency, the transients created when loads are switched on and off make it a rich source of high-frequency electromagnetic fields.

When any load is switched on, there is an initial current transient due to charging current in the stray capacitors (C1 and C2) of Figure 8.6.6. This is reflected back up the line by the motor inductance, and reflected further at all discontinuities along the way. There is nowhere for this unwanted energy to go, except into the environment. When a machine is switched off, the energy stored in the inductors and capacitors can only expend itself in the form of a burst of interference.

It is in such situations that the transformer-coupled resistance can be used to good effect. Figure 8.6.6 shows a filter whose sole function is to absorb unwanted transient energy. This energy can emanate from the motor switching current or from the transients generated by

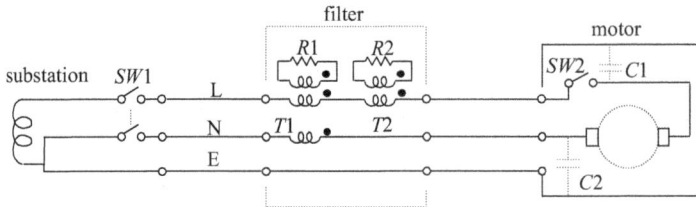

Figure 8.6.6 Transformer-coupled resistance.

other switched loads on the supply. One resistor absorbs differential-mode interference, while the other minimizes the common-mode current.

Transformer $T1$ is trifilar wound, and $T2$ is bifilar wound. The resistor $R1$ is approximately equal in value to the characteristic impedance associated with the common-mode loop, while $R2$ is the same as the characteristic impedance of the differential-mode loop. The mutual inductance of each transformer is determined by the lowest frequency at which the resistor is effective. Such a circuit can help to reduce the level of environmental pollution caused by switching transients in heavy-duty loads.

It could be used to prevent the nuisance tripping of the residual current detector in the mains switch box. Such tripping can occur at switch-on if there is a large value capacitance between the live and earth conductors at the load end.

8.7 Shielding

Since every book on EMC carries a section on shielding, there is no need for an extensive treatment of the subject in these pages. However, it is useful to apply the lessons learnt in the previous chapters to review the shielding properties of the structure.

The action of any electromagnetic field on a conductor is to induce a current along that conductor. If the frequency of the radiation is high, then most of this current will be concentrated on the surface of the conductor; the skin effect predominates.

With a dipole antenna, this current is routed to a receiver via a transmission line. If the conductor is a square, flat surface such as an equipment panel, then the current travels to the edge of the square. Since this represents a discontinuity in the current flow, some of it is reflected back. Some of it continues to flow, down the edge and back along the opposite face of the panel.

Any transient current flowing in a conductor will emit a radiated field. That is, the panel acts as a repeater and radiates a significant proportion of the transient energy back into the environment. Since current is flowing on both sides of the panel, radiation is in all directions. As far as shielding is concerned, such a panel is highly inefficient.

If the shield were to be spherical, then there would be no direct path between the inner and outer surfaces. At the outer surface of the conductor, part of any external field would be reflected back into the environment, but some would be absorbed into the material. At the inner surface of the conductor, some energy would be reflected back to the outer surface and some would propagate into the volume enclosed by the sphere. The ratio between external and internal power levels is described as the shielding effectiveness of the material.

This shielding action is equally useful in preventing internally generated field from radiating out into the environment. This means that a conducting surface can provide an effective barrier to the propagation of electromagnetic energy at all frequencies, provided that there are no gaps in the surface.

8.7.1 Equipment shielding

In practical situations, the construction of the enclosure is more akin to a box-like structure than a spheroid. There are joints between panels and framework, holes in panels for ventilation, gaps to allow visibility of meters, indicator lamps, and screen displays, and more holes for potentiometer spindles, switches, push-buttons, and keyboards.

This means that any current flow will be subject to multiple discontinuities and multiple reflections. Since current density will increase at the discontinuities, 'hot spots' can develop in the conducting surface immediately adjacent to slots. Under resonance conditions, these slots can act as dipole antennae.

Methods of dealing with gaps between structural components are to use crimped seams or conductive gaskets. Cut-outs used to provide visibility of meters and displays can be covered with conductive glass with the conductive surface of the glass RF-bonded to the shield all around the edge. Holes for access shafts can be protected by a waveguide below cut-off. All these techniques are described in detail in Reference 8.1.

The purpose of most units of electronic equipment is to process signals which arrive and depart via connectors and cables. If the cables are unscreened, then any interference they pick up will be routed directly into the shielded enclosure. This will render the shield ineffective.

One option would be to filter the signals at the shield wall. Another option would be to sheath the cable with a screen braid and RF-bond this braid directly to the shield wall, either directly or via the shells of suitable connectors. Such a construction ensures that most of the interference current flows via the cable screen to the outer surface of the enclosure.

Methods of maintaining the integrity of any shield are illustrated in Figure 8.7.1.

It is possible to design enclosures which have a shielding effectiveness of 100 dB or more. However, this can only be confirmed by testing. In the absence of such tests, it is better to assume a limit to the frequency at which shielding is effective.

This can be determined by following guidelines provided by the Ordnance Board [8.2]. Essentially the procedure is to examine the equipment and measure the length of the largest gap in the shielding, $Lgap$. The cut-off frequency is then:

$$f_{\text{limit}} = \frac{300 \times 10^6}{4 \times Lgap} \tag{8.7.1}$$

Below this frequency, it can be assumed that the shielding effectiveness is 20 dB, that is, a voltage or current ratio of 10 to 1. Above this frequency, it is better to assume that the shielding effectiveness is zero. When assessing the EMC of any particular signal link, the approach could be to implement the analytical techniques described in Chapters 2–6, and then to invoke the shielding factor as described above. This would give a worst-case figure.

In cases where the structure does not meet the rigorous guidelines of Figure 8.7.1, it still provides some measure of shielding by virtue of the fact that there will be circulating

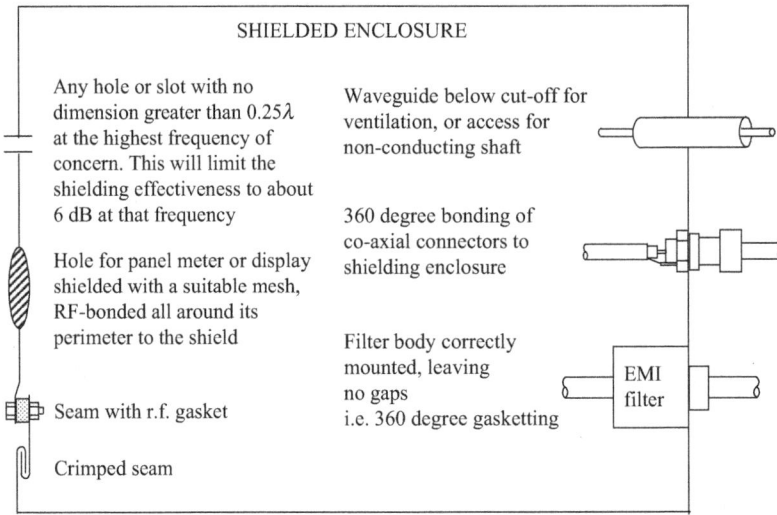

Figure 8.7.1 Maintaining integrity of shield.

currents in the conducting loops which absorb the energy of the incoming radiation, in much the same way that a breakwater protects boats inside a harbor. Chapter 3 describes a method that has been successfully used by researchers to locate regions in aircraft structures where the indirect effects of a lightning strike are least severe. This identifies routes along which cables can safely be laid. Such an approach can apply to any structural assembly.

Carrying out tests of actual performance, as described in Chapter 7, would allow a more accurate figure to be assigned to the shielding provided to any particular signal link.

Other techniques for calculating shield effectiveness can be found in Reference 1.5.

8.7.2 Shielding of buildings

Protecting buildings and structures from the effects of lightning presents a more daunting task, but the approach is essentially the same.

Figure 8.7.2 illustrates the protection measures which have evolved. Instead of a continuous conducting surface enclosing the building, the outer shield consists of a cage-like network of conducting links. Lightning conductors on the roof provide attachment points for any direct strike as well as a vertical conducting path down the outer walls to a grid-like network of conductors buried in the earth. Linking conductors provide horizontal paths between the verticals.

The nature of antenna-mode transient current is that it flows on the outer surface of conductors. The purpose of the horizontal links is to enable the downlinks to share the current. This minimizes the amplitude of the transient current in any particular conductor, and minimizes the level of pulsed magnetic field it creates. The presence of these links ensures that the lightning transient flows on the outer surface of the building. That is, the assembly of lightning conductors acts in the same way as a shield. Most of the current flows down the lightning conductors into the earth. It will also flow away from the building via any

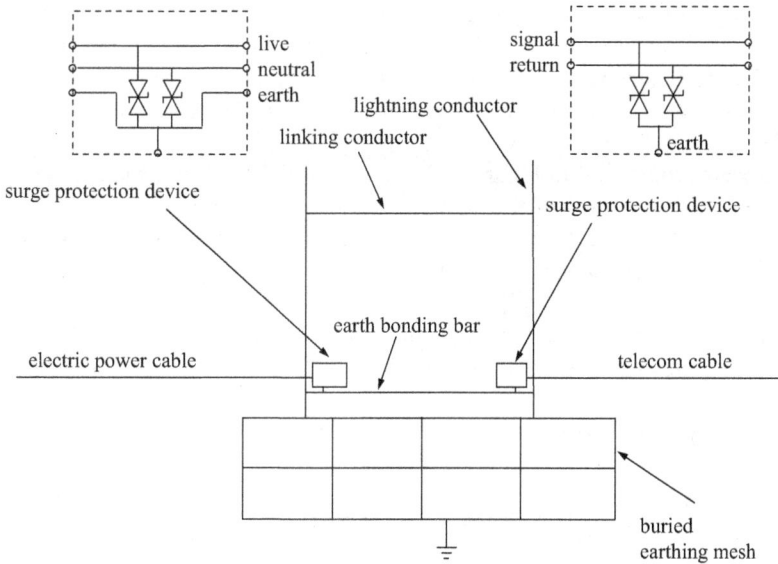

Figure 8.7.2 Lightning protection of building.

other conductor. So it can be expected that a proportion will be carried by the earth conductor of the power line.

High transient current in the conductors inevitably creates high transient voltages between these and any other conductors in the near vicinity, with the risk of arcing. This can happen at ground level as well as at height.

To prevent damage to telecommunications cables, surge protection devices can be fitted at the point of entry to the building. These are usually assemblies of metal oxide varistors.

At voltages below a pre-determined threshold, a varistor behaves like a pair of back-to-back Zener diodes. If the threshold voltage is exceeded, the current increases rapidly, and avalanche breakdown occurs. Some devices are capable of handling extremely high currents, of the order of 40 kA, albeit for a short period, but long enough to survive a lightning current surge. During a lightning transient, the reaction time is fast enough to prevent the voltage transient from exceeding the voltage withstand capability of the equipment being protected.

Current into the telecommunications line is limited by the inductance and resistance of the line when it is acting as an antenna. The cable acts to carry a small proportion of lightning current away from the building, and, hopefully, continues to serve its normal function after the event.

A similar surge protection device can be fitted at the point where the service power cable enters the building. Although the 'earth' conductor of the power cable will carry the larger proportion of the current flowing into the cable, electromagnetic coupling will create a high transient voltage between the earth and live conductors, and between the earth and neutral conductors. When this happens, the components of the surge protection device change into the conducting state. This has two effects; the live and neutral conductors now act to carry some of the transient current away from the building, and the transient voltages between the conductors are held to a level that the insulation can withstand.

If lightning strikes either the telecommunications cable or the power cable outside the building, the relevant surge protection device will act to ensure that the electrical equipment inside the building remains safe. A more detailed description of the method used to protect buildings from the effects of lightning is provided by a manufacturer of surge protection devices [8.3].

The measures illustrated in Figure 8.7.2 can provide protection from the worst effects of lightning, and injury to humans, fire, and structural damage. Preventing damage to electrical and electronic systems requires further measures, such as other surge suppression devices between the main switchboard and the equipment. Lightning protection zones, such as screened rooms, can also be installed.

It should be noted that surge protection devices can only provide protection from damage. They cannot prevent interference from upsetting the normal functioning of the equipment.

To prevent upset due to interference, one or more of the other techniques illustrated in this chapter need to be included in the system design.

8.7.3 Use of carbon

If the surface is highly conductive, some of the radiation is reflected, but most is conducted into spaces where its presence is undesirable. If the surface contains resistive material such as carbon, much of the unwanted electromagnetic energy is converted into heat.

Examples of carbon used for this purpose are as follows:

- Blocks of carbon composite material are used to line anechoic chambers.
- A bitumen layer is added to the walls of secure buildings.
- Carbon composite used in the structure of aircraft absorbs much radiated energy.
- Static dischargers have carbon tips to create a voltage drop between the structure and the environment. This limits the discharge current to safe levels.
- Fuel pipes made of carbon composite material prevent the buildup of static voltages.
- Carbon in vehicle tires, making contact with tarmac, prevents vehicles acquiring a static charge.

It is a useful exercise to investigate the materials planned for use in the equipment under development to check whether the properties of carbon can be exploited.

CHAPTER 9

System design

Chapters 2 and 3 have used the concepts of electromagnetic theory to create circuit models that allow electromagnetic coupling between adjacent circuits to be simulated using circuit theory. The resultant simplification in the mathematics comes at a price. Action and reaction anywhere in the system is assumed to be instantaneous. This means that the wavelength of the maximum frequency of the simulation must be greater than 10 times the maximum dimension of the assembly-under-review.

Chapter 4 shows how this limitation can be relaxed by invoking the concept of distributed parameters. Even so, the limitation is that the wavelength of the maximum frequency of the simulation must be greater than 10 times the maximum size of the cable section. However, this means that the maximum frequency of the simulation is no longer limited by the length of the cable.

Chapters 5 extends the application to analyze coupling between cable and environment. Here, the simplifications are achieved by assuming worst-case analysis. This is acceptable when analyzing EMC, where the criterion is to ensure that the actual interference is always less than that predicted.

The transient analysis of Chapter 6 reveals many features of electromagnetic coupling that are not generally known and which provide a much improved understanding of the phenomena.

The development of the modeling process was made possible by the fact that many tests were carried out on many cable assemblies. Chapter 7 describes some of them. Correlating the results of each test with those of the model allowed defects in the model to be corrected. This was a more unforgiving approach than one based purely on theory. Even so, every equation is traceable back to formulae derived in textbook theory.

Chapter 8 has identified several techniques which can be used to minimize the level of interference coupled into and out of a signal link. It is possible to glean a few simple guidelines that identify the basic concepts used in all these techniques, and these are listed in section 9.1. If implemented at the initial stage of a project, these guidelines should ensure that the EMC of the system can be achieved in a cost-effective way.

In all the previous chapters, the approach has been to build up circuit models from basic building blocks and to develop those models to deal with increasingly complex coupling mechanisms. Having developed these models to a point where they provide a reasonably accurate simulation of the performance of actual hardware, it is possible to reverse the process and describe a top-down approach. This is the objective of section 9.2.

Section 9.3 identifies the function of the interface circuits as buffers between the signal processing functions on printed circuit boards and the signal distribution function of the wires and cables. Chapter 8 provides details of a wide range of such interface circuits.

Having established a clear set of relationships between currents and voltages in the signal link and the electric and magnetic fields in the near vicinity, it is possible to extend those relationships to include the EMC requirements. That is, the design of the system can be tailored to meet those requirements. Section 9.4 establishes a relationship between the threat environment and the level of interference experienced by the victim circuit. Section 9.5 relates the design of a potential source of interference to the maximum level of radiated interference, as specified by the formal requirements.

One of the responsibilities of every engineer is to ensure adequate preparations are made before embarking on a design project. Section 9.6 identifies those aspects of the planning necessary to implement the approach described here. In particular, it recommends that enough bench test equipment be provided to meet the needs of the project and that mathematical software be installed on the personal computers. It goes on to describe methods of identifying those signal links that could pose problems and of characterizing those links during the initial development of the system. It concludes by stating that the best way of dealing with an EMC problem is to create a circuit model and review it in the light of the guidelines of section 9.1.

9.1 Design guidelines

Although the approach has been to create circuit models that simulate the behavior of electromagnetic field coupling, the fundamental parameters used in the construction of every model have been derived from electromagnetic theory and all the equations used in the analyses of Chapters 2–6 are traceable back to that theory. Chapter 7 shows that there is a clear correlation between actual performance of a physical assembly and the response of its circuit model.

It is possible to distil a few simple guidelines from the analyses and assessment of the previous chapters. If implemented, they will ensure that the system under development has the best possible chance of meeting all the EMC requirements.

9.1.1 Structure as a shield

Probably the most important conclusion of the previous chapter is that designers should avoid the temptation to use the structure as a convenient return path for signals and power.

Every conductor of the structural assembly possesses the property of inductance. Any changing current will develop a voltage along the length of that conductor. If voltages can exist between different locations on the structure, then that structure cannot be regarded as an equipotential surface. Every signal and every power line that uses the structure as a return conductor is guaranteed to interfere with every other signal in the system.

Section 8.1 explains this in more detail, and goes on to reason that it is better to design the structure as a network of conducting loops. Any interference field will cause current to circulate in each loop, and the field created by the circulating current will tend to balance the effect of the incoming radiation. Such a design will create a structure that behaves essentially as a shield. Section 8.7 provides more information on the design of shields.

9.1.2 Return conductors

Every conductor acts as an antenna and any electrical signal can be treated as a sequence of small (or large) step pulses. A transient step pulse traveling along any conductor will create an electromagnetic wave that spreads out in all directions. If another conductor is routed alongside, then the electromagnetic wave will create a current which flows in the opposite direction along that second conductor. That is, the second conductor acts as a receiving antenna. The closer the two conductors are to each other, the more current is captured.

If the second conductor is assigned to carry the return current, then the conductor pair is acting as a transmission line. It appears as though an electromagnetic wave is traveling along the intervening space, guided by the conductor pair. The tests described in sections 7.5 and 7.6 illustrate this effect.

If the return conductor follows a different route, then the coupling between send and return conductors is reduced while the coupling between differential-mode current and common-mode current is increased. The signal link will emit more interference and its susceptibility to interference will increase. Section 2.8 shows how the parameter values of the circuit model are affected by the relative positions of the conductors in the cross section of the cable assembly.

Two important objectives are achieved by keeping the send and return conductors of the signal link as close together as possible. The efficiency of the link is improved (there are fewer losses), and the interference level is minimized. Improved EMC and improved efficiency go hand-in-hand.

Since the transfer admittance of the cable as an emitter is the same as the transfer admittance that defines its susceptibility, then the action of routing the two conductors will also reduce the susceptibility of the signal link.

9.1.3 Ground loops

In the vast majority of signal links, it is simply not practical to use any of the techniques described in section 8.4 to achieve common-mode rejection. Both the output signal and the input monitor are referenced to the zero-volt conductor of their respective printed circuit boards. Rather than using the conducting path via the structure or the single-point ground of Figure 8.3.1, it is good practice to connect the return conductor to local structure at both ends of the line. This will form a multitude of ground loops (or earth loops) that each carry common-mode current. Electromagnetic field generated by common-mode current caused by external interference will link both conductors of any signal link and will balance out the interfering signal.

With the configuration illustrated by Figure 8.3.2, the field generated by the signal conductor will induce a current that flows in the opposite direction along the return conductor. The tests described in section 7.5 illustrate this fact. As the frequency increases, less and less current will flow in the common-mode loop.

In printed circuit boards, the ground plane acts as a set of return conductors, each located precisely under the associated signal conductor. Figure 8.2.2 illustrates the fact that current can flow in opposite directions in the ground plane. For this to happen, current must be flowing in a loop in the ground plane. The ground plane acts as a multitude of ground loops.

If the link-under-review is intended to carry a narrow-bandwidth, low-frequency signal, then it is sometimes better to isolate the return conductor from structure at the receiving end.

However, the floating configuration acts to amplify interference levels at the quarter-wave frequency. In such configurations, it is necessary to ensure that the interface circuitry does not respond to frequencies in this range. This can be done by the use of bandwidth filtering.

If the single-point ground configuration is implemented, then the system will exhibit the worst EMC characteristics possible. Section 8.3 identifies the essential difference between a configuration that uses ground loops and one that is based on the concept of the single-point ground.

9.1.4 Current balance

The existence of a return conductor adjacent to the send conductor provides an opportunity to balance the currents. If the current in the return conductor is equal in magnitude but opposite in direction to that in the send conductor, then the total current at the interface will be zero; that is, the common-mode current will be zero and there will be no interference.

Such a situation is never achieved in practice. But it is something to aim for. Circuit configurations described in section 8.4 help to minimize interference.

9.1.5 Differential-mode damping

It is good practice to minimize differential-mode reflections at the interface circuitry. If differential-mode energy is reflected back into the transmission line, the only way it can disappear is via electromagnetic radiation. In other words, matched transmission-line design is best for low emissions.

Since the transfer admittance for conducted susceptibility is the same as that for conducted emission, it follows that matched transmission-line design is best for low susceptibility.

Ideally, the terminations at each end of the line should be the same as the characteristic impedance of the line. This will generally be somewhere in the range 50–200 Ω. Open-circuit and short-circuit terminations will cause 100 percent reflections of all signals. Inductive or capacitive loads will cause 100 percent reflection of transient steps in any signal.

Section 8.5 describes various ways of damping differential-mode reflections.

9.1.6 Common-mode damping

Viewed from the perspective of conducted susceptibility, current in the common-mode loop is undesirable, since some of it will inevitably appear in the differential-mode loop. Viewed from the perspective of conducted emission, current in the common-mode loop is undesirable, since all of it will inevitably reappear as interference.

The only way of preventing unwanted energy from radiating out into the environment or coupling into the common-mode loop is to absorb it, that is, to convert it into heat energy.

Components capable of absorbing electrical energy are soft ferrites (especially when operated in frequency ranges where they are most resistive), steel-cored transformers and chokes, and other ferromagnetic materials that exhibit heating losses. Nickel is sometimes used to coat copper conductors to make them more lossy.

It is possible to insert a defined value of resistance in the common-mode loop or the antenna-mode loop using one of the techniques described in section 8.6. These can be especially useful in reducing the amplitude of high-current long-duration transients in power supplies, the kind usually dealt with by surge suppressors.

Carbon is a material worth considering when selecting components and materials. Sections 8.5.5 and 8.7.3 provide examples of its use.

9.1.7 System assessment

The modeling techniques described in the previous pages describe how to create a circuit model to simulate the interference coupling mechanisms of any signal link. In the majority of configurations, it is possible to assess the adequacy of the design merely by examining the circuit model from an engineering point of view. If any risks are identified, the design can be modified at an early stage in the development process.

Critical signal links can be identified at a very early stage of product development. One of the actions of such a feasibility study would be to identify tasks that should be carried out in the subsequent stages to define the detailed design and analyze the performance of these links. A method of performing such an assessment is described in section 9.2.

9.1.8 Bench tests

Having identified critical signal links and carried out a prediction of the performance of each, the next step is to check the actual performance by subjecting it to bench tests.

Chapter 7 is devoted to the subject of bench testing, in terms of the method of checking and in terms of the method of analysis. A representative circuit model can be created of the link-under-review, and this model can be used to predict the results of formal testing in an EMC test house.

Reports of such analyses will be a significant feature of design reviews.

9.2 Relating the diagrams

Any plan to control the EMC must necessarily involve a top-down view of the system, and this requires the systematic use of diagrams. Since the creation of block diagrams, circuit diagrams, and wiring diagrams is part of the normal design process, it is logical to extend the use of the technique to assess the EMC characteristics of the system.

The purpose of this section is to provide an overall view of the process and indicate some of the design considerations involved in meeting the EMC requirements. Since there are many books that provide detailed guidance on the design of electronic equipment, information on other aspects of design is not included here.

9.2.1 Circuit diagrams

One such diagram is the circuit illustrated by Figure 9.2.1. The push-pull emitter follower identifies the module as the output stage of an audio amplifier, the stage which supplies drive current into a loudspeaker. The focus is purely on the circuitry of the module, with minimal information about the interconnecting cables and no information about the circuitry in other modules.

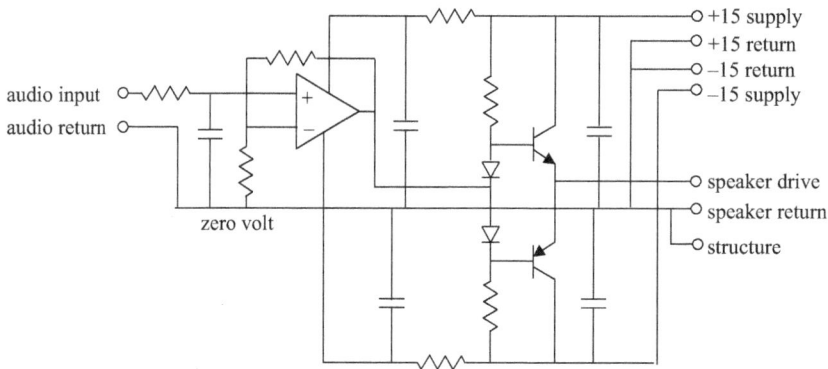

Figure 9.2.1   Circuit diagram – audio power amplifier.

No component values are shown in this particular figure because the diagram is for illustrative purposes only. If component values had been defined, it would have been possible to analyze the performance of this module. Simulation Programs with Integrated Circuit Emphasis (SPICE) are designed specifically to deal with such a task. All that the equipment designer needs to do is draw the circuit on a computer screen, define the component values, add a voltage source to the input terminals, and invoke one of the analysis options. Very significantly, SPICE software is capable of analyzing both transient behavior and frequency response of the system.

Although SPICE can be used to simulate the behavior of extremely complex circuitry on printed circuit boards, that facility comes at a price. The previous chapters have shown that the analysis of interference coupling is based on the assumption that voltages exist between 'reference' or 'ground' conductors. SPICE analyses of the functional performance of complex assemblies are based on the assumption that all points connected to the reference conductor are at zero volts. Given such an assumption, there is no point in attempting to use such software to analyze interference coupling in complex assemblies.

Section 8.2 shows that although voltages exist along the surface of a ground plane, the simultaneous existence of circulating currents allows that plane to be regarded as an equipotential surface. Hence, SPICE analysis of complex circuitry on a printed circuit board with a ground plane can provide a reliable simulation of the behavior of that circuitry. This makes the software a valuable analytical tool. The words 'integrated circuit emphasis' indicate the intended application of the software, the analysis of small sized modules.

In this particular example, the signals being processed are in the audio range, up to about 20 kHz. At this frequency the quarter-wavelength is more than 3.5 km. Since the largest dimension on the printed circuit board is of the order of 10 cm (corresponding to a quarter-wave frequency of 7.5 GHz), it is not likely that antenna-mode interference will have any direct effect on the board-mounted circuitry.

Although it is possible for high levels of radio frequency (RF) interference to arrive via the input (or output) connectors, the design of the interface circuitry should be such that all signals outside the operating range of the amplifier are severely attenuated. This is probably the most important precaution.

It is worth noting that non-linearity in the relationship between voltage and current in transistor and diode junctions can cause the modulation of an RF signal to be picked up. The same is true if the center frequency of the RF signal lies on the sloping part of the frequency response curve of a filter. Care should be taken to select transistors and diodes that are not capable of responding to high-frequency RF signals outside the bandwidth of the signal being processed.

It is also advisable to keep the signal loops on the board as small as possible, either by using a ground plane or by routing the tracks carrying each signal and its return current closely together.

Given such precautions, it is reasonable to assume that there will be no internal interference between components on the board. By ignoring effects that are judged to be irrelevant, the computational power of SPICE software can be put to good use in analyzing the behavior of extremely complex circuit boards.

Even though the main purpose of SPICE analysis is to simulate system function, it is still possible to assess and analyze the interference coupling characteristics of critical signal links on the board by reviewing them in the light of the guidelines of section 9.1.

9.2.2 Wiring diagrams

However, this assumption that all connections to the reference node are at zero voltage ceases to have any validity when the inter-module wiring of the equipment is considered. The wiring diagram of Figure 9.2.2 shows how this particular printed circuit board could be interconnected with the other modules of the equipment and how equipment units are assembled to create a functioning system.

In this system, the cables carrying signals and power between the various units are much longer than the printed circuit traces connecting different components on the printed circuit boards. Since the interface circuitry for each signal link is designed to provide a buffer between the interconnecting cable and the signal processing circuitry on the board, it is possible to assume that interference picked up by the cable can be treated separately to that picked up by the traces on the printed circuitry.

The analysis of section 5.3 shows that the maximum threat voltage which can be induced in a cable of defined length is proportional to the inverse of the frequency. Figure 5.3.7

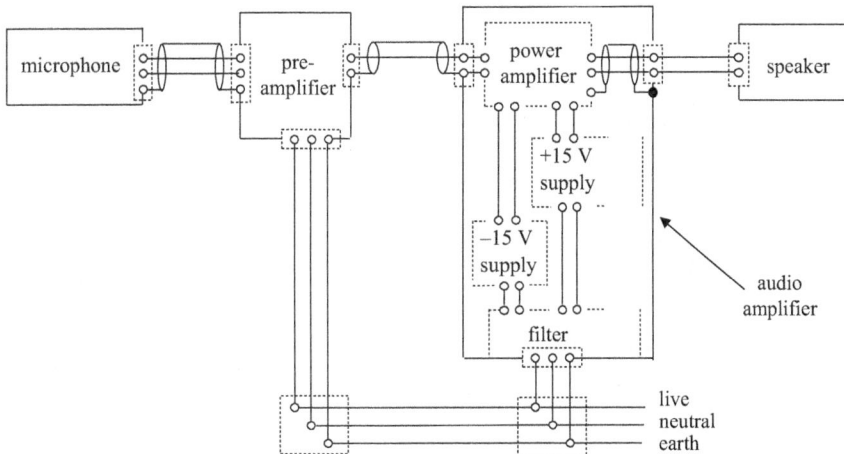

Figure 9.2.2 Wiring diagram – simple audio system.

illustrates the relationship. That is, the longer the cable, the higher the voltage which can be induced.

Figure 5.3.7 also illustrates the fact that the frequency at which maximum voltage can be induced reduces as the length of the cable increases. For a long cable, the design of the interface circuitry should ensure that the signal passed on the processing circuitry due to the threat voltage is within acceptable limits. This design would almost certainly include a low-pass filter, ensuring that the level of attenuation increases as the frequency increases.

This being so, the focus of attention shifts to the design of the interconnecting cables.

Any external electromagnetic field will link with the cables and will cause antenna-mode current to flow in the outer surface of every exposed conductor. Equally, any antenna-mode current created by differential-mode current in the cables will radiate interference away from the system.

For such a system, the assumption that every point on a conductor is at the same voltage is no longer valid. It becomes necessary to represent the coupling to and from the signal loop as a triple-T network, as illustrated by Figure 2.7.6.

9.2.3 Block diagrams

It is best to approach the problem in a systematic manner. Since most electronic systems are much more complex than the simple public address system depicted here, it is normal practice to illustrate the function of each equipment unit by means of a block diagram, as illustrated by Figure 9.2.3.

Such a diagram gives the designer an overall view of the relationships between the different units, and is usually created during the initial feasibility study of the project. At this stage of development, many details of the design have yet to be defined, and this includes the details that control the EMC of the system.

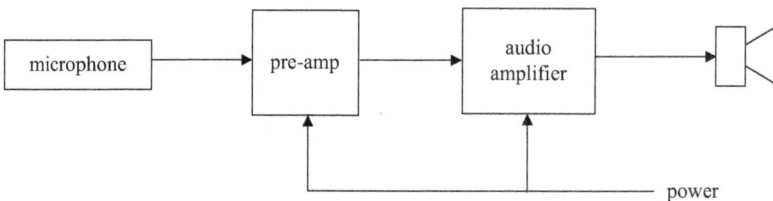

Figure 9.2.3   Block diagram – simple audio system.

9.2.4 Interface diagrams

Given the existence of a block diagram that identifies every signal transmitted between every processing module, it becomes possible to focus attention on each signal link, and this function is performed by an interface diagram. Figure 9.2.4 is just such a diagram. It identifies all the conductors and all the components involved in transmitting a signal from the pre-amplifier to the power amplifier. By limiting the scope of the diagram to the components involved in coupling interference between one signal and its environment, unnecessary complication is avoided.

The normal function of block diagrams is to allow attention to be focused on the modules (such as the audio power amplifier of Figure 9.2.1) that process the signals. By creating an

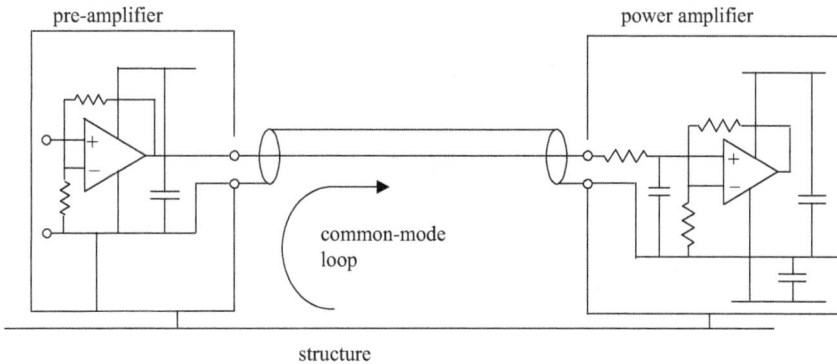

Figure 9.2.4   Interface diagram of link between pre-amplifier and power amplifier.

interface diagram for each signal link, the same block diagram can be used to focus attention on the signal links, that is, on the lines on the block diagram, rather than the boxes.

The interface diagram identifies the cable conductors, the connector terminals, and the circuitry at each end of the·cable, and indicates how the structure interconnects the 'ground' reference conductors at each end. It identifies the common-mode loop and the differential-mode loop. It defines precisely how the transmitter circuitry interfaces with the cable and precisely how the buffer at the receiving end is designed.

This diagram illustrates clearly that three conductors are involved in carrying a signal from one location in the system to another. In this case, it is the inner conductor of the co-axial cable, the screen of that cable, and the structure. Differential-mode current is carried by loop formed by the signal and return conductors (in this case, the core and screen of the co-axial cable); common-mode current is carried by the loop formed by the return conductor and structure. In this context, the structure represents the combined effect of all other conductors linking the two equipment units.

This diagram is arguably the most important from the point of view of EMC design. The designer is able to define what sort of cable to use – wire pair, co-axial cable, screened pair, multiconductor assembly, or something else. He or she is free to use any of the interface circuits described in Chapter 8, or to define a completely new interface. It is also possible to decide whether or not to enclose the cable with an overall screen or to route it along conduits.

Anyone viewing the interface diagram is provided with a complete picture of the features of that link which are relevant to EMC. If the information is contained in several drawings, each with a generous sprinkling of earth symbols, it is rather like peering into the mist.

9.2.5 Circuit models

In the case of the interface diagram of Figure 9.2.4, the signal carried by the differential-mode current is more likely to suffer from interference than to cause it. Hence, any current in the common-mode loop can be classed as an interference source. As far as EMC analysis is concerned, the task is to define how this will affect the signal. Assessing the coupling between these two loops can be accomplished by the use of the circuit model of Figure 9.2.5. So the creation of the circuit model becomes the next step in the process.

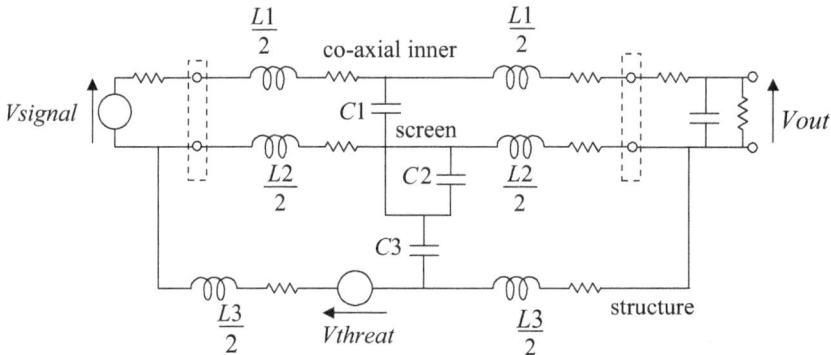

Figure 9.2.5   Circuit model of link between pre-amplifier and power amplifier.

A circuit model for a signal link can be defined as one that simulates the interference coupling mechanisms. It allows both the susceptibility and emission characteristics to be analyzed.

In many cases, the mere act of creating a circuit model allows the characteristics of the signal link to be assessed. It is possible to view such a model in much the same way that an experienced designer can look at the circuit diagram of a functional module and identify its strengths and weaknesses.

In the case of the link-under-review, the only coupling between differential-mode and common-mode loops is via the transfer impedance of the screen. It would be useful to know the frequency response characteristic of this impedance. If this information is available, then the component values for the circuit model can be deduced. Section 2.10 gives details on the method of modeling co-axial cables.

There may be problems when resonance conditions arise and the current in the screen reaches a high amplitude. If, as in this example, the bandwidth of the differential-mode signal is restricted to audio frequencies, then there should be no problems. There should be a low-pass filter at the audio amplifier interface to ensure that any high-frequency content is rejected.

9.2.6 Deriving component values

Initial values for the components of the circuit model can be derived from a review of the physical assembly of the wiring harness.

The common-mode loop is essentially made up of four sections: the co-axial cable between the pre-amplifier and the audio amplifier, the mains cable delivering power to the audio amplifier from the mains trunk cable, a short section of trunk cable, and the mains cable delivering power to the pre-amplifier. Figure 9.2.6 illustrates the magnetic properties of this assembly.

The partial inductance of each section of the loop can be derived by treating the section as a composite conductor made up of several conductors in parallel. Section 3.1 describes how to derive the inductance of a single composite conductor. Summing the values derived for the four partial inductors gives a value for the loop inductance ($L3$ in the circuit model of Figure 9.2.5).

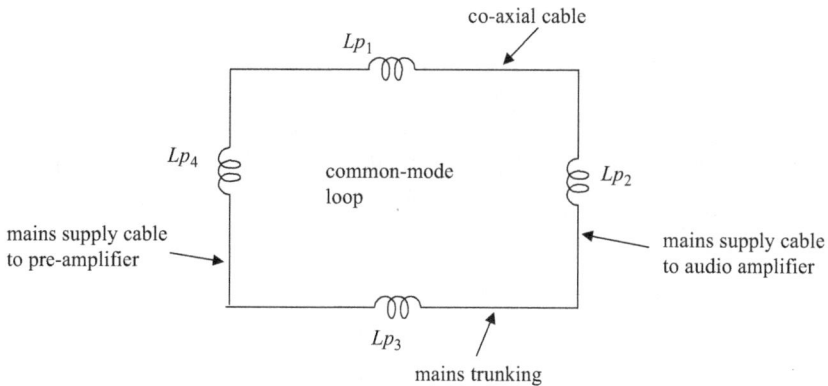

Figure 9.2.6   Inductance of common-mode loop.

Since the value of capacitance $C3$ is closely related to the inductance $L3$, this parameter can also be defined. Section 2.3, on the duality between L and C, provides the relevant formulae.

With properly terminated co-axial cable, the impedance common to the common-mode and differential-mode loops is the transfer impedance of the screen. As indicated previously, the model can be created if information on the frequency response of this parameter is available. Such information could be provided by manufacturer's data sheets.

Given knowledge of the diameter of the two conductors of the co-axial cable, the inductance per meter of the differential-mode loop, $L1$, can be defined. Since the characteristic impedance of the cable is (to a first approximation) equal to the square root of the ratio between inductance and capacitance, it is possible to derive a value for $C1$ in the circuit model.

If the signal link is more complex (for example, if the conductors form part of a multi-conductor cable with an overall screen), then the technique described in Chapter 3 can be used to determine the coupling parameters.

Alternatively, data from a set of bench tests can be used to construct the model. This could be provided by tests carried out on the prototype equipment, or by data from a library of 'representative circuit models' as described in section 9.6.6.

Values for the components at each interface are either defined by the designer or specified in the relevant drawings of each equipment unit.

9.2.7 Analyzing the signal link

The most appropriate EMC test for this particular configuration is the conducted suscept-ibility test. A voltage transformer is used to inject a known voltage at a set of known fre-quencies into the common-mode loop, and the amplitude of the differential-mode current is monitored. Figure 4.4.4 illustrates a setup that could be used.

Given knowledge of the values of the components of the circuit model, it is possible to predict the response of the signal link to such a test during the initial stage of development.

9.2.8 Testing the link

If it the results of such a simulation indicate that possible interference problems might occur, then bench testing of this link could be included in the development program for the product. Correlating the test results with the response of the model will allow an accurate circuit model to be created, one which can be used to define the performance of that particular link.

Such an approach will provide a high level of confidence in the ability of the system to pass the formal EMC tests on the first attempt. Significant cost saving can be achieved by avoiding the need to modify production-standard equipment and resubmit it to the qualification test procedure. Delays inherent in this process, which affect the time-to-market of a new product, can also be avoided.

9.3 Printed circuit boards

Detailed design of printed circuit boards is well covered in existing books, so there is no need for this section to reprise that information. However, there are a few points that are worth making.

In section 8.7, various ways of maintaining the integrity of an equipment shield were identified. In every case, the purpose was to provide a barrier against the incursion of external fields and to prevent the egress of internal fields into the environment. Figure 8.7.1 illustrates the concept.

The same concept can be applied to the design of printed circuit boards. If each interface between a board and the external wiring is limited to the transmission or reception of one signal, then that interface will act as a buffer between the board and interference carried by that wiring. Sections 8.4–8.6 describe a number of designs that can achieve this purpose.

A properly designed buffer will attenuate any interference signal arriving via the interconnecting cable to a level that is within acceptable limits. This could be a requirement that the normal function of the signal processing circuitry should not be upset, or a requirement that no damage should occur and that the system should be able to recover after the event. The buffer should include filter components to limit the signal bandwidth to that which is necessary for the intended signal function.

With multi-layer boards, the ground plane can provide the shielding against external fields. If the supply rail is also constructed of a plane surface, then the shielding can be very effective. Even if the board is single-sided, the presence of adjacent boards will provide a measure of protection.

Although it is still possible for interference to be picked up directly by the printed circuit tracks, the fact that buffers exist at every interface means that interference arriving via the cables can be dealt separately to the assessment of the performance of the board itself; the frequencies of the threat signal would be quite different. During such an assessment, the guidelines of section 9.1 continue to apply. Concepts that are applicable at system level are still relevant at board level.

The concept of 'composite conductors' can be used to analyze the coupling between adjacent tracks on a board, or the coupling between a signal link on a board and the local environment. Chapter 3 describes the technique and Figures 8.1.1 and 8.2.1 provide practical illustrations.

The concept of segregation of 'noisy' and 'quiet' circuits is often implemented on printed circuit boards. If one section of the board is used to mount logic circuitry and another section holds analog components, then it is not a good idea to mount each set of components on its own separate ground plane. The ground plane should cover the entire board, since it provides a return path for each signal trace at precisely the right location. Any separation of ground planes on a board will disrupt the paths for return currents, with an increased probability of interference.

Since the wiring on printed circuit boards is much shorter than that used externally, the frequencies at which problems occur will be much higher. Since the threat level reduces with frequency, interference from distant sources is unlikely (unless the board is completely exposed to the environment). This means that interference problems at very high frequencies are more likely to be due to emanations from nearby boards. Sniffers are available to detect sources of high-level emissions [9.1].

Sections 8.4–8.6 are devoted to describing ways of designing the interface circuitry. The most important consideration is that the design of the circuitry at both ends of the cable assembly should be considered at the same time. Books on circuit design [9.2–9.4] can provide guidance on the design of the functional circuitry, while the book *EMC Design Techniques* [1.10] has an excellent chapter on printed circuit boards.

9.4 Susceptibility requirements

Susceptibility requirements can be defined in terms of the threat environment and the acceptable functional behavior of the circuit interfered with. For example, $\pm1\%$ maximum error is harder to design for than $\pm10\%$.

Although there is a wide range of test specifications defined in a host of EMC regulations, it should usually be possible to define the threat environment in terms of a frequency response curve. Table 9.4.1 defines the frequency characteristics of the environment to which some U.K. military equipment can be subjected [9.5].

It is clear from the table that the most severe threat to military equipment lies in the range between 100 kHz and 100 MHz. So this is the range depicted in Figure 9.4.1.

Table 9.4.1   Frontline and operational support equipment field strength

Frequency	Average field strength (V/m)
10 kHz–100 kHz	10
100 kHz–500 kHz	10
500 kHz–1.6 MHz	10
1.6 MHz–5 MHz	560
5 MHz–10 MHz	380
10 MHz–30 MHz	200
30 MHz–100 MHz	200
100 MHz–200 MHz	60
200 MHz–700 MHz	70
700 MHz–1 GHz	60

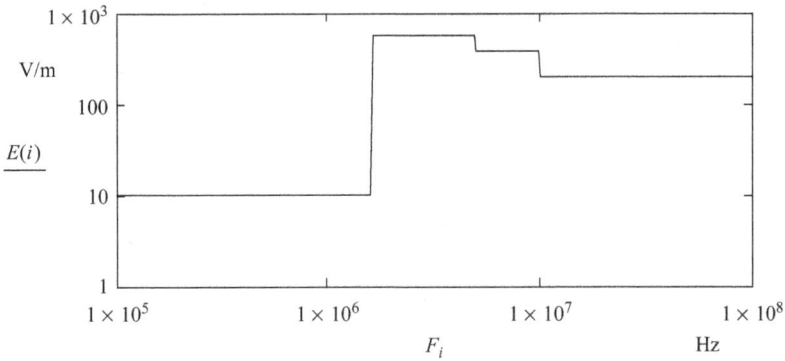

Figure 9.4.1 Example of military threat environment, 100 kHz to 100 MHz.

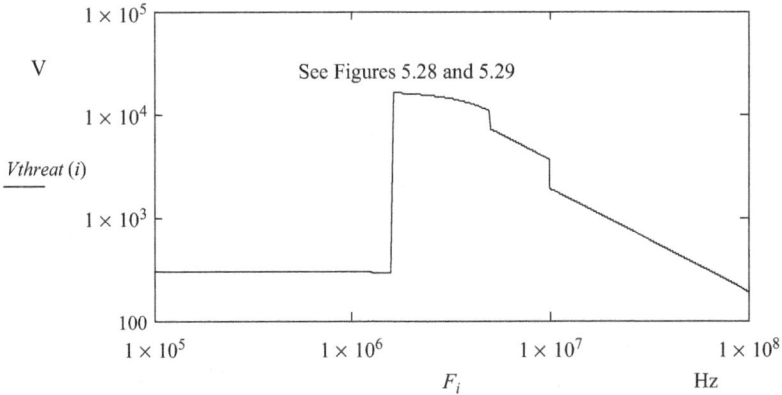

Figure 9.4.2 Threat voltage induced in the common-mode loop for a 15 m link.

In section 5.5, a method of relating the power density of the external interference to the differential-mode current induced in a signal link was derived. A Mathcad worksheet was used to calculate the frequency response of induced current in that link to a field of constant power density. Electric field strength is related to power density by:

$$E = \sqrt{Zo \cdot S} \qquad (9.4.1)$$

This being so, the response of Figure 5.5.9 could also be described as the response of the assembly to a field of constant electric field strength. Since the contractual requirements usually define the threat environment in terms of a field of varying intensity, it is necessary to include this parameter as a variable.

Given knowledge of the length of the assembly-under-review, it is a simple matter to define the threat in terms of the voltage that will be induced in the common-mode loop. Figure 9.4.2 illustrates the variation of that voltage with frequency for a signal link of 15 m. This is the same length as the link described in section 5.5, a twin conductor cable routed over a conducting structure. The computations are carried out by page 1 of Worksheet 9.4 and illustrated by Figure 9.4.3. It is worth noting that the threat voltage peaks at over 10 kV.

A representative circuit model was created in section 5.5 to simulate the coupling between the common-mode loop and the differential-mode loop, and a worksheet was derived to calculate the amplitude of the interference current in the signal circuitry. By applying the threat voltage *Vthreat* to the culprit loop, the current *Iout* can be calculated. The response is illustrated by Figure 9.4.4.

The computations were carried out by adding pages 1, 2, and 3 of Worksheet 5.5 to Worksheet 9.4. (See figures 5.5.3, 5.5.6, and 5.5.7)

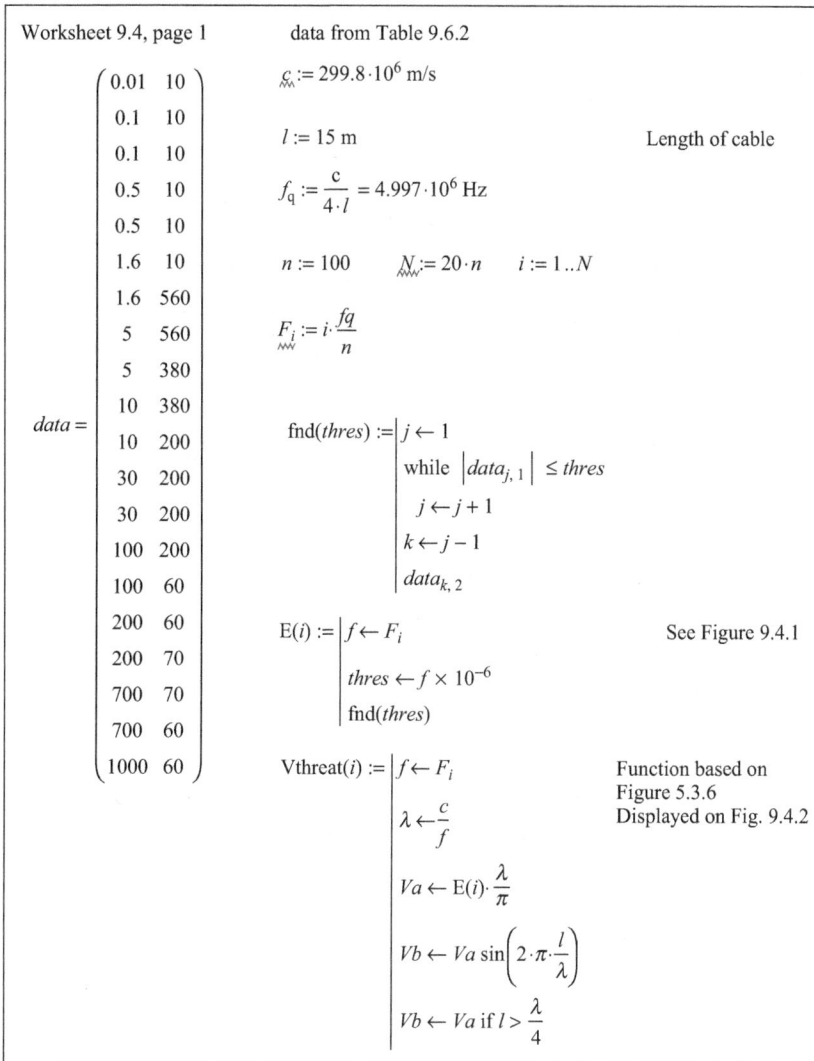

Worksheet 9.4, page 1   data from Table 9.6.2

$$data = \begin{pmatrix} 0.01 & 10 \\ 0.1 & 10 \\ 0.1 & 10 \\ 0.5 & 10 \\ 0.5 & 10 \\ 1.6 & 10 \\ 1.6 & 560 \\ 5 & 560 \\ 5 & 380 \\ 10 & 380 \\ 10 & 200 \\ 30 & 200 \\ 30 & 200 \\ 100 & 200 \\ 100 & 60 \\ 200 & 60 \\ 200 & 70 \\ 700 & 70 \\ 700 & 60 \\ 1000 & 60 \end{pmatrix}$$

$c := 299.8 \cdot 10^6$ m/s

$l := 15$ m   Length of cable

$f_q := \dfrac{c}{4 \cdot l} = 4.997 \cdot 10^6$ Hz

$n := 100 \qquad N := 20 \cdot n \qquad i := 1..N$

$F_i := i \cdot \dfrac{fq}{n}$

$\text{fnd}(thres) := \begin{vmatrix} j \leftarrow 1 \\ \text{while } \left| data_{j,1} \right| \leq thres \\ \qquad j \leftarrow j + 1 \\ k \leftarrow j - 1 \\ data_{k,2} \end{vmatrix}$

$\text{E}(i) := \begin{vmatrix} f \leftarrow F_i \\ thres \leftarrow f \times 10^{-6} \\ \text{fnd}(thres) \end{vmatrix}$ See Figure 9.4.1

$\text{Vthreat}(i) := \begin{vmatrix} f \leftarrow F_i \\ \lambda \leftarrow \dfrac{c}{f} \\ Va \leftarrow \text{E}(i) \cdot \dfrac{\lambda}{\pi} \\ Vb \leftarrow Va \sin\left(2 \cdot \pi \cdot \dfrac{l}{\lambda}\right) \\ Vb \leftarrow Va \text{ if } l > \dfrac{\lambda}{4} \end{vmatrix}$

Function based on Figure 5.3.6
Displayed on Fig. 9.4.2

Figure 9.4.3   Calculating the threat voltage for a 15 m cable.

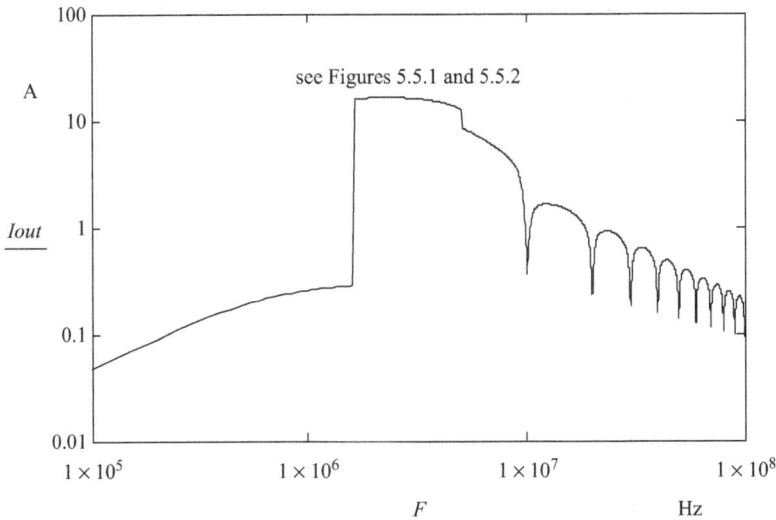

Figure 9.4.4   Frequency response of current in the differential-mode loop.

It is clear from this graph that the critical range of frequencies lies between 1.5 and 10 MHz. So the design of the interface circuits, the shielding, and the signal processing will need to focus on providing protection to the system function over this range of frequencies.

The above example is intended for illustrative purposes only, to show how system design can be related to the susceptibility requirements.

It is also worth noting that this is a worst-case analysis. In practice, there will be a significant reduction in threat level due to shielding afforded by the structure as well as re-radiation of the interference back into the environment. Although this is a limitation of the circuit modeling approach, any errors will be on the side of caution.

9.5   Emission requirements

As with susceptibility requirements, there is a host of formal requirements for controlling the emission characteristics of electrical equipment, and again it is possible to use a military requirement for illustrative purposes. Table 9.5.1 is a copy of the emission requirements defined in the U.K. Defence Standard for signal and secondary power lines [9.6].

Table 9.5.1   Emission requirements of DCE02B

Frequency	Current (dBμA)
20 Hz	130
2 kHz	130
50 kHz	50
500 kHz	20
150 MHz	20

The test involves the use of a current transformer clamped round the cable-under-test and connected to a current measuring device. The values in the second column of the table define points on the characteristic. Figure 9.5.1 illustrates that characteristic. It defines the upper limit for measured emissions.

Creating this graph was a simple matter of converting the values of dBμA in the second column of the table to values in amperes and plotting these values at the appropriate frequency. Figure 9.5.2 reproduces the first section of Worksheet 9.5 because some readers might be unfamiliar with the relationship between dBμA and A.

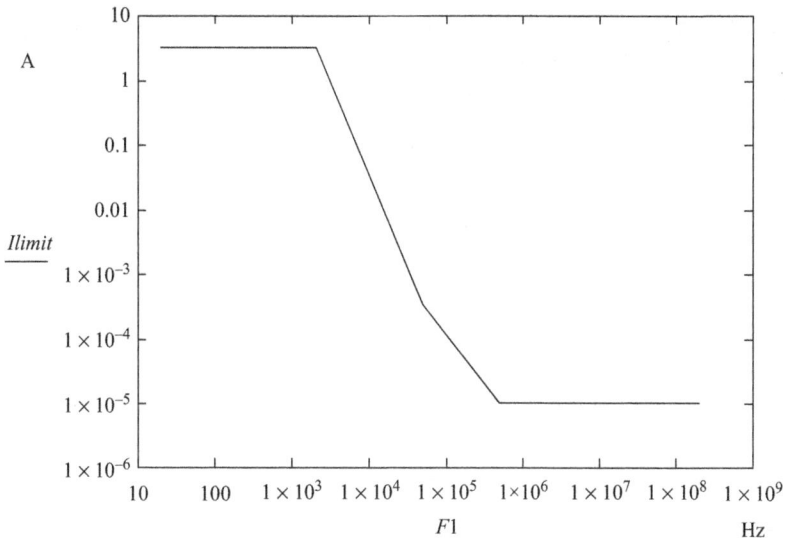

Figure 9.5.1 Maximum acceptable emission.

Worksheet 9.5, page 1

$$\text{Amps}(dBmicroA) := \left| \begin{array}{l} dBA \leftarrow dBmicroA - 120 \\ y \leftarrow \dfrac{dBA}{20} \\ A \leftarrow 10^{y} \end{array} \right.$$

$$data := \begin{pmatrix} 20 & 130 \\ 2000 & 130 \\ 50\cdot 10^3 & 50 \\ 500\cdot 10^3 & 20 \\ 200\cdot 10^6 & 20 \end{pmatrix} \qquad F1 := data^{\langle 1\rangle} \qquad dB := data^{\langle 2\rangle}$$

$$i := 1..\text{rows}(data) \qquad Ilimit_i := \text{Amps}(dB_i)$$

Figure 9.5.2 First page of worksheet 9.5.

The emission requirements illustrated by Figure 9.5.1 can be compared to the predicted response of any given signal link, for example, the link depicted by Figure 9.5.3.

This particular link is used as an example because it is the one which was subjected to a susceptibility analysis in the previous section.

Analyzing the emission characteristics of this link can be carried out by copying pages 1, 2, and 3 of Worksheet 5.5, (Figures 5.5.3, 5.5.6, and 5.5.7) adding them to Worksheet 9.4, and making a few minor changes to the program. Basically, these changes are:

- setting the value of the radiation resistance *Rrad* to zero (Inserting a 50 Ω resistor in series with the structure is only necessary when susceptibility is being analyzed.),
- setting the voltage source in loop 2 of the model to zero (When simulating emission, it is assumed that there is no voltage source in the common-mode loop.),
- setting the voltage source in loop 1, *Vout*, to unity (The voltage source is assumed to be in the differential-mode loop.).

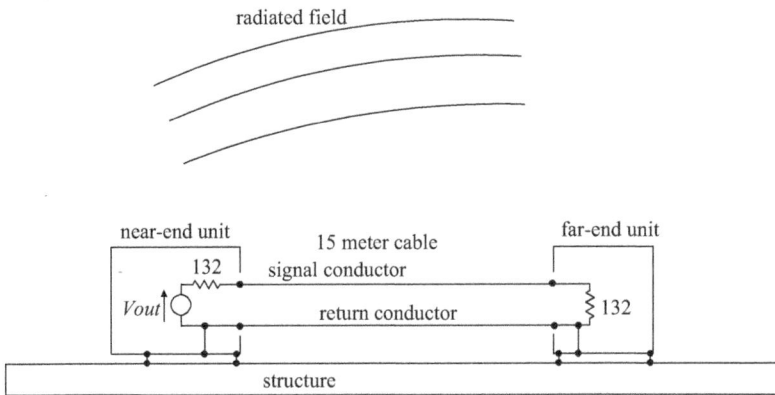

Figure 9.5.3   Estimating the level of the radiated field.

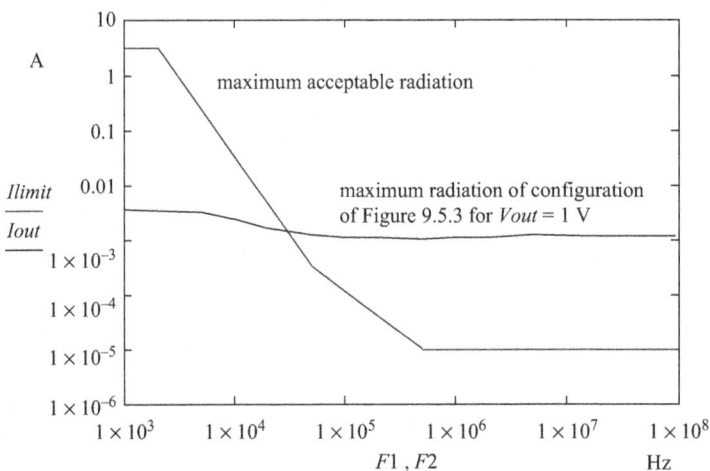

Figure 9.5.4   Comparing the response of the proposed design with the emission requirements.

It is clear from this analysis that the configuration-under-review is non-compliant with the requirements at frequencies above 50 kHz.

However, this analysis is of the worst-case conditions.

The design can be remedied by

- limiting the bandwidth and the amplitude of the signal to that which is essential for the intended function, and
- implementing one or more of the techniques described in Chapter 8.

9.6 Planning

9.6.1 Performance requirements

These usually take the form of a set of EMC test requirements for the particular system being developed. Several groups have developed independent sets of such requirements: military, aircraft, spacecraft, automotive, ships, commercial, etc. Different countries have different regulations. It is usually the case that the manufacturer involved in designing the product already has a full set of regulatory documents in place. If not, then there are several consultancies that specialize in offering advice. *The International Journal of Electromagnetic Compatibility* [9.7] identifies a number of such organizations.

It is sometimes the case that external EMC requirements are inadequate to meet the needs of safety or reliability. In which case, in-house EMC requirements may be called for.

Once the EMC requirements have been defined, the task is to design the equipment to meet those requirements. Normal procedure is to carry out formal EMC testing after all other tests have been completed. If the result of the formal test process is a number of non-compliances, then the equipment will need to be modified and resubmitted. This can be an expensive process, since at this stage all the manufacturing drawings will have been completed, all the manufacturing processes will have been written, and all the components will have been purchased. Also, there is not any guarantee that the modified equipment will be compliant.

The previous chapters of this book have shown that a clear correlation can be established between test requirements and the detailed design of the equipment. This allows EMC to be treated in the same way as requirements such as frequency response, functional performance, mass, size, reliability, and cost. This being so, analysis and test of the EMC requirements can be integrated into normal design process.

9.6.2 Bench test equipment

Two essential items of equipment are needed to supplement the general-purpose test equipment; the current transformer and the voltage transformer. Although these transducers are available commercially, they are tailored to meet the needs of the formal regulatory requirements rather than the needs of the equipment-under-development. They are also expensive.

Chapter 7 provides details of the design, manufacture, and calibration of simple, low-cost devices that can be constructed in-house. These particular devices were designed to cover a frequency bandwidth of about 20 kHz to 20 MHz. The coverage can be extended downwards by using larger devices or upwards by using smaller devices. The devices described in sections 7.1 and 7.2 use 10-turn windings on the toroids to ensure that the impedance reflected by the test equipment into the loop-under-test is as small as possible.

Some commercial transformers extend the frequency range by employing just one turn on the primary winding.

It would be a worthwhile exercise to manufacture a set of triaxial cables, as described in section 7.3. It is important that the test equipment does not create spurious coupling with the equipment-under-test.

The use of such devices need not be restricted to bench testing during the development process; they can also be used for in situ testing during trials and trouble-shooting. The availability of bench test equipment to check EMC enables newly purchased equipment to be assessed before it is integrated into the system.

9.6.3 Software

Any organization involved in the design of electronic equipment will already have a number of personal computers devoted to the design and manufacturing process. SPICE software will be installed on many of these to analyze the functional performance of the circuitry on the printed circuit boards.

Such software is based on the use of nodal analysis and the existence of a reference node at zero volts. If it is assumed that all connections to the reference plane are at zero volts, then extremely complex boards can be analyzed. However, this assumption carries with it the hidden assumption that there is no such thing as interference.

This does not mean that SPICE software is incapable of analyzing interference coupling. It is entirely possible for this software to analyze any of the lumped parameter models of Chapter 2. Moreover, such models can be combined with models of complex printed circuit boards. Provided the user is aware of the limitations of such software, valuable information can be gleaned from the exercise.

However, there will always be occasions where such an approach is inadequate, making it necessary to use mathematical software such as Mathcad. This type of software has the advantage that the computations are not restricted to circuit models, as many of the worksheets in this book can confirm. If both types of software are installed on the same computer, it becomes possible for them to exchange data.

9.6.4 Critical signal links

In any complex system, it is necessary to select the signal links that may be critical in terms of high susceptibility or high emission. So it is useful to identify the criteria used during the selection process.

The most obvious criterion is the length. Equation (5.1.15) indicates that the power received by a dipole receiver is proportional to the square of the wavelength. If the link-under-review is treated as one monopole, with the structure acting as the other monopole, as illustrated in Figure 5.3.1, then maximum power will be received when the actual length is equal to one quarter wavelength of the interfering signal.

If the cable is routed along the structure, then the response of Figure 5.3.7 should provide guidance on the level of interference to expect at different frequencies.

As far as susceptibility is concerned, the most critical links are often, but not always, those that carry very low level signals, such as those from remote transducers.

As far as emission is concerned, the most critical links are those that carry the highest transient currents. Power supplies fall into this category, since they almost always contain switch-mode converters. The transmission of fast digital signals between separate items of equipment can also create high levels of emission.

It is normal practice to create a block diagram at the early stage of a project to identify the main items of equipment and to define their function. Since the links between blocks at this level of definition are likely to be the longest, as well as carrying the most important signals, such a diagram can be used to identify the most critical links.

9.6.5 Critical frequencies

Experience has shown that both emission and susceptibility are characterized by a series of peaks and troughs. Since EMC requirements are usually defined in terms of limits which should not be exceeded, it is invariably at peaks in the response that these levels are exceeded. Any analysis of the interference coupling mechanisms needs to focus on these peaks, on the frequencies at which they occur as well as on their amplitude.

Since the peaks occur at frequencies corresponding to multiples of the quarter-wave frequency of the line and since the quarter-wave frequency is a function of the length of the signal link, it is not difficult to make a first estimate as to what the critical frequencies are. Figure 5.5.8 illustrates this for the configuration of Figure 5.5.1.

The relationship between power and frequency ensures that there is less power available at frequencies higher than that corresponding to a quarter-wavelength. If it is assumed that the threat level is constant with frequency, the frequency at which the first resonance occurs is likely to be the most critical. Even so, some EM environments have very much higher levels at very high frequencies. Hence, it is important to consider the entire range of frequencies when assessing the requirements.

9.6.6 Characterization

Although the methodology described in this book can be used to create circuit models of any particular type of cable coupling using manufacturer's data, there are always uncertainties. For example, the relative permittivity and the radiation resistance are usually unknown variables (unless one has to hand a field solver that can accurately predict permittivity and radiation resistance from dimensions and material characteristics).

If the link-under-review is constructed from different types of cable, or if the effect of the terminations is uncertain, tests on a representative rig can be used to create a representative circuit model that can be used with confidence.

However, it is not necessary to characterize every signal link in the system. The representative circuit model of a particular type of cable can be applied to every cable of that type. Since the L, C, R, and G parameters of a cable are proportional to length, tests on a long cable can be used to define the characteristics of a short cable.

Tests of long cables over a low frequency range can be used to predict the response at high frequencies. When carrying out such a test, care should be taken that the cross section is uniform along the entire length and that the effect of the terminations is minimal. The test described in section 7.5 is an example.

One observation made during these tests was that it did not seem to matter how the cable was laid out, or whether or not there were tight bends. These features might cause local hotspots in the field distribution, but they did not seem to have any effect on the overall response.

Another observation was that when two conductors are twisted together to create a twin-core cable and subsequently separated, the lengths of the two conductors were different. The more turns per meter, the greater the difference. If the lengths are different, then so are the impedances. If the impedances are different, then the coupling between common-mode and differential-mode signals must increase. Specifying a high number of turns per meter is not a good idea. There are other ways of achieving close spacing between conductors, for example, ribbon cable.

Capacitors are really open-circuit transmission lines coiled into a small volume. Just like transmission lines, they have inductive properties. Unlike transmission lines, there is only one resonant frequency, as the response of Figure 7.7.3 illustrates. This could be due to capacitive coupling between adjacent turns. Interference within the component seems to cancel out the higher frequency resonances that would occur with a normal transmission line.

Tests can be devised to measure the value of each 'component-within-a-component' as illustrated by Figure 7.7.4. By applying an input signal via a 50 Ω cable and monitoring the output via another 50 Ω cable, the component can be measured over a range that includes a peak or a null in the frequency response. A circuit model can be postulated to simulate the component. Analysis of this model allows the theoretical curve to provide a faithful simulation of the measured response. Adjusting component values of the model allows the actual and simulated responses to coincide. The end result is a representative circuit model of the component, valid over the frequency range of the tests carried out.

By creating a library of representative circuit models to simulate signal links and components, it becomes possible to simulate any signal link in the system. This is essentially the approach used with SPICE software, but adapted to meet the needs of EMC.

Several examples of representative circuit models are provided in the previous pages, and these are listed in Table 9.6.1.

These have been derived from the small set of general circuit models listed in Table 9.6.2.

9.6.7 General approach

Since there is a multitude of regulatory test specifications, there is no point in attempting to compile a list of such regulations. In any case, the relevant set of regulations is dependent on the type of equipment being assessed. However, these requirements have one common factor; they all specify the use of electronic equipment to carry out the tests. Previous chapters of this book have shown that it is always possible to create a model of the coupling mechanism between two independent circuits. So it is always possible to relate the performance of the equipment under test to that of the formal test setup.

Formal requirements are not the only factor to consider when assessing the EMC of the product-under-review. Cross-coupling within the system will inevitably cause problems. Nor is there any guarantee that a product that meets the formal requirements will experience no interference problems during its lifetime.

Whether or not it is standard design procedure it is useful to carry out a failure mode, effects and criticality analysis (FMECA) on the system-under-review; it is a worthwhile

Table 9.6.1   Representative circuit models

Description	Figure
Screened pair example	3.3.4
Cross-coupling example	4.3.3
Conducted emission test setup	4.4.3
Conducted susceptibility test setup	4.4.5
Exposed cable – simplified	5.4.3
Radiation coupling via the structure	5.5.4
Transient emission from cable	6.6.5
Voltage transformer	7.1.6
Current transformer	7.2.7
Cable under test	7.5.12
Murata 100 nF capacitor	7.7.4

Table 9.6.2   General circuit models

Description	Figures
Three-conductor signal link	4.2.4
Isolated conductor	5.2.2 and 7.4.4
Isolated cable	5.2.8 and 7.5.5
Exposed cable and structure	5.3.2
Exposed cable and structure – simplified	5.4.1
Transient emission	6.6.1

exercise to spend some time assessing the possible problems that could arise due to EMI. The section 'Banana Skins' in every issue of the *EMC Journal* [9.8] provides some hair-raising examples of what can go wrong.

When dealing with any problem relating to electromagnetic interference, the approach of the circuit designer should always be to create an interface diagram of the link-under-review such as the one illustrated by Figure 9.2.4, and assess its characteristics using the guidelines of section 9.1. This task should take less than a day for even the most convoluted link.

If this does not immediately identify the cause of the problem, the diagram can be used to create and assess a circuit model of the link. The results of this initial assessment can determine whether there is a simple solution or whether more detailed analysis is necessary. Tests on a prototype assembly can be devised to check that the solution is viable before implementing the modification.

The objective should be to acquire a high degree of confidence that the system-under-review will meet the defined requirements without over-designing the protection measures. It is much easier to do this when the design is based on the results of circuit modeling than when it is based on the consensus of opinion at a design review meeting.

If further development is needed to improve the design of a critical assembly, then the information acquired by the circuit modeling approach can be used as a baseline for more sophisticated analyses using three-dimensional field solvers.

Mathcad worksheets

As far as circuit modeling is concerned, Mathcad software eliminates most of the tedium involved in developing programs. It can handle tasks that include the calculation of values of circuit components, analysis of the response of the circuit model, analysis of test data to define the response of an assembly under test, and display a single graph showing the responses of model and hardware. This can all be done on the same worksheet; that is, without the need to invoke subroutines that are recorded on separate files. Moreover, the equations used in the computations are displayed in exactly the same way as they would appear in a textbook on circuit theory.

In a programming language, equations look something like:

$$x = (-B + SQRT(B^{**}2 - 4^*A^*C))/2^*A$$

With Mathcad software, the same equation looks the same way as it does in a reference book:

$$x = \frac{-b + \sqrt{b^2 - 4 \cdot a \cdot c}}{2 \cdot a}$$

This makes the programs much easier to read and understand. Mathcad software also avoids the need to use a rigorous procedure to write a program. If the programmer gets the syntax wrong, the software refuses to accept the entry; and provides a message indicating what is wrong.

On a worksheet, the equations can be set out from left to right, top to bottom. Text can be included anywhere on the page.

Doubtless there are other software packages which provide the same sort of facility. However, for consistency in presentation, Mathcad is the one used here.

Since this software has been available for over a decade, there is a fair probability that the reader is already using it. If not, it should be fairly easy task to modify the programs in the worksheets of this document to convert them into programs which work with other software packages.

A few symbols in Mathcad have a special meaning:

$a := b + c$ 'a' is defined as the sum of b and c

$a = 2$ the value of 'a' is 2

If the 'equals' sign is in bold typeface, the software interprets this as the 'boolean equal' and returns a zero or a one.

The programming operator looks like that shown on Figure A.1:

$$\underset{\sim\sim\sim}{\text{root}}(a, b, c) = \begin{array}{|l} discr \leftarrow b^2 - 4 \cdot a \cdot c \\[1em] num \leftarrow -b + \sqrt{discr} \\[1em] denom \leftarrow 2 \cdot a \\[1em] \dfrac{num}{denom} \end{array}$$

Figure A.1 Simple Mathcad program.

The programming operator behaves like a function, taking input variables and returning an output. This output is the last variable to be declared. In the example above it is the value of the ratio of '*num*' divided by '*denom*'. The program can return a single variable, a vector, or an array.

Local variables defined in the function are not visible outside. However, variables declared in the worksheet above the program function are visible within the function.

Included in the software are a number of built-in functions. As far as circuit modeling is concerned, the most important is 'lsolve(M, **v**)'. The argument 'M' is a square array, while '**v**' is a vector. 'M' contains as many rows as there are elements in '**v**'. This function returns a solution vector '**x**' such that M**x** = **v**.

Another useful characteristic of the software is that it can distinguish between the 'j' operator as used with complex numbers and the variable 'j'. It can be defined as an imaginary number by typing the characters '1j'. It appears on the worksheet as 'j'.

It is also possible to intermingle text with variables. This facility is put to good use in defining the units appropriate to each variable. To avoid cluttering up the worksheets, this is usually restricted to places where a variable is defined or a final result is displayed.

Armed with this information, there should be no real problems in understanding the programs contained in the Mathcad worksheets of this book. Since the figures in the book which depict Mathcad worksheets have been hand-copied from copies of the actual worksheets, it is possible that errors have been introduced in the transcription process. Further errors could be introduced by the reader hand-copying the text onto his or her own computer. To prevent the chance of such errors creeping in, copies of the original Mathcad files are available in a zip folder which can be downloaded from www.designemc.info. These files can be run on any computer which has Mathcad version 15.0 (or higher) software installed.

The techniques described in the previous pages can be used to simulate the interference-coupling characteristics of any signal link. Every electrical or electronic system will have its own particular set of interference problems. It should be possible for individual designers to carry out tests and create circuit models of their own critical links. Such information can be shared. Mathcad users have access to PlanetPTC, a mix of dynamic channels that enables PTC customers and product development professionals to actively participate in exchanging ideas.

However, it does not really matter which software is used to carry out the calculations. The key feature of any circuit model is the fact that it is a shorthand method of defining a set of equations. It becomes possible to describe the hardware, the tests, the model, and to illustrate the results in a single report. There are many forums and many communities which can be used to publish such reports.

MATLAB®

Translating Mathcad worksheets into MATLAB® m-files was a fairly simple task. That being so, it should not be difficult for MATLAB users to understand the function of the Mathcad worksheets replicated in this book. There are some differences which could be puzzling at first.

With MATLAB, there are five windows; Command Window, Command History, Current Folder, the Workspace, and the Editor. The program is really a text file in the Editor Window, and needs to be saved to a file before it is run, a task performed by typing and entering a command in the Command Window. All variables and their computed values are accessible in the separate Workspace. The Current Folder holds the files which are accessible to the Command Window, and the Command History keeps a record of all the actions taken to date. If a computation is carried out to produce a graphical picture, the picture is displayed in a separate Figure Window.

With Mathcad, there is only one window; the Worksheet. Variables and expressions can be defined anywhere on the page and intermingled with text, in the same way that calculations are recorded in a laboratory notebook. There is no list of numbers down the left-hand side of the page. Of course, there is a basic rule; the sequence of commands must be left to right; top to bottom. Graphs are created and modified by invoking menu commands. The 'run' command is invoked by clicking anywhere on an empty portion of the worksheet. This means that everything; definitions, expressions, functions, numerical results, graphs, and explanatory text, is displayed on the worksheet. Any part of the worksheet can be copied and pasted onto a Microsoft word document. Copies of the Mathcad worksheets are available in rich text format (.rtf) at www.designemc.info.

Both Mathcad and MATLAB work with arrays and vectors, and the related operations are called up by similar expressions. For example; $A_{i,j}$ in a Mathcad worksheet can be translated into $A(i, j)$ in MATLAB, $V(k) := 0$ into $V = \text{zeros}(k,1)$, lsolve(Z, V) into linsolve(Z, V), etc. Similarly, $\ln(f)$ can be translated into $\log(f)$, $|x|$ into abs(x), $\sqrt{x}$ into sqrt(x), etc.

In Mathcad, all vectors are column vectors; in MATLAB, vectors can be defined as row vectors or column vectors.

In Mathcad function statements, the parameters on the worksheet are 'visible' to the computations in that function. In MATLAB, they are not; every variable used by a function must be included in the set of input variables.

In Mathcad, the function statements form part of the worksheet. In MATLAB, these are stored as separate Function Files. This means that those MATLAB m-files which invoke the use of special functions must have the relevant Function Files available in the Current Folder.

If in doubt about the method of computation employed by any worksheet, the reader who is familiar with MATLAB can easily clarify its purpose by examining the text of the equivalent m-file. Each Mathcad worksheet has been translated into one or more MATLAB files and the set of files stored in a zip folder. This folder can be downloaded from www.designemc.info.

In the preceding pages of the book, each page of a worksheet is presented in the form of a figure, and these figures are interspersed with diagrams and descriptive text. To ease the task of relating the MATLAB files to the Mathcad worksheets, the text of each worksheet is available in its entirety in portable document format (pdf) file, and these files are also available at the above website.

The hybrid equations

It is entirely possible that the hybrid equations of section 4.1 are unfamiliar to engineers, even those who are experienced in the application of electromagnetic theory. If the starting-point equations are viewed with scepticism, then designers are unlikely to have any confidence in the succeeding mathematics. That being so, it is necessary to provide proof of the validity of (4.1.4) and (4.1.5).

It is even more likely that the average circuit designer has never seen them before. This section provides a succinct derivation of these equations. It is also made clear that both conductors of the line possess the properties of inductance, resistance, capacitance, and conductance. By defining the variables in terms of loop parameters, it avoids the use of the concept of the equipotential ground. It is basically a set of lecture notes copied from a blackboard at Glasgow University in 1959.

Figure C.1 illustrates how voltage and current can vary along the length of a transmission line.

For the purpose of deriving the transmission line equations, the following definitions apply:

R = resistance per unit length: Ω/m

L = inductance per unit length: H/m

C = capacitance per unit length: F/m

G = conductance per unit length: S/m

Voltage and current are functions of both x and t. For the section dx:

$$(R + j \cdot \omega \cdot L) \cdot dx \cdot I = V - \left(V + \frac{\delta V}{\delta x} \cdot dx \right) = -\frac{\delta V}{\delta x} \cdot dx \tag{C.1}$$

$$(G + j \cdot \omega \cdot C) \cdot dx \cdot V = I - \left(I + \frac{\delta I}{\delta x} \cdot dx \right) = -\frac{\delta I}{\delta x} \cdot dx \tag{C.2}$$

Hence:

$$\frac{\delta V}{\delta x} = -(R + j \cdot \omega \cdot L) \cdot I \tag{C.3}$$

273

$$\frac{\delta I}{\delta x} = -(G + j \cdot \omega \cdot C) \cdot V \tag{C.4}$$

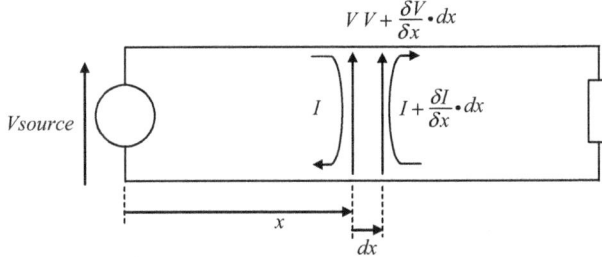

Figure C.1   Transmission line voltages and currents.

Differentiating (C.3) gives:

$$\frac{\delta^2 V}{\delta x^2} = -(R + j \cdot \omega \cdot L) \cdot \frac{\delta I}{\delta x} = (R + j \cdot \omega \cdot L) \cdot (G + j \cdot \omega \cdot C) \cdot V \tag{C.5}$$

Differentiating (C.4) gives:

$$\frac{\delta^2 I}{\delta x^2} = -(G + j \cdot \omega \cdot C) \cdot \frac{\delta V}{\delta x} = (R + j \cdot \omega \cdot L) \cdot (G + j \cdot \omega \cdot C) \cdot I \tag{C.6}$$

Substituting γ^2 for $(R + j \cdot \omega \cdot L) \cdot (G + j \cdot \omega \cdot C)$ in (C.5):

$$\frac{\delta^2 V}{\delta x^2} = -\gamma^2 \cdot V \tag{C.7}$$

This relationsip can be derived from:

$$V = C \cdot e^{\gamma \cdot x} + D \cdot e^{-\gamma \cdot x} \tag{C.8}$$

This can be confirmed by differentiating (C.8) twice. Another way of defining this relationship is:

$$V = A \cdot \cosh(\gamma \cdot x) + B \cdot \sinh(\gamma \cdot x) \tag{C.9}$$

This is because:

$$\begin{aligned}
V &= A \cdot \left(\frac{e^{\gamma \cdot x} + e^{-\gamma \cdot x}}{2} \right) + B \cdot \left(\frac{e^{\gamma \cdot x} - e^{-\gamma \cdot x}}{2} \right) \\
&= \frac{A + B}{2} \cdot e^{\gamma \cdot x} + \frac{A - B}{2} \cdot e^{-\gamma \cdot x} \\
&= C \cdot e^{\gamma \cdot x} + D \cdot e^{-\gamma \cdot x}
\end{aligned}$$

From (C.3) and (C.9):

$$I = -\frac{1}{R + j \cdot \omega \cdot L} \cdot \frac{\delta V}{\delta x} = -\frac{\gamma}{R + j \cdot \omega \cdot L} \cdot [A \cdot \sinh(\gamma \cdot x) + B \cdot \cosh(\gamma \cdot x)]$$

Now, since:

$$\frac{(R + j \cdot \omega \cdot L)}{\gamma} = \frac{R + j \cdot \omega \cdot L}{\sqrt{(R + j \cdot \omega \cdot L) \cdot (G + j \cdot \omega \cdot C)}} = \sqrt{\frac{R + j \cdot \omega \cdot L}{G + j \cdot \omega \cdot C}} = Zo$$

then:

$$I = -\frac{1}{Zo} \cdot [A \cdot \sinh(\gamma \cdot x) + B \cdot \cosh(\gamma \cdot x)] \tag{C.10}$$

Figure C.2 illustrates the boundary conditions of the line.

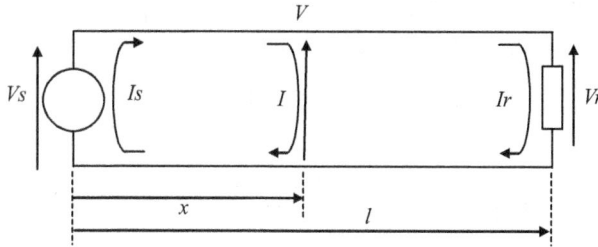

Figure C.2   Transmission line – boundary conditions.

From Figure C.2, boundary conditions are:

$$V = Vr \quad \text{at } x = l$$
$$I = Ir \quad \text{at } x = l$$

At the receiving end, x can be replaced by l in (C.9) and (C.10):

$$Vr = A \cdot \cosh(\gamma \cdot l) + B \cdot \sinh(\gamma \cdot l) \tag{C.11}$$

$$-Zo \cdot Ir = A \cdot \sinh(\gamma \cdot l) + B \cdot \cosh(\gamma \cdot l) \tag{C.12}$$

Multiplying (C.11) by $\sinh(\gamma \cdot l)$ and (C.12) by $\cosh(\gamma \cdot l)$:

$$Vr \cdot \sinh(\gamma \cdot l) = A \cdot \cosh(\gamma \cdot l) \cdot \sinh(\gamma \cdot l) + B \cdot \sinh(\gamma \cdot l) \cdot \sinh(\gamma \cdot l) \tag{C.13}$$

$$-Zo \cdot Ir \cdot \cosh(\gamma \cdot l) = A \cdot \sinh(\gamma \cdot l) \cdot \cosh(\gamma \cdot l) + B \cdot \cosh(\gamma \cdot l) \cdot \cosh(\gamma \cdot l) \tag{C.14}$$

Subtracting (C.14) from (C.13) gives:

$$Vr \cdot \sinh(\gamma \cdot l) + Zo \cdot Ir \cdot \cosh(\gamma \cdot l) = B \cdot (\sinh^2(\gamma \cdot l) - \cosh^2(\gamma \cdot l)) = -B$$

Hence:

$$B = -Vr \cdot \sinh(\gamma \cdot l) - Zo \cdot Ir \cdot \cosh(\gamma \cdot l) \tag{C.15}$$

Similarly:

$$A = Vr \cdot \cosh(\gamma \cdot l) + Zo \cdot Ir \cdot \sinh(\gamma \cdot l) \tag{C.16}$$

substituting A and B in (C.9):

$$\begin{aligned}
V &= [Vr \cdot \cosh(\gamma \cdot l) + Zo \cdot Ir \cdot \sinh(\gamma \cdot l)] \cdot \cosh(\gamma \cdot x) \\
&\quad - [Vr \cdot \sinh(\gamma \cdot l) + Zo \cdot Ir \cdot \cosh(\gamma \cdot l)] \cdot \sinh(\gamma \cdot x) \\
&= Vr \cdot [\cosh(\gamma \cdot l) \cdot \cosh(\gamma \cdot x) - \sinh(\gamma \cdot l) \cdot \sinh(\gamma \cdot x)] \\
&\quad + Zo \cdot Ir \cdot [\sinh(\gamma \cdot l) \cdot \cosh(\gamma \cdot x) - \cosh(\gamma \cdot l) \cdot \sinh(\gamma \cdot x)]
\end{aligned}$$

Hence:

$$V = Vr \cdot \cosh[\gamma(l - x)] + Zo \cdot Ir \cdot \sinh[\gamma(l - x)] \tag{C.17}$$

Similarly, substituting A and B in (C.12) and reducing the hyperbolic products gives:

$$I = Ir \cdot \cosh[\gamma \cdot (l - x)] + \frac{Vr}{Zo} \cdot \sinh[\gamma \cdot (l - x)] \tag{C.18}$$

At the sending end, $x = 0$, $V = Vs$, $I = Is$. Substituting these values in (C.17) and (C.18) gives:

$$\begin{aligned}
Vs &= Vr \cdot \cosh(\gamma \cdot l) + Zo \cdot Ir \cdot \sinh(\gamma \cdot l) \\
Is &= \frac{Vr}{Zo} \cdot \sinh(\gamma \cdot l) + Ir \cdot \cosh(\gamma \cdot l)
\end{aligned} \tag{C.19}$$

where the propagation constant is:

$$\gamma = \sqrt{(R + j \cdot \omega \cdot L) \cdot (G + j \cdot \omega \cdot C)} \tag{C.20}$$

and the characteristic impedance is:

$$Zo = \sqrt{\frac{R + j \cdot \omega \cdot L}{G + j \cdot \omega \cdot C}} \tag{C.21}$$

It is worthwhile emphasizing the fact that, in this derivation, the parameters R, L, G, and C are defined in terms of Ω/m, H/m, S/m, and F/m. The concept of 'per-unit-length' parameters is inherent in all the transmission line equations to be found in textbooks. Even so, this appendix is the only place in this book where such parameters are invoked. Section 4.1 shows that the analysis of transmission line behavior can be related to the actual resistance, inductance, capacitance, and conductance of the conductors.

Definitions

These definitions are those used in this book. They are not necessarily the same as those appearing in other documents.

antenna-mode current	Unidirectional current in the conductors of a cable. That which flows between the cable and the environment.
buffer circuit	The circuit which forms an interface between the conductors of a signal link and the processing circuitry of the equipment unit.
circuit equations	A set of equations which relate voltages to currents in a circuit model.
circuit model	A model which obeys the rules of circuit theory and which simulates the behavior of the assembly under review.
circuit parameter	A parameter used in a circuit model.
common-mode current	Current which flows in the loop formed by the cable and the structure.
common-mode gain	The ratio of the output voltage of a buffer circuit to the common-mode input voltage when the differential-mode input is zero.
common-mode rejection	The ratio of the differential-mode gain to the common-mode gain. Usually quoted in terms of decibels.
composite conductor	A set of elemental conductors, aligned in parallel, which enables the distribution of currents or voltages in the actual conductor to be simulated.
conducted emission	The current induced in the common-mode loop by a voltage source in the differential-mode loop. See Figure 4.4.1.
conducted susceptibility	The current induced in the differential-mode loop by a voltage source in the common-mode loop. See Figure 4.4.5.
culprit loop	The loop in which the interfering voltage source is located.
current transformer	A transducer which monitors the magnetic field surrounding a group of conductors and generates a voltage proportional to the sum of the currents in those conductors.
delay-line model	A circuit model which simulates the differential-mode behavior of a transmission line.
differential-mode current	Current which flows in the loop formed by the signal and return conductors.
differential-mode gain	The ratio of the output voltage of a buffer circuit to the differential-mode input voltage when the common-mode input voltage is zero.

distributed parameters	Parameters which are derived from the use of per-unit-length parameters.
earth	The conductor(s) designated to carry fault current in an AC distribution network. Always connected to the conducting structure.
elemental conductor	A conductor which represents a small segment of the surface of a composite conductor.
EMC	Electromagnetic compatibility. The ability of a device, unit of equipment, or system to function satisfactorily in its electromagnetic environment without introducing intolerable electromagnetic disturbances to anything in that environment.
EMI	Electromagnetic interference.
EUT	Equipment under test.
floating configuration	A wiring configuration where the signal link is isolated from the structure at one (or both) ends.
general circuit model	A model which includes interface circuit components as well as cable coupling components, but which does not specify component values.
ground	Another name for the conducting structure. The terms 'ground', 'earth' and 'structure' are interchangeable in the analyses described in this book.
grounded configuration	A wiring configuration where the return conductor of the signal link is bonded to local structure at each end.
loop equations	A set of equations, derived from the primitive equations, which relate loop voltages to loop currents.
loop parameter	Parameter which has been derived from groups of primitive parameters or partial parameters. Loop parameters can be measured by electrical test equipment.
lumped parameter	A resistor, capacitor, or inductor which represent the relevant properties of a defined length of conductor. The term also applies to circuit components.
partial capacitance	The capacitance of a composite conductor, or the capacitance of a loop segment.
partial current	A portion of the current flowing in a conductor.
partial inductance	The inductance of a composite conductor, or the inductance of a loop segment.
partial inductor	Inductor associated with partial current or partial voltage.
partial parameters	Those parameters associated with the behavior of composite conductors as antennae. The term is also used to distinguish between incident and reflected currents in transmission lines.
partial voltage	A portion of the voltage in a circuit loop.
per-unit-length parameter	Resistance, inductance, capacitance, or conductance which is defined in terms of Ω/m, H/m, F/m, or S/m.
primitive capacitance	A component which relates the voltage on an isolated conductor to the energy contained by the electric field in which it is immersed.
primitive current	Unidirectional current flow in a single conductor of a multi-conductor assembly.
primitive equations	A set of equations relating primitive voltages to primitive currents.
primitive inductance	A component which relates the current in the conductor to the energy contained by the magnetic field in which it is immersed.
primitive parameters	Parameters associated with conductors which are treated as antennae.

primitive voltage	Voltage along a conductor due to electromagnetic coupling.
radiated emission	A type of test where the emission is measured by equipment connected to a receiving antenna located in the vicinity of the EUT.
radiation resistance	The apparent resistance of an antenna when it is carrying maximum current during resonance conditions.
radiation susceptibility	A type of test where the interference source is an antenna located in the vicinity of the EUT.
relative permittivity	The ratio of the average permittivity of the environment of the signal link to the permittivity of free space.
relative permeability	The ratio of the average permeability of the environment of the signal link to the permeability of free space.
representative circuit model	A circuit model which simulates the interference coupling mechanisms of a specific signal link.
return conductor	The conductor designated to complete the loop carrying the differential-mode current. (It also carries the common-mode current in the return/structure loop.)
signal conductor	The conductor which is exclusively allocated to carry the differential-mode current.
signal link	The conductors and interface circuitry involved in transmitting a signal from one location to another in a particular system.
SPICE	Simulation Program with Integrated Circuit Emphasis.
structure	The conducting elements of the structure. Can be referred to as 'ground' or 'earth'.
threat environment	The frequency response of the power density of the worst-case external radiation to which the equipment can be subjected. It can also be defined in terms of the waveform of a transient current or voltage in the structure.
threat voltage	The amplitude of the voltage in the culprit loop.
time-step analysis	A method used to predict currents and voltages in the circuit model a discrete time after a step change has occurred in any current or voltage in that model.
transfer admittance	The ratio of the current in the victim loop to the source voltage in the culprit loop, when there are no other voltage sources.
transfer impedance	The ratio of the amplitude of the voltage appearing in the output loop to the amplitude of the current in the input loop. With screened cable, this is the impedance of the screen.
victim loop	The loop carrying the signal which is regarded as being susceptible to interference.
virtual conductor	An imaginary conductor which enables the coupling between cable and environment to be simulated.
voltage transformer	A transducer which uses magnetic field coupling to induce a voltage in the loop-under-test and which allows the induced voltage to be measured.

References

Chapter 1

1.1. Williams, T. *EMC for Product Designers* (Section 1.1.1). 2nd edn. Jordan Hill, Oxford: Newnes; 1996. p. 4. ISBN: 0-7506-2466-3.

1.2. Europe EMC guide. *The International Journal of Electromagnetic Compatibility*. Retrieved from http://www.interferencetechnology.eu

1.3. Armstrong, K.: 'EMC design of SMP and PWM power converters'. *EMC Journal*. 2011, March: p. 28.

1.4. Williams, T. *EMC for Product Designers*. 4th edn. Jordan Hill, Oxford: Newnes; December 2006. ISBN: 0-750-68170-5.

1.5. Tesche, F., Ianoz, M., Karlsson, T. *EMC Analysis Methods and Computational Models*. New York, NY, USA: John Wiley & Sons, Inc., 1997. ISBN: 0-471-15573-X.

1.6. Defence Standard 59-411, Part 5, Issue 1, Amendment 1. Electromagnetic Compatibility. Part 5. Code of Practice for Tri-Service Design and Installation. (Section 8.9. Single Point Reference Connection). Glasgow, UK: Ministry of Defence; January 2007. p. 29.

1.7. Shepherd, J., Morton, A.H., Spence, L.F. *Higher Electrical Engineering* (Section 7.28. Equivalent Phase Inductance of a Three-Phase Line). Pitman, London, UK: Pitman Publishing Limited; 1985. pp. 234–235. ISBN: 0-273-40063-0.

1.8. Shepherd, J., Morton, A.H., Spence, L.F. *Higher Electrical Engineering* (Section 7.16. Equivalent Phase Capacitance of an Isolated Three-Phase Line). Pitman, London, UK: Pitman Publishing Limited; 1985. pp. 216–219. ISBN 0-273-40063-0.

1.9. Burrows, B.J.C. 'The computation and prediction of induced voltages in aircraft wings. CLSU memo 18'. April 1974. Culham Lightning, Units 13/15, Nuffield Way, Abingdon.

1.10. Armstrong, K. *EMC Design Techniques for Electronic Engineers*. Armstrong/Nutwood, UK: Nutwood UK Limited, Cornwall, UK; 2010. ISBN: 978-0-9555118-4-4. Retrieved from http://www.emcacademy.org/books.asp

1.11. Paul, C.R. *Introduction to Electromagnetic Compatibility*. 2nd edn. Hoboken, NJ, USA: Wiley-Interscience; January 2006. ISBN: 978-0-471-75500-5.

Chapter 2

2.1. Skitek, G.G., Marshall, S.V. *Electromagnetic Concepts and Applications* (Section 2.5. Electric Field Intensity of a Line of Charge). Englewood Cliffs, N.J., USA: Prentice Hall; 1982. ISBN 0-13-248963-5.

2.2. Page, L., Adams, N. I., *Principles of Electricity: Inductance of Straight Conductors*. New York, USA: D Van Nostrand; 1958. p. 325.

2.3. Skitek, G.G., Marshall, S.V. *Electromagnetic Concepts and Applications* (Section 8.3. Magneto-static Field Intensity from the Biot-Savart Law: Magnetic Field due to a Filamentary Current Distribution of Finite Length). Englewood Cliffs, USA: Prentice Hall; 1982. ISBN 0-13-248963-5.

2.4. Skitek, G.G., Marshall, S.V. *Electromagnetic Concepts and Applications* (Section 12.12. Skin Effect, and High and Low Loss Approximations). Englewood Cliffs, USA: Prentice Hall; 1982. ISBN 0-13-248963-5.

Chapter 3

3.1. Skitek, G.G., Marshall, S.V. *Electromagnetic Concepts and Applications* (Section 7.4. Image Solution Method: Capacitance between two Cylindrical Conductors). Englewood Cliffs, USA: Prentice Hall; 1982. ISBN 0-13-248963-5.

Chapter 4

4.1. Skitek, G.G., Marshall, S.V. *Electromagnetic Concepts and Applications* (Section 12.2. General Equations for Line Voltage and Current). Englewood Cliffs, USA: Prentice Hall; 1982. ISBN 0-13-248963-5.

Chapter 5

5.1. Skitek, G.G., Marshall, S.V. *Electromagnetic Concepts and Applications* (Section 14.4. The Half-Wave Dipole). Englewood Cliffs, USA: Prentice Hall; 1982. ISBN 0-13-248963-5.

5.2. Ordnance Board Pillar Proceeding P101 (Issue 2). 'Principles for the design and assessment of electrical circuits incorporating explosive components. (Annex. E. Appendix 1. The Radio Frequency Environment)'. p. E1-3. Bristol, UK.

Chapter 6

6.1. Savant, C. J, Jr., Roden, M.S., Carpenter, G. L., *Electronic Design – Circuits and Systems* (Appendix A. SPICE. Section A.2.4.3 Transient Analysis). 2nd edn. Redwood City, California: The Benjamin-Cummings Publishing; 1991. ISBN 0-8053-0292-1.

Chapter 7

7.1. Ediss Electric Ltd. Totton, Hampshire, UK: Ediss Electrical Ltd., Retrieved from http://www. ediss-electric.com

Chapter 8

8.1. Gnecco, L.T. *The Design of Shielded Enclosures*. Woburn, MA, USA; Newnes: 2000. ISBN 0-7506-7270-6.

8.2. Ordnance Board Pillar Proceeding P101 (Issue 2). 'Principles for the design and assessment of electrical circuits incorporating explosive components. (Annex. E. Appendix 1. Section 28: Shielding Assessment)'. p. E1–14. Bristol, UK.

8.3. Thomas & Betts Limited. *A Guide to BS EN 62305:2006. Protection against Lightning.* Nottingham, UK: Thomas & Betts Limited; 2008.

Chapter 9

9.1. 'EMC probes'. *Magnetic Sciences*. Retrieved from http://www.magneticsciences.com/ EMCProbes.html

9.2. Horowitz, P., Hill, W. *The Art of Electronics*. 2nd edn. Cambridge, CB2 1RP, UK: Cambridge University Press; 1989. ISBN 0-521-37095-7.

9.3. Savant, C. J, Jr., Roden, M.S., Carpenter. G. L., *Electronic Design – Circuits and Systems*. 2nd edn. Redwood City, California. The Benjamin Cummings Publishing; 1991. ISBN 0-8053-029-1.

9.4. Ludwig, R., Bretchko, P. *RF Circuit Design – Theory and Applications*. Upper Saddle River, New Jersey: Prentice Hall; 2000. ISBN 0-13-122475-1.

9.5. Defence Standard 59-411, Part 2, Issue 1, Amendment 1. 'Electromagnetic Compatibility. Part 2. The Electric, Magnetic & Electromagnetic Environment. Table 18. Front Line and Operational Support Equipment Field Strength'. Ministry of Defence; January 2008. p. 28.

9.6. Defence Standard 59-411, Part 3, Issue 1, Amendment 1. 'Electromagnetic Compatibility. Part 3. Test Methods and Limits for Equipment and Sub Systems. Appendix B.2. DCE02.B Conducted Emissions Control, Signal and Secondary Power Lines. 20 Hz–150 MHz. Figure 51. DCE02 – Limit for Air Service Use'. Ministry of Defence; January 2008. p. 84.

9.7. *The International Journal of Electromagnetic Compatibility*. 1000 Germantown Pike, F-2 Plymouth Meeting, PA 19462, USA: ITEM[TM]. www.interferencetechnology.com

9.8. *The EMC Journal*. Eddystone Court, De Lank Lane, St Breward, Bodmin, Cornwall, UK. Nutwood UK Ltd. www.theemcjournal.com

Index

admittance
  of current transformer, 187
  of isolated conductor, 190, 192, 194
  of open circuit line, 203
antenna gain, 104, 130
antenna-mode current
  definition, 277
  monitored on scope, 210
  propagation velocity, 205
  simulation, 210

block diagram, 23, 249, 252, 253, 265
broadband over power lines, 233
bubble plot, 71, 74, 78

characteristic impedance
  definition, 83
  of delay line model, 144, 149, 156
  derivation, 276
  of free space, 105
  inner & outer braids, 189
  measurement, 211, 212
  of virtual conductor, 168
characterization
  of cable, 197
  of capacitor, 213
  of components, 176
  of voltage transformer, 178
circuit capacitances
  of composite pair, 70
  of screened pair, 79
  of virtual conductor, 112
circuit capacitors
  for conductor pair, 40
  for three conductors, 48

circuit equations
  capacitive coupling, 40
  definition, 277
  inductive coupling, 38
  and loop equations, 26
  for three conductors, 46
circuit inductance
  composite pair, 70
  screened pair, 79
  three-conductor assembly, 47
  virtual conductor, 112
circuit parameters
  capacitance, 39
  cross-coupling, 91
  definition, 6, 26, 277
  exposed cable, 117
  inductance, 38
  and loop parameters, 33
  resistance, 41
common-mode choke, 223, 228, 229, 235
common-mode current
  definition, 163, 277
common-mode filter, 238
common-mode rejection
  definition, 277
common-mode resistor, 235, 236
composite conductor
  concept, 7
computational electromagnetics, 2, 3
conducted emission
  circuit model, 12, 13, 97
  damping, 231, 248
  definition, 277
  test setup, 96, 267

conducted susceptibility
circuit model, 13, 98
definition, 277
test setup, 98
conductor resistance
general formula, 41
crossover frequency, 41, 89, 119, 193
current transformer
assembly, 184
calibration, 185
circuit model, 186
definition, 277

delay line model, 147, 156
differential amplifier, 226
differential analogue driver, 228
differential logic driver, 227
differential logic receiver, 227
differential-mode current
definition, 50, 277
distributed parameters
concept, 6
definition, 278
maximum frequency, 86
single-T model, 9, 83
transformation, 85
transformation equations, 86
triple-T model, 10, 87

earth
definition, 278
earth conductor, 122, 183, 232, 240, 243
earth loops, 3, 21, 122, 247
electric field strength
at a point, 28
and power density, 105, 115, 258
and threat voltage, 18, 101, 114, 258
electromagnetic compatibility
definition, 1, 278
electromagnetic interference, 1, 4, 7,
267, 278
elemental conductors
concept, 7
primitive equations, 64
radii, 63
spacing, 63
envelope curve, 116
equipment shielding, 218, 241
equipotential ground plane, 7, 217, 218

floating configuration
definition, 278
floating transformer, 230
frequency domain, 4, 137, 175

general circuit model, 168, 266, 267
general-purpose software, 3
general purpose test equipment, 14, 79, 96,
176, 263
ground
definition, 278
grounded configuration
definition, 278
ground loops
guidelines, 22, 247
ground plane
concept, 57
on printed circuit board, 222
half-wave dipole
radiated power, 102
hybrid equations, 81, 82, 85, 273

in situ, 2, 264

Line Input Simulation Network,
233
loop capacitance
composite pair, 71
conductor pair, 37
loop equations
definition, 278
in derivation process, 7, 11
general formula, 67
for triple-T model, 87
loop impedances
and circuit impedances, 46,
47, 79
and primitive impedances, 46, 67
loop inductance
co-axial cable, 56
conductor pair, 36
three-conductor assembly, 77
loop parameters
definition, 278
in derivation process, 6
and test equipment, 6, 25
lumped parameters
definition, 278
and distributed parameters, 9

in single-T model, 82
in triple-T model, 87

magnetic field strength
  and electric field strength, 105
  at monitor antenna, 133
  at a point, 30
  and power density, 105, 133
  and radiated emission, 133
mesh analysis, 7, 49, 125, 137
mesh equations, 7, 79

nodal analysis, 49, 137, 264

opto-isolator, 230

partial capacitance
  in composite conductor, 66
  definition, 278
partial current
  antenna-mode current, 163
  in composite conductor, 65, 77, 78
  definition, 278
  due to reflections, 6, 19, 136, 144
  and mesh analysis, 7
partial impedance, 65, 70
partial inductance
  definition, 278
  composite conductor, 65, 66
  screened pair, 78
  section of loop, 254
partial voltage
  in composite conductors, 62, 65, 77
  definition, 278
  in transient analysis, 19, 136, 144
permeability
  definition, 33
permittivity
  definition, 29
power density
  and field strength, 105
  and frequency, 115
  and power received, 106
  and radiated emission, 132, 133
  and radiation susceptibility, 130
  and threat voltage, 120, 128
  and transmitted power, 104
power density vector, 103, 105
power received, 102, 106, 264

power supply
  loops, 220, 224
  returns, 228
  transient damping, 231, 239
primitive capacitance
  definition, 278
  general formula, 29
  of virtual conductor, 107, 112
primitive equations
  definition, 278
  derivation process, 7
  for elemental conductors, 64
  for three conductors, 45
  for two conductors, 34, 109
primitive inductance
  definition, 278
  general formula, 33
  of virtual conductor, 107, 112
propagation constant, 83, 276
proximity effect, 8, 220

radiated emission
  of cable and model, 204
  definition, 279
  test setup, 132
radiated power, 16, 102, 103
radiated transient
  test setup, 155, 206
radiation resistance
  of dipole, 103
  measured value, 201, 205
radiation susceptibility
  definition, 279
  test setup, 130
reflection coefficient, 144
relative permeability, 33, 71, 279
relative permittivity
  definition, 30, 279
  measurements, 205
representative circuit model
  concept, 14
  table, 267
resonant frequency, 35, 116, 189, 266

screened pair
  bubble plot, 78
  with differential analogue driver, 228
  with differential logic driver, 227
  representative circuit model, 78

series LCR circuit, 138, 139, 214
shielding
    of buildings, 242
    of equipment, 241
signal link
    basic building block, 23
    circuit model, 13, 89
    wiring diagram, 11
single-point ground, 3, 217, 224, 225, 247, 248
skin effect
    formulae, 41
    and proximity effect, 8
SPICE
    definition, 279
structure as a shield, 246

threat voltage, 18
    definition, 279
    and electric field, 101, 114
    and frequency, 117
    and threat environment, 258
three conductor signal link
    general circuit model, 89
time constant, 34, 143
transfer admittance
    of critically damped cable, 95
    definition, 26, 279
    duality, 54

of open-circuit cable, 94
of short-circuit cable, 95
transfer function
    of voltage transformer, 179, 182
transfer function analyzer, 213
transfer impedance
    of capacitor, 214, 215
    of co-axial cable, 27, 57, 254, 255
    of current transformer, 186, 187, 188
    definition, 279
transformation equations, 86
transformer driver, 230
transient emission, 168, 174, 267
triaxial cable, 188, 237

virtual conductor
    capacitance, 112
    concept, 16
    definition, 279
    inductance, 112
voltage transformer
    assembly, 177
    definition, 279
    representative circuit model,
        180
    transfer function, 179

wiring diagram, 249, 251